THE ECOLOGY AND EVOLUTION OF INVASIVE POPULATIONS

THE ECOLOGY AND EVOLUTION OF INVASIVE POPULATIONS

'The author has spared himself no pains in his endeavour to present the main ideas in the simplest and most intelligible form'

Einstein, on relativity, (1920)

'No efforts of mine could avail to make the book easy reading'

Fisher, on population genetics, (1930)

The Ecology and Evolution of Invasive Populations

Ben Phillips

OXFORD

UNIVERSITY PRESS

Great Clarendon Street, Oxford, OX2 6DP,
United Kingdom

Oxford University Press is a department of the University of Oxford.
It furthers the University's objective of excellence in research, scholarship,
and education by publishing worldwide. Oxford is a registered trade mark of
Oxford University Press in the UK and in certain other countries

Published in the United States of America by Oxford University Press
198 Madison Avenue, New York, NY 10016, United States of America

British Library Cataloguing in Publication Data
Data available

Library of Congress Control Number: 2024946954

ISBN 9780192898630
ISBN 9780192898647 (pbk.)

DOI: 10.1093/9780191924910.001.0001

Printed and bound by
CPI Group (UK) Ltd, Croydon, CR0 4YY

Links to third party websites are provided by Oxford in good faith and
for information only. Oxford disclaims any responsibility for the materials
contained in any third party website referenced in this work.

The manufacturer's authorised representative in the EU for product safety is
Oxford University Press España S.A. of El Parque Empresarial San Fernando
de Henares, Avenida de Castilla, 2 – 28830 Madrid (www.oup.es/en or product.safety@oup.com).
OUP España S.A. also acts as importer into Spain of products made by the manufacturer.

Contents

Introduction

1.1 Why is this book interesting?

Invasive populations are everywhere, from the lichen growing on your roof tiles to the flux of populations across continents. Not only are they everywhere, but they often carry consequences. A new virus emerges and rapidly spreads around the world; a tumour quietly grows in a loved one's body; an introduced species marches across a continent. All of these are invasive populations. In all cases, we are interested in the rate at which these populations spread. Sometimes we seek to speed up an invasion; often we would like to slow it down.

Invasive populations are not just a management problem, of course. Invasions are also threaded deep into our own evolutionary history. *Homo sapiens*, having spread out of Africa and across the entire globe, is an incredibly successful invasive species. Understanding the process and consequences of invasion shed light on our collective past. Most people reading this book carry stories of historical invasion written not only in their culture, but also in their DNA.

Our collective desire to understand invasive populations has inspired a rich body of theory. This theory develops from foundations laid in physics and statistics early in the twentieth century and has been elaborated to a wide array of ecological circumstances (see Chapter 2). Despite the elegance of the basic theory, however, it has often failed to explain real patterns in nature. Something important has been missing. In the last 20 years, it has become clear what that something was: evolution.

The last few decades have seen a growing awareness among those studying invasive populations that evolution matters: evolution plays out on timescales that matter to ecological systems. There has also been a reawakening among evolutionary biologists that ecological context matters. It is now clear that there is an interplay between ecology and evolution that is subtle, complex, and only ignored at our peril. Eco-evolutionary thinking is the new black, but in many ways it is simply a reconciliation of fields that were never really separate. Amid this broader backdrop, it is probably inevitable that evolutionary thinking would make its way into our understanding of invasions. The last 20 years has seen a suite of new and important ideas around invasive populations; ideas that greatly enrich the original ecological theory and that, I think, offer new perspectives on ecology and evolution more broadly.

My own interest in invasive populations began accidentally, with an introduced species in northern Australia: the cane toad. I was a field biologist, employed to study the impact of toads on native snakes. I was planning some field-based experiments to measure that impact, and this required me to have

The Ecology and Evolution of Invasive Populations. Ben Phillips, Oxford University Press. © Ben Phillips (2025).
DOI: 10.1093/9780191924910.003.0001

some idea of when toads would arrive at my study site. My efforts to work out their expected time of arrival revealed something astonishing: cane toads were spreading across the country at nearly 55 km per year. This implied a new land-speed record for amphibians, and also that the toad invasion had been steadily accelerating over time. An invasion that had initially spread at around 10 km per year was now, 70 years later, going more than five times faster than that. What was going on?

This book begins with the basic ecological theory of invasive populations (Chapter 2). Had I known this basic ecological theory at the time, I probably would have reached the wrong conclusion about my accelerating toads. But as it was, I was spectacularly ignorant about invasions, though I did know a little about evolution. Using my one hammer, then, I speculated that there may have been rapid evolution of dispersal rate during the toad invasion. This led me to dream up a mechanism that is now called spatial sorting—one of the fascinating and important phenomena that we now know happens on invasion fronts (see Chapter 3). I wasn't the first to imagine this mechanism, nor the first to find empirical evidence for it, but something about toads and timing caused people to notice this time.

Following a large community effort, it transpired that the evolution of dispersal only partly explains the dramatic acceleration of the toad invasion. There are, it turns out, several evolutionary processes afoot in any invasive population. Some very familiar (e.g., natural selection); others with exotic and exciting names like 'mutation surfing' and 'the Olympic village effect'. As we grapple for understanding, and place names on the things we see, there is inevitably some nomenclatural confusion. But one of the most fascinating insights emerging from this struggle to classify process is that these exotic evolutionary processes are, in fact, just our well-known evolutionary processes—natural selection and drift—but they are playing out through space rather than time. Natural selection has a shy younger sibling—spatial sorting—which is natural selection operating through space rather than time. Drift also has a shy younger sibling—serial foundering—which is just genetic drift operating through space. If this all seems wildly confusing, Chapters 3 and 5 go to careful lengths to explain.

The notion that well-known evolutionary processes have spatial counterparts is intriguing. Evolutionary biologists and ecologists have long been interested in spatial processes, and there are several well-developed fields that treat space explicitly: areas of study such as hybrid zones, metapopulations, and range edges, to name a few. Invasion fronts, complex though they are, are in some ways simpler than these other systems. This relative simplicity has allowed us to see more clearly that there are spatial counterparts to our classic evolutionary processes. This insight at once simplifies our language, and allows us to see links between fields that we might not have seen before. While there remains much to be done on the evolution of invasive populations, there is also much to be done translating insights from invasive populations back to older and more extensive bodies of theory.

One of the most alluring aspects of working on invasive populations is their very obvious dynamism. Populations spread in a travelling wave across space and suddenly appear in places where they were absent before. An ecological perspective looks outwards from there, towards interactions with the invaded community: a new predator arrives and alters trophic interactions; a virus arrives and impacts its host population; a tumour grows and invades vital organs. An evolutionary perspective looks inwards and sees waves within waves: gene frequencies rapidly shifting, and advantageous alleles sweeping backward and forward making their own travelling waves within the invasive population. All together, it is a rich and humblingly complex dynamism, and there is endless detail to be explored. The latter parts of this book explore some of these complexities. While there is a sense that the notions explored in the first six chapters are beginning to solidify into coherent theory, much of the rest of the book looks down on the wings of the house that are still under construction.

The invasion of cane toads took over my world for a while. We learned much from the spread of toads, and many of those lessons have relevance beyond toads, to invasive populations in general. If nothing else, toads focused people's attention and made a strong case that interesting things happen in invasive populations. It is astonishing to me that lessons learned from toads are now used to understand the growth of tumours (and this has become one of my favourite stories about the value of curiosity-driven research). The truth is, however, that evolution in invasive populations is an idea whose time has come. In hindsight, there were scientists all around the world suddenly, and independently, working on evolution in invasive populations. It took some time for everyone to notice everyone else's work, but despite (or perhaps because of) this independence a remarkable coherence has emerged. We now have a suite of empirical systems to draw from, and the evolution of invasive populations is now a very active field of research, with wide-ranging implications. I am lucky enough to have been caught up in this surge of interest and, to an extent, carried along with it. It has been a busy and productive time, and I have learned a lot.

The ecology and evolution of invasive populations have attracted interest from research groups across the world. There is now a large body of work on the topic, and the field is sufficiently mature that a synthesis is definitely in order. Certainly, the rate at which research has accumulated over the last several years surpassed my capacity to keep up with all of it without some dedicated effort. This book attempts this synthesis (and also furnished me an excuse to play catchup). The broad sweep of the narrative is science at its best: a wonderful interplay between empirical and theoretical work.

The ecological theory of invasive populations was initially inspired by observations of invasions in nature: invasive species or populations shifting as a consequence of climate change. Arguably, the theory is now ahead of the empirical work. The evolutionary theory of invasive populations has largely followed the same course (although much faster): observations of accelerating invasions and trait shifts from natural invasions demanded new theory. Within a decade, the theory has raced ahead of the empirical work. To address this gap, empiricists have now moved away from natural invasions, instead constructing invasions in the lab, often in close association with theoreticians. Lab-based invasions are replicable and manipulable, allowing rigorous experimental tests of theory. These lab-based invasions are providing the kind of data that a field biologist could only ever dream about. They are inspiring new theory, and new analyses; understanding that can be applied back towards the invasive populations we seek to manage and understand.

There is still much to be learned, but enough has solidified that a synthesis is in order. The field is an exciting one, and I have no doubt that much will be added to this synthesis in the next decade or so. The subject matter is, at times, necessarily complex, but I have made substantial effort to make it as broadly accessible as possible. I started my career as a field biologist, so I know that mathematical notation can, for the uninitiated, be as difficult to parse as hieroglyphics. I also know that mathematics is concise, beautiful, and powerful. I use mathematical notation throughout the book, but I use it more often for its concision than for its power. I use it to clarify the narrative more often than I use it to demonstrate a point. If a picture says a thousand words, a formula will often say more. This information density leads me to suggest that when you see a bunch of pronumerals coming, slow down; there is more information per line for your brain to process. For me, numerical simulation was a powerful route to understanding the math, so I provide examples of numerical simulation as we go along. My hope is that this book will provide a rapid route for the reader in understanding the fascinating processes that are set in motion when populations spread through space. This understanding is important, because invasive populations are everywhere, and they carry consequences.

1.2 Some examples, and thought tools

This book is essentially designed to be read from left to right. Ideas are built upon each other. I provide back-references for those who wish to skip ahead, though reading Chapters 6–9 before the others is likely to leave you feeling a bit lost. Chapter 4 gets a little intense at times, but you can skim over that one and come back to the detail later, if that helps.

This is not a book about invasive species, necessarily, but about invasive populations generally. As such, we can draw insight from invasions across a range of settings: from invasive species through to tumours and the growth of a bacterial colony. As an entrée, we will spend the rest of this chapter looking at several examples of invasive populations: two invasive species (as it turns out), and an experiment with replicate invasions of beetles in a lab. We will use these case studies to introduce a few simple thought tools and to cement an idea mentioned earlier: the idea of invasion speed, and how it varies.

1.2.1   The muskrat, that old workhorse

Charles Elton was an English ecologist based at Oxford. He was a man who clearly enjoyed contemplating nature as much as he enjoyed the company of his family, playing Haydn on his piano, savouring his wife's lyric poetry, and (presumably when his wife was not looking) escaping into trashy Westerns replete with gunslingers and outlaws. In 1958, he wrote a landmark book, *The Ecology of Invasions by Animals and Plants* (Elton, 1958), which is widely attributed as establishing invasion ecology as a sub-discipline. The book is a Kerouacian read, full of acute observations and thoughtful insight, and it contains a wealth of empirical examples. The book is also noticeable for a near complete absence of mathematics: Elton was intolerant, it seems, of too much theory. But he was a fine observer and a careful thinker. One of the very first examples[1] Elton chose to illustrate the phenomenon of biological invasion was an inglorious and muddy rodent.

The muskrat is a medium-sized semiaquatic rodent, native to North America. It is something like a small beaver, a tiny capybara, or an enormous rakali.[2] It builds distinctive mound-shaped dens in and around swampy areas, so its presence is easily detected. In 1905, an enterprising Czech decided that muskrat fur might be a nice way to diversify his farming income and so he brought five muskrats from North America to Czechoslovakia. The inevitable escapes happened, and what followed was a spectacular invasion, noticed by the locals (those mounds!), and beautifully documented by a local ecologist. In around 22 years, five muskrats in a Czech farm rapidly became hundreds of thousands of animals spread across 200,000 km^2 of Europe (Figure 1.1). It proved a profitable move by the farmer, and subsidiary introductions were later made into Finland and Russia to establish what became a thriving fur trade.

[1] The fourth example, in fact.

[2] Or water rat: this last one for the Australians.

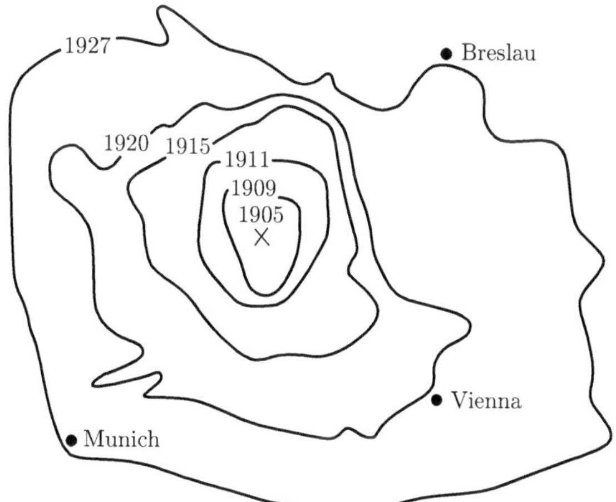

Figure 1.1 *The spread of muskrats in Europe as reported by Skellam (1951).*

Interestingly, seven years prior to Elton's book, the muskrat invasion had been noticed by a much more mathematically sophisticated character, John Skellam. Skellam was a zoologist by training (almost certainly having sat through some of Elton's lectures along the way), who worked through several teaching posts after graduation. During this teaching period, his early interests in embryology developed somehow into an interest in ecology, and mathematics. A self-taught mathematician, he was eventually employed by The Nature Conservancy (an institution Elton helped establish) as a biometrician. He was, by all accounts, a kind and thoughtful man who was more interested in helping others than in publishing his own work. While Elton's book undoubtedly popularized the study of invasions among ecologists, it was Skellam's mathematical treatment of invasions that gave ecologists their most powerful thought tools for understanding invasions (see Chapter 2), and it is striking that Elton does not mention Skellam's foundational (though quite mathy) work once in his entire book. This omission is even more striking when we consider that they were contemporaries, they worked nearby each other, and they must certainly have known of each other's work.

Elton's treatment of muskrats was a salutary example of how invasions can be kicked off by small numbers of introductees and have spectacular consequences. Elton was interested in illustrating a story about the explosion in numbers and area occupied that can occur when populations are introduced to new areas. Skellam, by contrast, was more focused: he was interested, quite specifically, in the rate at which the invasion spread; he wanted to turn that map into a graph that showed how fast the invasion front was moving.

To do this, Skellam did what modellers always do: he treated the real world as a messy approximation of some ideal form, some Platonic process. He looked at Figure 1.1 and said, 'well, the distribution of these things seems to be roughly a circle that is getting larger over time.' Looking at Figure 1.1, this does seem a halfway reasonable, if slightly strained, claim. We all know that the area of a circle is a function of its radius:

$$A(t) = \pi r(t)^2, \tag{1.1}$$

where $A(t)$ is the area and $r(t)$ is the radius and both are a function of time t.[3]

Anyway, Skellam reorganized equation (1.1), so he had

$$r(t) = \sqrt{\frac{A(t)}{\pi}}. \tag{1.2}$$

And then, for each of several time points, he measured the area that the invasion occupied at that time, $A(t)$. If we set 1905 (the year of introduction) to $t = 0$, his data look something like the values in the observed area column of Table 1.1.[4] By plugging these values of $A(t)$ into equation (1.2), he gets an estimate of $r(t)$, the 'effective radius', for each time point.

In this way, he converted the area occupied into the effective radial extent of the invasion; the average distance the invasion front had travelled from the

[3] Note that while the formula for the area of a circle is very familiar to us, this notation for the area and radius, $A(t)$ and $r(t)$, may not be. It should be read as 'A as a function of t' and it simply means that A depends upon another variable, called t: area changes with time. A variable is a quantity that might change over time (or space), as is clearly the case with the average radius (and so area) of the muskrat invasion. Some functions may depend upon more than one variable, and it becomes helpful at some point to specify these dependencies, so we start doing so now with the familiar formula for the area of a circle.

[4] Skellam's actual numbers are never reported. I measured these values off the map in his paper. I used a computer to measure the area inside each year; Skellam probably used a piece of graph paper.

Table 1.1 *The area occupied by muskrats at varying times since introduction, and the circle radius that would yield that area.*

Time (t)	Observed $A(t)$ (km^2)	Effective radius, $r(t)$ (km)
$t = 0$	0	0
$t = 4$	5,362	41.3
$t = 6$	14,068	66.9
$t = 10$	38,281	110.4
$t = 15$	80,904	160.5
$t = 22$	209,229	258.1

introduction point. We can see that in 22 years, the invasion had, on average, travelled about 258 km from where muskrats were introduced, and that they now occupied about 209,000 km^2 of Europe. Not a bad effort for a muddy rat. He then made a plot of $r(t)$ against t. This allowed him to get a sense of $r(t)$: he knew that the radius was a function of time, but what kind of function? To his (no doubt) great delight, Skellam discovered that it was essentially a straight line (Figure 1.2).

It is worth drilling a little deeper into this figure, because we can use it to illustrate a simple point, and to remind ourselves about a little bit of physics. This is a plot of distance (in km) against time (in years). You will probably recall that

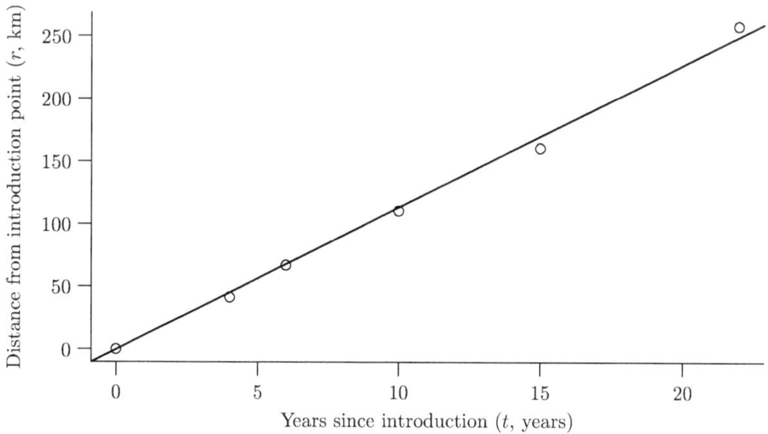

Figure 1.2 *The muskrat invasion in numbers. Skellam (1951) found that the relationship between radial distance from the introduction point and time since introduction forms a strikingly straight line.*

the slope of a straight line can be calculated as the rise over the run. The rise in this case is the change in distance, and the run is the change in time. So the slope of this line, b, can be calculated from any two points along the line as

$$b = \frac{\Delta r}{\Delta t} = \frac{r_2 - r_1}{t_2 - t_1}, \tag{1.3}$$

where Δr is the change in radius between our two points ($r_2 - r_1$), and Δt is the change in time ($t_2 - t_1$) of our two points. The units of r are units of distance (km) and the units of t are time (years). So the units of b are distance per time (km/y), which is a measure of speed. So the slope of this line tells us the speed of our invasion: the rate at which it is spreading across the landscape.

If the slope of the line gives us our speed, and it is a straight line (i.e., it has only one value for the slope), then the straight line relationship Skellam uncovered tells us that the muskrat invasion progressed at a more or less constant speed. It neither accelerated nor decelerated, but simply moved forward at the same speed between 1905 and 1927. I have calculated the speed of the muskrat invasion from Figure 1.2 as 11.3 km/y. If I were standing in Vienna in 1920 or Munich in 1926, was very patient, and had very good eyes, I would see a tidal wave of damp muskrats advancing towards me across the landscape at about 11.3 km/y. Terrifying stuff.

1.2.2   Cane toads catch us by surprise

In 1935, primed by circumstance to follow poor advice, Reginald Mungomery and Arthur Bell brought 101 toads to Australia. The sugar industry in north-eastern Australia had experienced a string of bad years, and there was enormous pressure on the industry's scientists to find a solution to the crop damage being done by beetles. Arthur and Reg's colleagues in Puerto Rico and Hawaii were claiming (with nearly no useful data to back these claims) that toads were a success story in their jurisdictions; crop yields were up and surely toads would work just as well in Australia. It was an easy decision to make. Wrong, but easy.

Since their introduction in 1935, cane toads have spread to encompass more than 1.5 million square kilometres of the Australian continent. It is not clear that they ever had an effect on the cane beetles, and the development of organochlorine pesticides a few years later rendered them redundant in that role anyway. What is clear is that the toads have done very well for themselves on this new continent, and from those initial 101 animals they now number well into the billions. It is a biological invasion on a vast scale—more like what we would expect from a flying insect than from a vertebrate—with toads having hopped more than 2,100 km from the sugar-growing districts where they were first introduced (Figure 1.3).

If we apply Skellam's idea of the 'effective radius' to the toad invasion and modify it slightly by drawing imaginary semicircles rather than circles[5] (a semicircle is a better choice given the dispersal barrier of the Pacific Ocean), we

---

[5] Which implies $r(t) = \sqrt{2A(t)/\pi}$ in this case.

can once again plot $r(t)$ against t for the entire toad invasion (Figure 1.4 lower left-hand side panel). The result is not the nice straight line that Skellam saw with muskrats. First, we see a very rapid increase in range area between 1935 and 1938. This is caused by the deliberate and well-catalogued introduction of toads to sugar-growing areas around Mackay and Isis during this time interval; this is clearly an effect of human intervention, so we can ignore that. From 1938 to 1975, however, we see a curved trajectory: the slope gets steeper as time goes on. If we recall that the slope is equivalent to the speed of the invasion, the toad invasion appears to be accelerating. Then, in 1975, we see a funny kink in our trajectory, and the invasion appears to accelerate again.

The 1938 and 1975 kinks are an indication that we have pushed Skellam's imaginary (semi)circle idea too far. We were willing to accept this simplification with muskrats because their final map looked like a fried egg; roughly circular in shape. The range of the toad as of 2006 (Figure 1.3) looks nothing like a fried egg, or even half a fried egg. Even if we look at the map as of 1975, a semicircle is a bit of a stretch.[6] Luckily, we have other ways of working out invasion speeds.

One obvious way to calculate invasion speed is to start at the introduction point, and draw an imaginary straight line in some direction outwards from that point. We then walk along this line, and every time we trip over one of those imaginary contours that marks where the invasion was at a particular time in history,

[6] A semi-ellipsoid, perhaps?

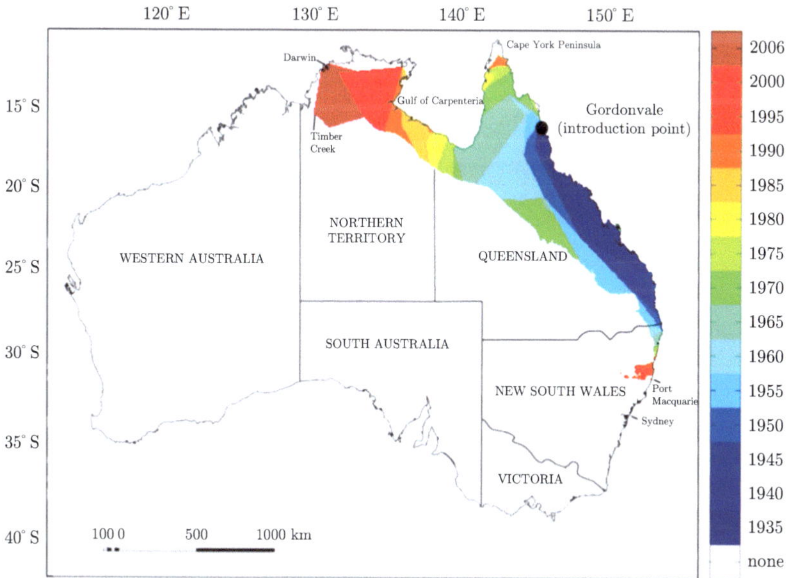

Figure 1.3 *The spread of cane toads in Australia from 1935 to 2006. From Urban et al. (2008).*

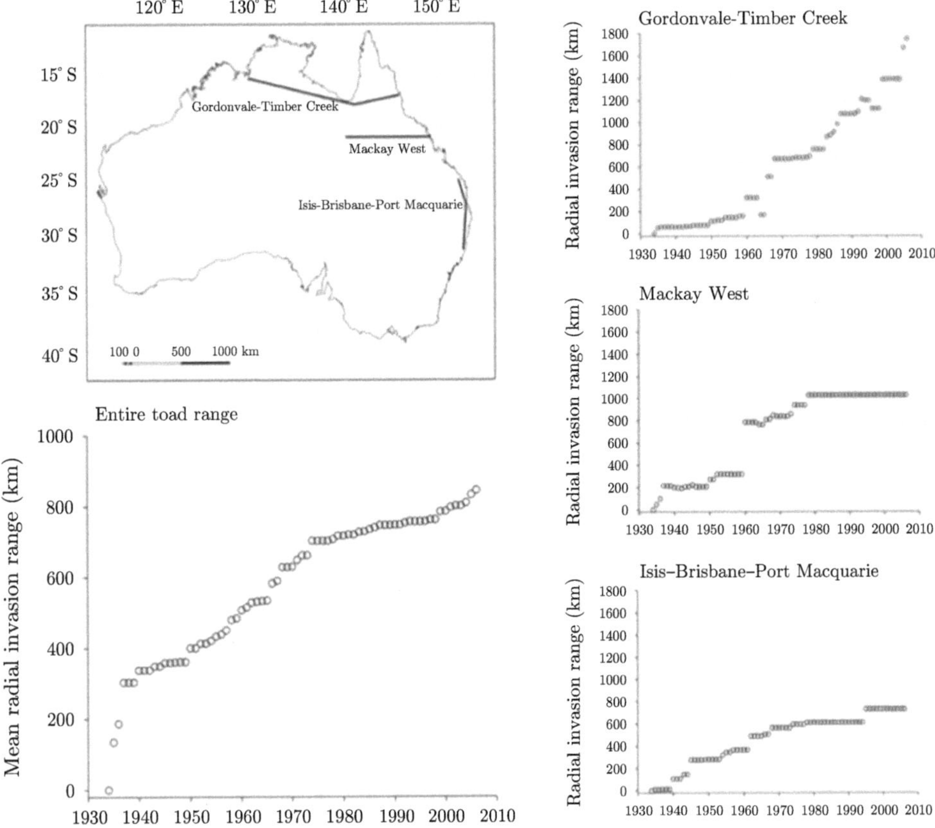

Figure 1.4 *Transects through the toad invasion. From Urban et al. (2008).*

we write down the distance we have walked and the historical time associated with the contour line. The result is a set of points recording {displacement, time} (it is displacement, rather than distance because we have been walking in a particular direction). We can plot these points up in the same way as we did for our radius versus time data, and in the same way the slope of the line is the velocity of the invasion (velocity because we now have a speed in a particular direction).

With the toad invasion, it makes sense to make three imaginary lines, one from each of the sugar cane growing districts in north-eastern Australia: Gordonvale, Mackay, and Isis (Figure 1.4). Toads were introduced to these latter two areas from Gordonvale in 1937. We draw these lines in directions that maximize the breadth of range covered, while still meeting the range edge at approximate right angles and avoiding interference from the other introduction points. We then walk along each line and tick off contour lines. The results are shown in the right-hand side column of panels in Figure 1.4, and we see very different stories, depending on which of our three transects we examine.

From Gordonvale, we see a very unambiguous signal of steady acceleration as the invasion progressed westward. There are no straight lines here; instead, there is a steadily steepening curve. As of 2006, the toad invasion was still spreading westwards (one consequence of which was, indirectly, this book). From Mackay, we see something complicated: the invasion appears to have been decelerating, then there is a sudden jump in displacement followed, once again, by a deceleration, and then—flatline—the invasion stops spreading in this direction from about 1980 onwards. In the south, from Isis, we see what looks like a steady deceleration to a near standstill; the invasion southwards has also largely stopped by around 1980.

In cane toads, it is clear that interesting things are happening: we have quite a lot of variation in the behaviour of the invasion front. In some directions, it has been steadily accelerating; in others, it has been steadily decelerating, and there are various patterns in between. If you go back to the muskrat map (Figure 1.1), you will notice that the same is true of muskrats. They did not spread equally fast in all directions: their spread to the north appears slightly slower than their spread in other directions, for example. Skellam asked us to quietly ignore that when he laid out his imaginary circles, and we happily did so. It is much harder to ignore the variation in spread rate that we see in toads, however.

1.2.3   Where you bean beetles?

All this variation within a single invasion is interesting, but it has many possible explanations. Maybe the farmers in northern Czechoslovakia just hated muskrats and spent every spare moment hunting them down; maybe the environment in northern Australia just gets better for toads as you head west. Maybe this variation is just due to sampling effort or random noise. Wouldn't it be great if we could replicate our invasion and watch it happen again? Better still if we could replicate it 10 or 20 times. Then we would really get a handle on what is driven by the environment, and what is just random noise. Excitingly, in recent years, people have started doing just that: bringing invasions into the lab, replicating them, and manipulating them.

You might recall from the introduction (Section 1.1) my claim that the evolution of invasive populations was an idea whose time had come. Well, in 2018, two labs independently conceived the exact same experiment, ran it independently, and then submitted their independent papers to the same journal and had them published in the same issue of that journal (Ochocki and Miller, 2017; Weiss-Lehman et al., 2017). Evidence, surely, of an idea whose time had come. For now, we will skip over the full details of their experiment, and we will simply look at results from one of their control populations.

The setup was very simple: they set up arrays of Petri dishes with dried beans in them, and they introduced to one end of each array a population of small beetles whose sole food is those beans. They then allowed the beetles a small opportunity to disperse between dishes every generation (about 30 days), and watched what happened over about 8 generations. The result, of course, was a biological invasion across these bean patches. Figure 1.5 shows us what one of those invasions looks like. It is a messy kind of affair, but you can see quite clearly that the population has spread from left to right over time.

It is messy, but it is worth considering for a moment how complete our knowledge of this invasion is. Literally every individual in the population, across every patch, was counted every generation. Anyone who has ever attempted ecology in the field will know that this is simply not possible outside. Most examples of natural invasions are built off datasets in which a tiny fraction of the individuals are observed, and typically the observation effort is patchy across time and space. There is a very large amount of (often unacknowledged) guesswork that goes on when invasions are studied in natural settings.

The other great advantage of these lab-based invasions is that they can be replicated. In this case, the invasion was replicated nine times. These replicates were run all at the same time and across near-identical and near-homogeneous environments (Petri dishes with beans in them). We would expect these invasions to all track along at more or less the same speed, but they really didn't. Figure 1.6 shows that over only ten generations there was already considerable variation in invasion speed: the slowest moving invasion had travelled less than three quarters of the distance of the fastest moving invasion. The other thing that is obvious is that this discrepancy grows over time; there is considerably more variation in invasion distance at time 10 than there is at time 5. Some of these invasions just seem to be going faster than others. Remember, these invasions are all happening at the same time across a homogeneous environment, with all invasions starting

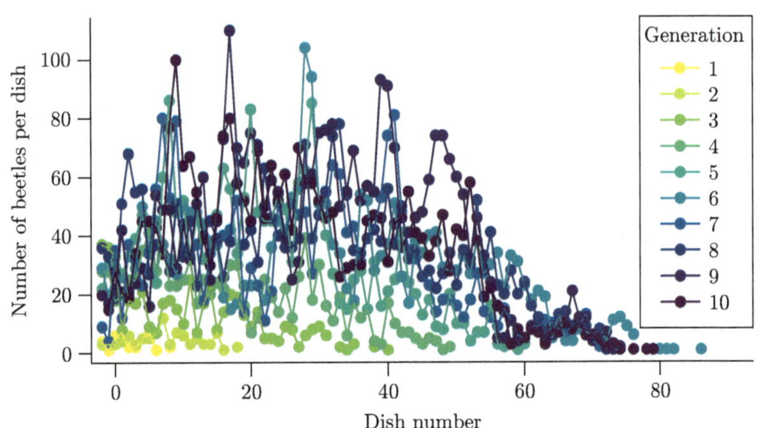

Figure 1.5 *The invasion of a population of bean beetles through a linear array of Petri dishes. Data from Ochocki and Miller (2017).*

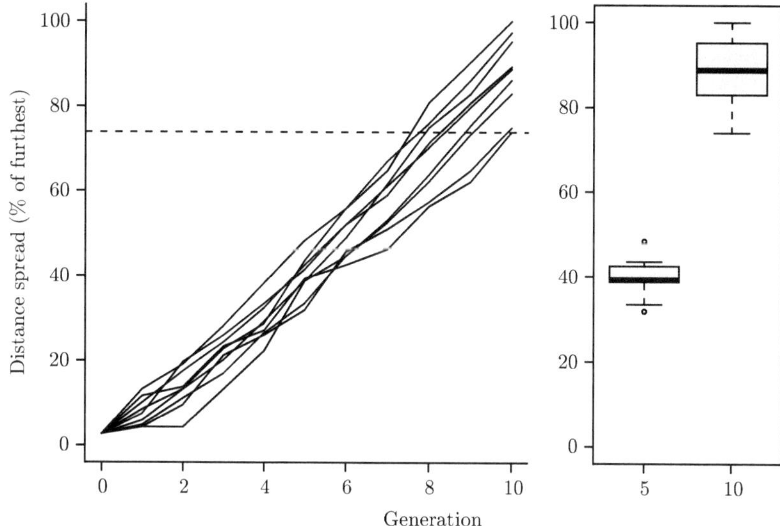

Figure 1.6 *Spread distance over time for replicate invasions of bean beetles. Distance spread is expressed as a percentage of the furthest distance attained by any invasion (188 patches). The right-hand panel shows variation in distance spread at 5 and 10 generations, respectively. Data from Ochocki and Miller (2017).*

with individuals sampled from the same beetle population. Where is this variation coming from? Chapters 4 and 5 will give us a fairly good idea.

1.3   Where are we off to?

The above three examples give us a sense of how we might observe invasions, and some of the dynamics that are possible. Some invasions progress at constant speed, some accelerate, some decelerate. Some invasions do all three of these things at different times and places. Even when we try to provide invasions with a perfectly homogeneous laboratory environment, replicate invasions differ in their patterns of spread.

Clearly, interesting things are happening, but we can't say much more than that at this stage. The rest of this book works to unpack our understanding of invasions. We start with the basic theory of why populations invade new areas (Chapter 2), and we discover that invasion happens because of dispersal and population growth. We also discover that there are many ways in which dispersal and growth can be described, and this description makes a big difference to our expected invasion speeds.

In Chapter 3, we unpack the basic theory around the evolution of invasive populations. We find out that there is strong and unambiguous evolutionary pressure for invasive populations to evolve. This evolutionary pressure comes from natural selection and spatial sorting; with spatial sorting being a spatial analogue of natural selection. These forces often conspire to cause invasions to accelerate.

Chapters 2 and 3 describe how we might expect things to go if we lived in a clockwork universe. They give us powerful insights into the ecology and evolution of invasions, but they don't really help us understand all the variation in invasion speeds that manifest in the real world. Chapters 4 and 5 look at how we might fold chance and uncertainty into our understanding. We discover that uncertain outcomes are a fact of life with invasions, we find out where this uncertainty springs from, and we discover that random evolutionary events, in particular, can profoundly affect our deterministic expectations derived in Chapters 2 and 3. Random evolutionary outcomes can slow invasions down, quite dramatically. In Chapter 5, we also find another evolutionary process (drift) with a spatial analogue (serial foundering) that matters on invasion fronts.

In Chapter 6, we reveal that there is a whole other class of invasions that we have (so far) been ignoring. These are invasions that are pushed forward by dispersal rather than pulled forward by population growth. We see that such invasions are probably quite common, and that this dynamic can change many of our expectations derived in the previous chapters.

The first six chapters outline most of what we know (or think we know) about invasive populations. They certainly give us a much deeper understanding of what might be going on when we see invasions play out in nature. Next, we go on to explore parts of the house of theory that are still under construction (or at least, places where the paint is still wet). Chapter 7 takes us out to some of these edges, and explores a few interesting wrinkles in the story. We then cleave back to more established theory in Chapter 8 where we look at how biotic interactions might change the ecological and evolutionary outcomes of invasion. Finally, we end with a chapter focused on management. Chapter 9 looks at a small selection of diverse case studies, from toads to tumours, and draws out some common threads; new avenues for management that come from integrating an evolutionary perspective into the dynamics of invasions.

1.4 The wrap

Each chapter will end with this wrap section, giving us a quick summary and a pointer to other resources.

In this chapter, we found out that biological invasions are everywhere. Situations as seemingly diverse as the growth of a tumour or the spread of cane toads are both just biological invasions. We set out the basic rationale for this book and how it is structured. We also looked at three case studies—muskrats, toads, and beetles—that between them give us a small taste of the ecological dynamics that can emerge from invasions. With these case studies, we looked at the phenomenon of invasion speed and how it might be measured, and we introduced a little mathematical notation.

For those interested in deepening their understanding of mathematical modelling, it is hard to go past Otto and Day's (2007) *A Biologist's Guide to Mathematical Modelling in Ecology and Evolution* for a beautifully explained and carefully developed grounding in the basic tools. For those interested in a broader introduction to the ecological field of invasion biology, the work of Lockwood et al. (2007) is a nice place to start: easy to read, broad coverage of topics, and a suite of empirical examples.

<table>
<tr><td>

2

</td><td>

Why do populations invade new areas?

</td></tr>
</table>

When first confronted with an invasive population, many people feel that the invasion must require some kind of motivation or organization, some group-level agreement, as in the way an army invades territory. That is, of course, one way to do an invasion, but it is a relatively uncommon way. Most invasive populations do not have a Genghis Khan or an Alexander the Great at the helm. The spread of a pathogenic fungus across a wheat region does not require a leader; it requires no consensus among the fungal cells to head west. But as we have seen, such leaderless invasions happen routinely. And 'happen' is the correct word. Biological invasions happen in the same way that bush fires and chemical reactions happen: given the correct conditions, invasions just spontaneously execute, generating a travelling wave of advance that baffles our intuition into invoking a leader and a plan.

Take a simple Petri dish, coat the bottom of the dish with sterile, nutrient-filled jelly, and carefully place a drop containing a few thousand bacteria into the very centre of the dish. Place the lid on, and watch what happens over the next 24 hours. What happens is that the population of bacteria grows rapidly, and invades the space of the Petri dish. Given enough time, it will expand to the very edges of the dish and run out of habitat (Figure 2.1). This just happens. It is easy to accept that this just happens when we talk about bacteria in Petri dishes, but when the invasion involves more complex organisms (birds, toads, or rats, for example), we easily forget that a malevolent plan is not really necessary.

Given the seemingly orchestrated way in which populations spread, it takes a leap of insight to realize that a leader, or group-level agreement, is not necessary. That leap of insight happened somewhere between 1937 and 1950. As is often

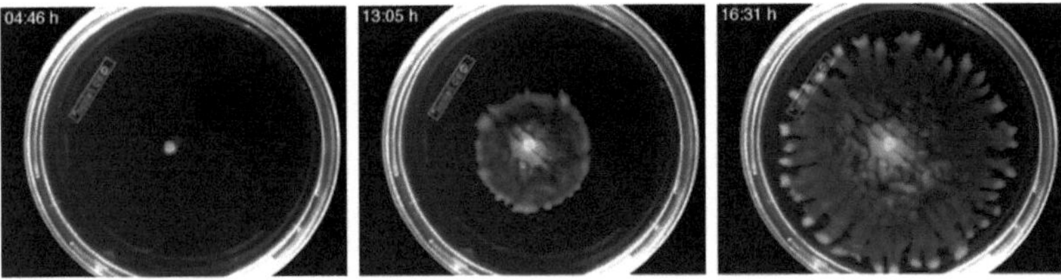

Figure 2.1 *Bacteria spreading in a Petri dish. The centre of the dish was inoculated with a small drop of bacteria,* Pseudomonas aeruginosa, *at 00:00 hrs. From van Ditmarsch et al. (2013).*

the case, the insight built off work that had been done in another field. The key insight did not come from ecology. Instead, and rather poetically in the context of this book, that key insight came from evolutionary biology.

Ronald Fisher, by all accounts, was a witty, mildly irascible, and (in his later years) curmudgeonly genius. He was a giant of modern evolutionary synthesis, and a founder of modern statistics. He was not, however, one of those kindly, zany geniuses as personified by Einstein: Fisher was spectacularly wrong on a number of counts; most notably, he was a staunch defender of eugenics, and did not appear to resile from this position even after the horror of the Holocaust. It is good to remember that even intellectual geniuses can be very wrong indeed.

Not content with inventing, among other things, quantitative genetics, ANOVA, and statistical likelihood, Fisher (1937) wrote a surprising paper entitled 'The wave of advance of an advantageous allele', which caused, well. . ., waves. Fisher's 1937 paper dealt with the situation in which a favourable mutation occurs in a population. A great deal of theory had been done by that point describing how, in a large well-mixed population, we might expect this favourable mutant (Fisher's 'advantageous allele') to rapidly increase in frequency until it replaced all other versions of that gene (alleles) in the population. These earlier models had largely ignored space, or treated it very simply. Indeed, nearly all basic models in ecology and evolution start by ignoring space and only following a population through time. They assume, for simplicity, that all individuals (or cells, or atoms, or whatever) in the population have an equal chance of interacting with all other individuals.[1] These spatial representations are undoubtedly the correct place to start for understanding systems, but it is *very* important to remember that space has been temporarily set aside. Fisher, of course, knew that space had been put on the backburner, and in his 1937 paper, he brought it back into play. The question he was interested in was not so much whether the favourable mutant would increase in frequency (this had already been well settled), but how fast it might spread through space.

Fisher's paper showed that his favourable allele would indeed increase in frequency, and it would spread through space. Most surprisingly, however, he showed that (following a number of generations for the pattern to establish) his allele would spread through space as a travelling wave with a constant shape and velocity. Physicists love waves, and in 1937 physicists had not imagined that waves of this type were possible. This was a whole new type of wave, previously unimagined. Interestingly for our story, Fisher also showed that the speed of this wave was entirely determined by only two things: the rate of dispersal, and the growth rate of the mutant allele.

One of the emergent themes from this book is that ideas have their moment in history. Advances in understanding cause ideas to bubble up, where they are often seen independently by multiple people. Fisher was not the only person to find this model and discover this new form of wave. A team of physicists in the USSR, Kolmogorov, Petrovskii, and Piskunov (Kolmogorov et al. (1937); and 'KPP' henceforth)—working entirely independently from Fisher—hit on essentially the

---

[1] Whenever you read the words 'large, well-mixed population' in a theoretical treatment, you can safely interpret this as 'we imagine space does not exist' or 'we imagine a zero-dimensional space', both of which are actually quite hard to imagine.

same model and derived essentially the same results as Fisher. Their work was just as rigorous (perhaps more so than Fisher's) and was published in the same year. Another set of physicists in the USSR also seem to have hit on the idea, publishing their results in 1938 (Frank-Kamenetskii and Zeldovich, 1938). Clearly, this was an idea whose time had come.

Fisher's model not only established theoretical waves, but also set off waves of interest and research that continue to this day. The model was mathematically interesting, so mathematicians and physicists took immediate interest, but it also applied to a wide range of systems, from chemical reactions to the propagation of fires, and of course to the spread of populations. The model that Fisher, Kolmogorov, Petrovskii, and Piskunov dreamt up is now known as the 'reaction-diffusion model' or the 'Fisher–KPP model', and since its inception it has been endlessly extended, polished, and applied to a wide array of circumstances, only one of which is ecology. Its first application to ecology was made by a character we have already met: Skellam, in his 1951 paper.

2.1   The Fisher–KPP model

It is worth going a little deeper on the Fisher–KPP model because it is foundational to our understanding of why populations spread: a few definitional tweaks, and this model became the first ecological model of an invasive population. Below we will step through Fisher's model, but for simplicity I will give it an ecological (rather than genetic) spin: rather than tracking an advantageous allele, we will track a population of animals (or cells, or plants). And because it is tedious to continually talk about a population of arbitrary organisms, I'm going to call it a population of cats. You will need to remember that cats are just a placeholder for [insert your organism of interest here].[2]

You will recall that space is typically set aside as an opening simplification. Space pretty rapidly makes things complicated, so Fisher's first trick was to introduce only a little space: one dimension, to be precise. Fisher placed his population of cats on an infinitely skinny, and infinitely long, string. Taking a lead from Cartesian geometry, he denoted location along this skinny string as x. Essentially, he placed a population of one-dimensional cats on a continuous number line. This is a thing that is possible in imaginary worlds, but in the real world we might find approximations to this scenario: a population of fish in a river; a population of bacteria inside a capillary; a population of aphids on the stem of your favourite plant; anything where the habitat is a lot longer than it is wide, and where there is a large number of individuals that are 'well-mixed' across the width of the habitat.

Fisher then picked an arbitrary point on his string (at location x) and thought about how the density of the population at this location might change in an instant of time (denoted t).[3] We don't know the density at x, so we introduce another pronumeral, $N(t,x)$, to describe this unknown density.[4] Maybe $N(t,x) = 0$,

[2] For the record, I am not a cat person, and possibly that is why I chose them as a reference case, considering what is about to happen to these cats.

[3] Density of a population is the number of individuals per unit space: so the number of cats per metre in our case, the number of cats per square metre of savanna, or the number of whales per litre of water.

[4] $N(t,x)$ should be read as 'N as a function of t and x', which is a concise way of saying that 'N might vary over time and space in ways that depend on time and space'.

maybe $N(t,x) = 42$, maybe $N(t,x) = 10^{10}$, we don't know and we don't care (though $N(t,x)$ can't be negative, obviously). Taking a lead from the calculus playbook, we then call the change in density at location x per instant of time, $\frac{\partial N(t,x)}{\partial t}$.

If you think about it, there are really only two processes that will cause N to change. The first process is a spatial process: movement, the coming and going of cats. In our instant of time, some cats might leave x and some other cats might arrive at x from somewhere else along the number line. We can call this $m(t,x)$: movement as a function of time and space. The second process is a temporal process: population growth, the balance of births and deaths. Some cats might be born, and some may die. We can call this process $g(t,x)$: growth as a function of time and space. So we can express the change in density over time as follows:

$$\frac{\partial N(t,x)}{\partial t} = m(t,x) + g(t,x), \tag{2.1}$$

which simply says the density at any arbitrary location changes according to the balance of immigration/emigration and the balance of births/deaths at that location and time. Considerable progress can be made in understanding invasive populations simply by understanding the truth encapsulated in this equation, so it is worth repeating: the change in density of cats at any place and time depends only on immigration/emigration ($m(t,x)$) and births/deaths ($g(t,x)$). When you think about it, it's just true. Setting this truth down as a mathematical object, however, allows further insights to be deduced, and this is what Fisher did. He was rather good at such things.

It is fair to warn you though that the Fisher–KPP model doesn't look like equation (2.1). Equation (2.1) is actually too general to be very useful. We actually need to replace $m(t,x)$ and $g(t,x)$ with functions that describe movement and population growth in relevant ways. Fisher chose quite specific descriptions of movement and growth. Fisher's model (tweaked slightly) actually looks more like this:

$$\frac{\partial N(t,x)}{\partial t} = D\frac{\partial^2 N}{\partial x^2} + rN\left(1 - \frac{N}{K}\right), \tag{2.2}$$

which is a substantially scarier looking beast. If equation (2.2) hurts to look at, just go back to equation (2.1) and give your eyes a rest. Once you have recovered, we can dart between the two. I have aligned the two equations carefully so that the two processes (immigration/emigration and births/death) are in the same order. That first ghastly term in equation (2.2), $D\frac{\partial^2 N}{\partial x^2}$, accounts for movement: it is the balance of comings and goings, $m(t,x)$. The second term, $rN(1 - \frac{N}{K})$, is just growth, the balance of births and deaths, $g(t,x)$. So the two equations say essentially the same thing: the density of cats at any arbitrary location changes according to movement and population growth. Fisher's model expresses this truth in a way that is less general but that affords considerable insight.

Fisher's model (equation (2.2)) describes dispersal as a random process called 'diffusion', and population growth as a density-dependent process called 'logistic growth' in which population growth stops at a given density called the 'carrying capacity'. For those interested in the details of diffusion and logistic growth, I describe diffusion in more detail in Appendix A, and logistic growth gets unpacked in section 6.1. The model describes the change in population density at an arbitrary location, but when you apply it to all locations, and you give the population a starting point, you get a travelling wave of advance (see Figure 2.2). The model has three parameters.[5] The parameters are: the 'diffusion coefficient', D, which determines the rate at which cats move through space; the 'intrinsic growth rate', r, which is the per capita balance of births and deaths when the population is at very low density; and the carrying capacity, K, which determines the local population size at which per capita population growth goes to zero.

Once we have this system set up, we can set the initial density at $x = 0$ to be some positive value, and have the density everywhere else set to 0. The partial differential equation that is the Fisher–KPP equation then determines what happens.

What happens is an invasion plays out. One of the really cool things about the Fisher–KPP analysis is that they took this mathematical object, this equation describing change in density at a single arbitrary place and time, and used it to work out how fast an invasion would spread through space. They showed that, ultimately, a travelling wave of invasion would develop and that this wave would

[5] Parameters are parts of a model that don't change over space and time; they are fixed numbers that affect the way the model behaves but that are not, in turn, affected by the behaviour of the model.

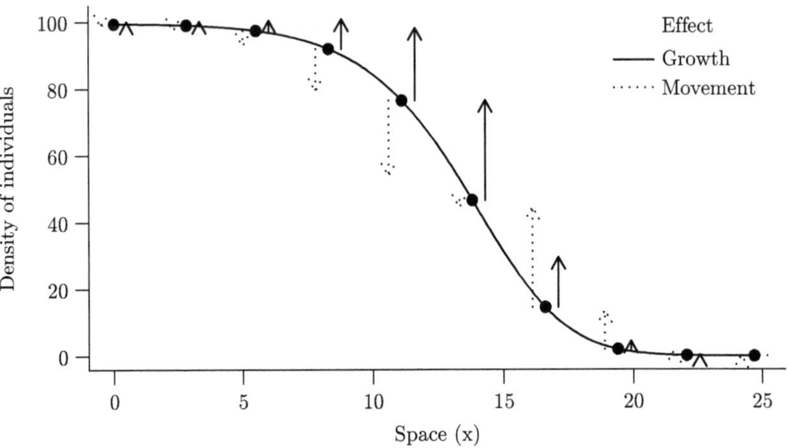

Figure 2.2 *Effects of population growth and movement along an invasion front, according to Fisher's model (equation (2.2)). Effects have been scaled such that, for each effect, the magnitude of the arrow reflects the strength of that effect relative to other locations along the wavefront. In this figure, r = 1.05, D = 1.05, and K = 100.*

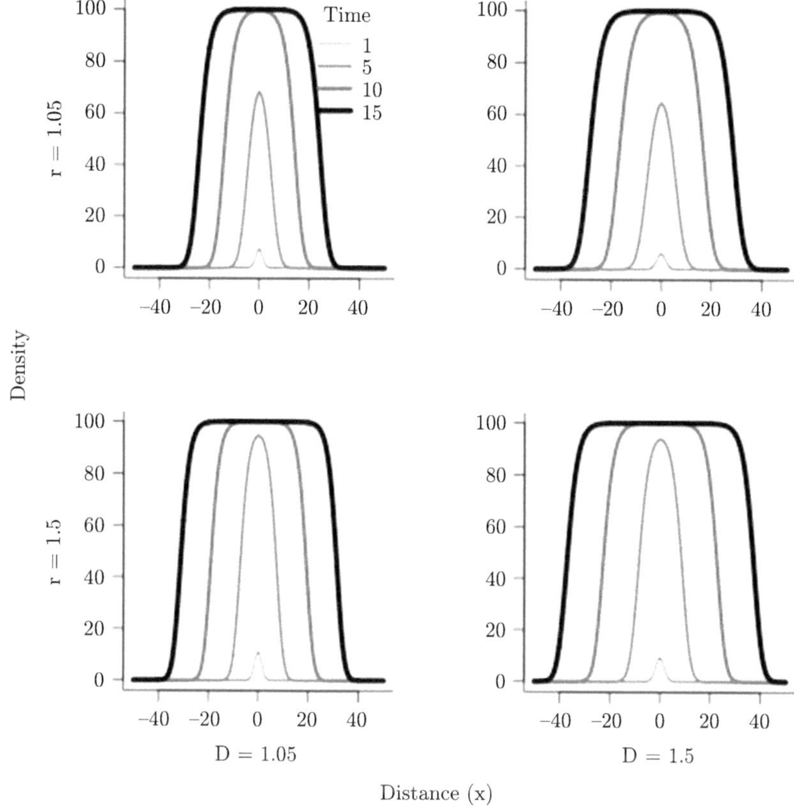

Figure 2.3 *Numerical realizations of the Fisher–KPP model at different values of r and D. The speed of spread is described by $v_F = 2\sqrt{rD}$, so invasions at bottom left and top right advance at the same speed despite differing in growth rate and diffusion.*

rapidly approach a constant speed. Following later authors, we will denote this speed as v_F, the 'Fisher velocity'. Both Fisher and the Soviet team showed that $v_F = 2\sqrt{rD}$.

This result is worth thinking about for a few moments. First, they showed that dispersal and population growth would cause a wave of invasion. Second, they showed that this wave would spread at a rate that was entirely determined by the rates of growth and dispersal. That third parameter, the final density of the population—the carrying capacity, K—was irrelevant as far as the speed of invasion was concerned. All that matters is how fast the population grows at a low density, r, and how fast individuals spread out through space, D. Figure 2.3 shows some numerical realizations of Fisher's model at different values of r and D. When both r and D are small (top left), the invasion progresses slowly, increasing as both r and D increase.

2.2 Skellam's model

As I mentioned earlier, Skellam (1951) was the first to point out that this kind of model could be applied to invasive populations. Fisher was interested in the spread of favourable mutations rather than populations necessarily (I just recast Fisher's model in terms of a population, for simplicity). Skellam was clearly motivated by invasive populations, and it was he who brought reaction–diffusion equations to ecology.

Skellam made a couple of small adjustments to the Fisher–KPP model. First, he allowed his cats to inhabit a two-dimensional landscape. Instead of cats on a string, we had, for the first time, flat cats on a map. The second thing Skellam did was that he allowed his population to grow without limit. Skellam's cats had the same rate of dying and being born regardless of how many cats were on the map. This type of growth is called 'exponential growth', and it rapidly leads to very large population sizes. Skellam's model was more spatially realistic, but less realistic about population growth, because (despite recent developments in human population size) no population can grow exponentially for very long; it eventually runs out of resources. Skellam's model looks like this:

$$\frac{\partial N}{\partial t} = D \left(\frac{\partial^2 N}{\partial x^2} + \frac{\partial^2 N}{\partial y^2} \right) + rN. \tag{2.3}$$

Compared to Fisher's model (equation (2.2)), you can see that the diffusion term (the balance of immigration and emigration) just got a little more complicated, and the growth term (the balance of births and deaths) just got simpler.

Skellam showed, among other things, that the radial speed of invasion in this model also approached a constant speed, and that speed was identical to the speed attained under the Fisher–KPP model. That is, $v_F = 2\sqrt{rD}$ again.

2.3 Reaction–diffusion and invasion speed

We now have two theoretical models that, starting from a simple description of how local density changes over time (through dispersal and population growth), show that invasions just happen. Moreover, not only do invasions just happen, but they propagate forward in beautiful waves that tend towards a constant speed. The speed of these waves is determined entirely by the rates of dispersal and population growth. All very nice.

These models are known as 'reaction–diffusion' equations because their application in chemistry preceded their application in ecology. If they had been used in ecology first, they may have been called 'growth-diffusion' models. Given their very first conception was actually in evolution, they probably should be called 'selection-diffusion' models, but history is seldom fair, and the chemists somehow got naming rights. A really cool example of invasion waves in chemistry is the Belousov–Zhabotinsky reaction (look up a video of it on the internet). Here, we have a reaction that starts with a bromine, forms an acid, and spontaneously goes back to a bromine again. It is a chemical oscillator. Start this reaction somewhere on a Petri dish with an indicator and you can watch the reaction propagate across the Petri dish, just like an invasive population (Strogatz, 2012).

Many biological invasions also seem to follow the predicted pattern of constant speed rather nicely. One of Skellam's first examples was the muskrat invasion, which we have seen earlier (section 1.2.1). Plotting the average extent of the invasion over time, Skellam found a strikingly straight line relationship, indicating a near constant spread velocity, such as we might expect from $v_F = 2\sqrt{rD}$ (see Figure 1.2). There are numerous other examples of biological invasions that proceed at near-constant speed ranging from bacteria to trees. There are, however, also many examples of biological invasions (such as cane toads) that do not propagate at constant speeds. How might theory deal with these cases?

2.4 Lagging, not flagging

Invasion ecologists often describe invasions that take quite a while to get going, and then suddenly take off like wildfire. Typically what happens is someone suddenly notices that these introduced crabs have become a feature of their landscape, and they dig up historical records and find that these things were introduced twenty years ago and have been seen sporadically since then. Now they seem to be everywhere and invading quite fast. What's going on?

Sometimes, perhaps very often, we can see that such experiences are probably just the result of the surprising nature of exponential growth. If a population of crabs starts with ten individuals and doubles in size every year, there will be more than 10,000 crabs by ten years. But in the year before, there were just over 5,000 crabs, and there will be more than 20,000 by the following year. If crabs are only detected sporadically, as a consequence of chance encounters with individuals, we very rapidly go from a situation of not seeing many crabs to suddenly seeing them all the time.

Sometimes, however, there seems a genuine case for lags. Sometimes, quite well monitored invasions clearly accelerate either steadily (as we see in cane toads) or they seem to suddenly switch gears and begin to travel faster. It turns out that there are quite a few possible explanations for why this might happen, and we will come back to this phenomenon several times throughout this book. One explanation, however, follows directly from the reaction–diffusion theory.

The mathematical result of $v_F = 2\sqrt{rD}$ is sometimes called the 'asymptotic' or 'equilibrium' spread rate. The reason for this terminology is that, in the mathematical model, the wave velocity takes a while to reach this speed. This 'gathering momentum' process can be thought of as the invasion front taking a while to develop its shape. An intuitive way to grasp this is to imagine a population with a high dispersal rate (i.e., a large value of D) but a small growth rate (i.e., small *r*, see top right of Figure 2.3, for example). Following introduction, the primary process we observe is diffusion, as individuals move through space and the population only grows slowly. Eventually, however, population growth catches up and the invasion front assumes its equilibrium shape. At that point, diffusion and reproduction are perfectly balanced across the invasion front, and the shape of the front stays almost the same, but steadily moves forward through space.

Depending on the relative values of *r* and *D*, this gathering momentum phase may last a number of generations before the invasion is, for all intents and purposes, bouncing along at its equilibrium spread rate (Shigesada and Kawasaki, 1997). This effect may explain some invasions in which there is a short 'lag' observed. This mechanism would also be compounded if there is a systematic change in detection effort as an invasion becomes more apparent. If we go from nearly no effort to systematically looking for crabs, the apparent range of crabs will increase in line with our extra effort.

While the effects of momentum gathering and detection might explain many short-term lags, there is a third class of invasion that is more difficult to explain through these mechanisms. These are the invasions that seem to steadily accelerate. The cane toad is one among many invasions that exhibit this pattern, and it was difficult to explain using reaction–diffusion models. To see how these might happen, ecologists needed to think hard about dispersal.

2.5 Accelerating waves, and long-distance dispersal

One of the limitations of reaction–diffusion models is that dispersal is treated as diffusion. You will recall I said earlier that Fisher chose very particular descriptions of both dispersal and population growth (the $m(t, x)$ and $g(t, x)$ terms in equation (2.1)). Diffusion is mathematically convenient (all of those comings and goings captured in a single term!), and has a strong mechanistic basis, but it results in a distribution of dispersal distances that is normally distributed. Every time. The reason for this is that, with diffusion, dispersal is the result of lots of random movements, and when we sum lots of random outcomes together, we tend to get a normal distribution (see Appendix A).

Ecologists have, of course, been out there measuring dispersal distances, and very often they find that dispersal is in fact not normally distributed (e.g., Clark, 1998). Instead, they often find a small number of individuals who move a surprisingly large distance. Could such long-distance dispersals explain accelerating invasions? The answer, of course, is yes.

2.5.1 Stratified dispersal

Millions of people around the world have lizards foraging on their windows. The Asian house gecko, *Hemidactylus frenatus,* is a cute little lizard that has the impressive ability to walk up vertical panes of glass as easily as you or I might walk across the living room. It is native to Java and, although capable of living in forests, does very well for itself in human settlements where it shelters in and around houses, and eats insects attracted to lights at night. Its impressive climbing ability, nocturnal habits, and small size mean that it regularly ends up in removalist boxes, vehicles, and shipping containers. Its Indonesian name Cecak (pronounced 'chechuk') is an abbreviation of its distinctive call ('Che chuk chuk chuk...') with which it readily announces its presence.

In Australia, Cecak was first noticed in the northern city of Darwin, around 1967. From there, it rapidly spread to other human population centres across northern Australia—not in a continuous fashion like we saw in cane toads, but by simply hitching a lift between towns. The lizard is largely absent from undisturbed habitat, but around towns and cities it has been steadily invading native bushland (Barnett et al., 2017). This looks very much like a two-speed invasion. On one hand, we have humans transporting the animals long distances by road, rail, and air, and on the other hand we have individuals reproducing and moving of their own accord.

Cecaks are a great example of stratified dispersal. Most of the population moves only a short distance in their lifetime; a small proportion of the population, through no particular fault of their own, takes great leaps. Theoretical ecologists confronted with such examples took the simple expedient of sticking with diffusion for describing movement, but imagining that the population is composed of two classes of individuals, where the classes differ in their diffusion coefficients. These 'stratified dispersal' models essentially have some proportion of the population moving according to D_1 and another proportion of the population moving according to D_2. This modelling strategy can be used to build a 'coalescing colony' model (Nichols and Hewitt, 1994; Shigesada and Kawasaki, 1997), in which a small number of individuals moves a long way and establishes satellite colonies, each of which grows according to standard reaction diffusion. As the colonies spread, they bud off new colonies (in proportion to their size) and coalesce with other colonies. Such a model can, it turns out, yield invasions that accelerate over time. Under particular settings, these models give invasions with speeds that are initially characterized by the smaller value of D (say, D_1) and that transition ultimately to the speed we would expect from the larger diffusion value, D_2 (Shigesada and Kawasaki, 1997). By allowing two dispersal speeds, theoretical invasions can be made to accelerate.

2.5.2   Fat kernels

Cecaks are a great example of two processes resulting in vastly different dispersal distances. But ecologists also regularly find distributions of dispersal distance that are more or less continuous, but are not more or less normal. Take, for example, a classic experiment on the dispersal of *Drosophila* (Dobzhansky and Wright, 1943). Fruit flies of the genus *Drosophila* are one of the most well-studied

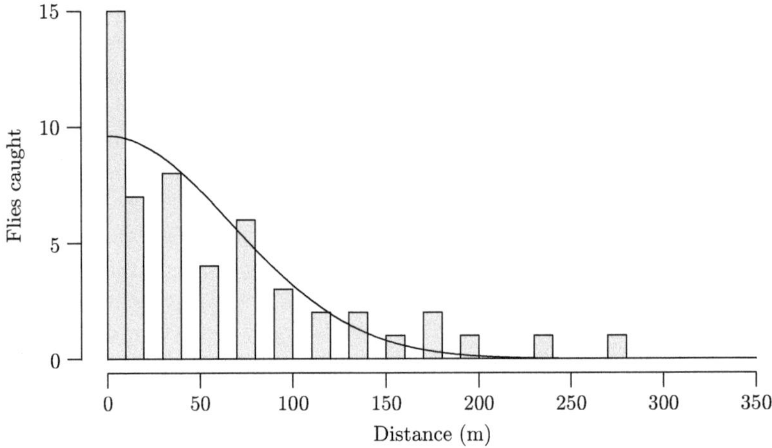

Figure 2.4 *Dispersal distances of orange-eyed mutant fruit flies after four days, compared with the best-fit normal distribution. There is an excess of observations in the centre and at large distances. Data from Dobzhansky and Wright (1943).*

organisms on the planet, not so much because they eat fruit, but because they are easy to culture in captivity, and their salivary glands have four large and easily observed chromosomes; they are a model organism in genetics, and we have learned a boggling amount from them.

Following Fisher's wave paper, there was a big focus among the cognoscenti of evolution on gene flow, and so the *Drosophila*-botherers got to work. Chief among these was Theodosius Dobzhansky[6] with some analytical help from Sewell Wright.[7] Dobzhansky decided that he would attempt to measure dispersal distances in this tiny, winged insect. To do this, he selected a mutant line with bright orange eyes, bred a very large number of these flies, and then released them from a single location. He then set up a large array of traps at varying distances from this release point, and hoped hard that they would recapture some of their flies. It worked surprisingly well, though the experiment had to be aborted after a few days because the flies moved too far too fast (Figure 2.4).

When we plot these data, and compare them to a normal distribution, however, it is pretty clear that they are not normally distributed. There are many more observations close to the release point than there should be. Less obvious (but also true) is that there are many more observations a long way from the release point than there should be (Figure 2.4). This property—an excess of observations close to the centre and far from the centre—is measured by the *kurtosis* of a distribution.[8] And the distribution of dispersal distances in these mutant flies has a higher kurtosis than the normal distribution. In practice, this means that there are many more individuals moving a long way than we would expect from a model—such as diffusion—in which dispersal distances are normally distributed.

[6] Famous for the quote that 'nothing in biology makes sense except in the light of evolution'.

[7] The only person smart enough to find disagreement with Fisher and be proven correct, so this was serious help.

[8] Strictly, kurtosis is the fourth moment of a distribution, the first three being, respectively, the mean, the variance, and the skewness.

Another way to think of this excess kurtosis is that the distribution of dispersal distances (often referred to as the 'dispersal kernel') has a fatter tail than the normal distribution. Fat-tailed dispersal kernels are regularly observed in nature, both in animals and in plants (e.g., Clark, 1998). How might fat-tailed dispersal kernels change our predictions about spread rate?

2.5.3   Integrodifference models of spread

To examine how the distribution of dispersal distances (the 'dispersal kernel') affects spread rate, it is simplest to use a different mathematical model to the reaction–diffusion model we saw earlier. This model—the integrodifference model—looks prettier than the reaction–diffusion model and is also (I think) more immediately understandable. It is prettier because it is expressed as an integral, and easier to understand because there is no diffusion term (with that confusing second derivative).

The integrodifference model was developed for invasion biology (in the context of epidemics) in the 1970s by Denis Mollison (1972b; 1972a), and extended further by Weinberger (1982). It was popularized in ecology (i.e., translated from the heavy math) by Kot et al., (1996) who (using the earlier theoretical developments) pointed out to ecologists that the shape of the dispersal kernel matters. The integrodifference models of Kot and colleagues were proposed in direct response to the observation that dispersal distances very often looked like Dobzhansky and Wright's data: dispersal distributions very often exhibited fat tails.

The discrete time integrodifference model (in one dimension) looks like this:

$$N_{t+1,x} = \int G(N_{t,y})k(y-x)\,dy. \tag{2.4}$$

This is quite different from the Fisher–KPP model. Let's step through it. First, we are no longer defining a change in N; we are tracking N directly. The particular N we are tracking is $N_{t,x}$, the density at location x at time t. To do this, we go to some other location, y, and look at $N_{t,y}$. We first work out how much this population at y is going to grow in the interval between t and $t+1$ as a consequence of the balance of births and deaths. We do this by applying a growth function, $G(N_{t,y})$ to $N_{t,y}$. We then work out how many of the individuals at y will disperse to the location we really care about, x. We do this by applying a dispersal function, $k(x-y)$, which specifies the probability that an individual moves from y to x. This probability depends upon the displacement between y and x: $d = y - x$. We then make this calculation for all values of y (i.e., all space) and sum the result (which is what the integral is doing). The result is the number of individuals at x at $t + 1$: $N_{t+1,x}$.

We could make this model approximately the same as the reaction–diffusion model (equation (2.2)), by setting $G(N) = rN(1 - N/K)$, and $k(d)$ equal to the probability density function for the normal distribution, with a mean of zero and variance given by the diffusion coefficient (see Appendix A). Making the integrodifference model equivalent to equation (2.2) would, in full, look something like this:

$$N_{t+1,x} = \int \frac{N_t K e^r}{K + N_t(e^r - 1)\sqrt{4\pi D}} e^{-\frac{(y-x)^2}{4D}}\, dy. \qquad (2.5)$$

Look away, dear reader. The point is only that the Fisher–KPP formulation is rather more elegant.

The real value of the integrodifference model is not that we get to use that beautiful integral symbol, but that we have a lot of control over how dispersal is described. By using different functions for the dispersal kernel, $k()$, we can imagine various kinds of dispersal, including asymmetric dispersal, and dispersal in which there is a fat tail to the distribution, such as what we saw in *Drosophila* above. Kot et al. (1996) introduced the idea of integrodifference models to invasion biology, and their key result is that the shape of the dispersal kernel matters—a lot.

For dispersal kernels whose tails decay exponentially (as happens, e.g., with a normal distribution), we see once again that a constant rate of spread emerges. Essentially the Fisher–KPP result, recapitulated. If we make the tails decay slower than exponential, however, by using a 'fat-tailed' distribution, we find invasions that accelerate over time.

2.5.4   Spread in fat-tailed *Drosophila*

Let's go back to those data on *Drosophila* collected by Dobzhansky and Wright and compare the spread rates we get from different descriptions of the dispersal data. We have already noticed that the normal distribution did not look like a good fit. What happens to the predicted spread rate if we choose a different function to describe dispersal? Figure 2.5 fits two alternatives to our dispersal data. The first alternative is essentially a normal distribution. Again, we can see that it doesn't do a good job of describing frequencies close to and far away from the release point. The shoulders of the normal distribution are just a little too broad for these data. The second alternative has much narrower shoulders, and (relative to the normal distribution) a fatter tail. It seems to do a slightly better job of describing the data at long distances.

We can use a rescaled version of either of these functions as our dispersal kernel, $k()$, in equation (2.4). Let's call the first one $k_1()$, and the second, fat-tailed kernel, $k_2()$. The rescaling we need here is to modify our functions such that the area under their respective curves is 1. This allows the function to describe probability density (for a continuous function), or probability mass (for a discrete function, see Appendix B).[9] Anyway, after rescaling,

$$k_1(x) = \frac{\sqrt{b}}{e^a \sqrt{\pi}} e^{a-bx^2} \tag{2.6}$$

$$k_2(x) = \frac{b^2}{4e^a} e^{a-b\sqrt{x}}, \tag{2.7}$$

where a and b are just parameters that adjust the shape of each curve. We already estimated these from the data to produce Figure 2.5. The only thing we need now is a growth function, $G()$—some function that describes how a *Drosophila* population might grow in discrete time. A useful model for population growth in discrete time is the Beverton–Holt function (Beverton and Holt, 1957),[10] which we will use here as our function G:

$$G(N) = \frac{R_0 N}{1 + ((R_0 - 1)/K)N}, \tag{2.8}$$

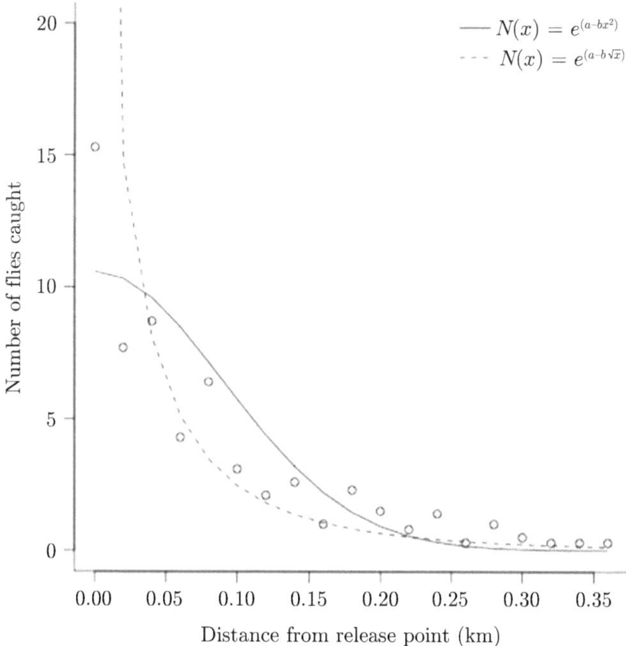

Figure 2.5 *Two alternative functions for describing the distribution of dispersal distances in* Drosophila. *The first function has tails that decay at the same rate as the normal distribution. The second function has a tail that decays more slowly than the normal distribution—it is a fat-tailed distribution. Functions are from Kot, et al. (1996).*

[9] To do this, I took each of these functions reported in Figure 2.5 and (1) worked out their indefinite integral between ∞ and 0 (i.e., the area under their curve between 0 and ∞), (2) imagined that the function was reflected around x = 0 such that it was symmetrical, so multiplied the indefinite integral by 2, and (3) then divided the entire function by the result. The resulting new, rescaled, function has an area of 1 under the curve between $-\infty$ and ∞.

[10] This, in turn, is a particular version of the more general Hassell–Comins growth model (Hassell and Comins, 1976).

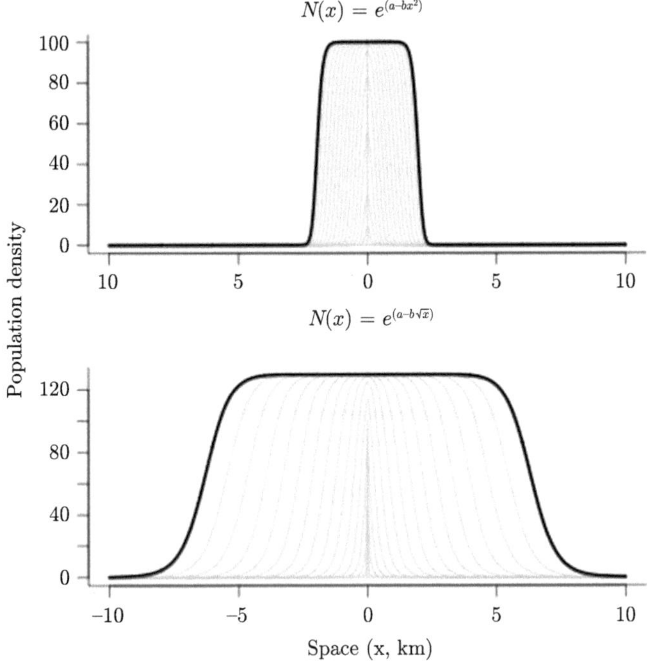

Figure 2.6 *Spread of* Drosophila *over fifteen generations under two alternative descriptions of dispersal. The top panel shows results from a dispersal kernel with exponential decay, and the bottom panel results from a fat-tailed kernel (with less than exponential decay).*

where R_0 is the proportional growth rate of the population, and K is the carrying capacity. *Drosophila*, no doubt, have their particular values for these parameters, but in what follows it doesn't matter too much what we choose: we are interested in how changing the dispersal kernel affects spread, so we will leave the growth function constant in our comparison. In what follows, we will set $R_0 = 5$, and $K = 100$.

Keeping $G()$ identical then, let's now plug our two different kernel functions into equation (2.4) and crank the handle (i.e., iterate the function) over fifteen generations. The results are shown in Figure 2.6. We can see immediately that our description of dispersal matters a lot. When dispersal has an exponentially bounded tail, the invasion soon settles into a constant rate of spread. When dispersal has a fat tail, however, this does not happen. When we use fat-tailed kernels to describe dispersal, spread rate accelerates over time.

2.5.5 Long-distance dispersal, light speed, and statistical structure

So basic ecological theory has a way to understand invasions that accelerate. It does this by describing dispersal in a way that admits occasional long-distance dispersal. Using the reaction–diffusion framework, we might build a model that has stratified dispersal (i.e., more than one diffusion coefficient), which can cause populations to exhibit jumps and coalescence, resulting in accelerating spread. Another method is to use the integrodifference model—which allows us to admit dispersal kernels with fat tails—which also generates invasions that accelerate over time. Whether we choose a stratified reaction–diffusion model, or an integrodifference model with fat-tailed kernels, we are using a model that allows us to mathematically describe long-distance dispersal. And long-distance dispersal can have dramatic effects on invasion speed.

While integrodifference models are nice, one criticism that might be levelled at them is that they allow invasions that accelerate forever. With a sufficiently fat-tailed kernel, it seems, it is possible for invasions to reach relativistic speeds and ultimately breach the speed of light. We haven't witnessed such an invasion yet; perhaps because it was moving too fast for us to see. But jokes aside, it would be nice to have a model in which invasion speed is naturally bounded.

In this quest, it is sensible to go back and think about our stratified dispersal model. Here, we had a model in which individuals dispersed according to one of two dispersal distributions. Some proportion, p, of the population dispersed according to a normal distribution described by diffusion coefficient, D_1, $N_1(x)$; the rest of the population $(1 - p)$ dispersed according to the normal distribution described by D_2, $N_2(x)$ (note that N here is now referring to the normal distribution, not population size). This model would be a good way to describe dispersal in our window-walking Cecaks, and can cause accelerating invasions.

What would the dispersal kernel look like in this model? Well, it would simply be the weighted average of our two normal distributions: $k(x) = pN_1(x) + (1 - p)N_2(x)$. Figure 2.7 gives us a sense of how this kernel might look. We can see that adding the possibility of occasional long-distance dispersal has made the base of the kernel's tail fatter than it had been. Ultimately, however, this kernel, built from two normal distributions, has an exponentially bounded tail. While we might see acceleration from $v_1 = 2\sqrt{rD_1}$, we can expect our invasion to maintain a speed at or below that set by the larger diffusion coefficient, $v_2 = 2\sqrt{rD_2}$.

In 2009, a pair of mathematical biologists, Petrovskii and Morozov, generalized and formalized this idea (Petrovskii and Morozov, 2009). They imagined that, instead of a kernel resulting from a mix of two values for D, we could actually have an infinite number of values for D. In fact, there could be a distribution of D in a population. This could arise for many reasons: variation between individuals in traits (some have larger wings than others), or in environments (some inhabit removalist boxes, others shoe boxes). Whatever its genesis, this variation in D, this statistical structure in the population, would generate an interesting population-level dispersal kernel.

Formally, they said, let's introduce a function that describes the distribution of D in the population. Let's call that function $\phi(D)$. So, just as we can have a distribution of dispersal distances that depends upon the diffusion coefficient, $k(x, D)$, we can also have a distribution of diffusion coefficients. We have seen in Appendix A that diffusion always results in a normal distribution of dispersal distances. If we examine this distribution at $t = 1$, it can be captured as follows:

$$k(x) = \frac{1}{\sqrt{4\pi D}} e^{-\frac{x^2}{4D}}. \tag{2.9}$$

We have learned that this distribution has 'exponentially bounded' tails, and that such tails generate invasions that travel at constant speeds.

Petrovskii and Morozov showed that if we introduce a distribution of values for D, $\phi(D)$, then we end up with kernels that are fat-tailed relative to a normal distribution. Formally, they defined:

$$k(x) = \frac{1}{\sqrt{4\pi}} \int \frac{1}{\sqrt{D}} e^{-\frac{x^2}{4D}} \phi(D) dD. \tag{2.10}$$

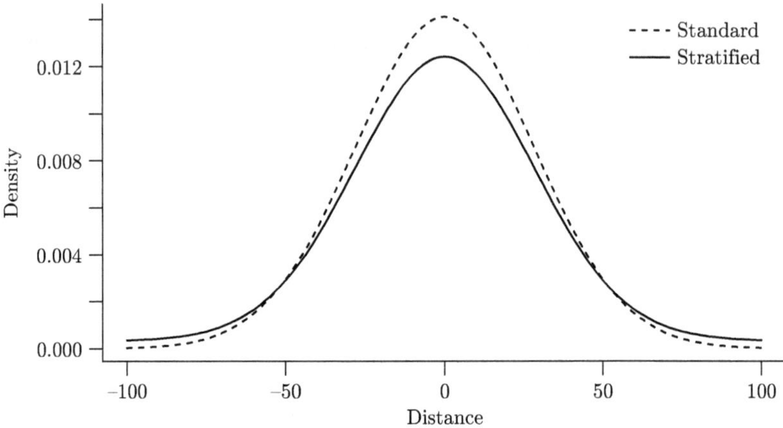

Figure 2.7 *A standard normal dispersal kernel, and a stratified kernel in which we add a small proportion of individuals with higher dispersal rates.*

Solving this expression for various functional forms of $\phi(D)$, they showed that sensible functions for $\phi(D)$ generally caused tails that are fatter than we would observe with a normal distribution. The implication is that within-population variation in D may well yield invasions that naturally accelerate. That is, if individuals in a population vary in their dispersal propensity, then we can expect dispersal distributions with fat tails.

As with our earlier example in which we considered only two values of D, however, they argue that these fat tails must be exponentially bounded at some point, because there must be some natural upper limit to D in any given population (toads can't fly; *Drosophila* don't have rocket boots). The implication of their argument is that by admitting variation between individuals (but some natural bound to that variation), we can naturally have invasions that accelerate but eventually asymptote at some constant speed.

2.6   The wrap

In this chapter, we have discovered that, given the right conditions, invasions can just happen, in much the same way that a chemical reaction can just happen. We have learned that invasions happen as a consequence of two processes: population growth and dispersal.

We were also introduced to an elegant mathematical model—the reaction–diffusion or the Fisher– KPP model—that has become a bedrock piece of theory for understanding invasions. Under the Fisher–KPP model, following an initial introduction, an invasion wave rapidly develops, taking on a constant shape and a constant invasion speed. This model, developed to describe the spread of an advantageous allele through a population, was first brought to ecology by Skellam in 1951. Skellam pointed out that the Fisher–KPP model could also describe the spread of invasive plants and animals.

While the Fisher–KPP model predicts constant invasion speed, it is clear from Chapter 1 that not all invasions behave this way; many invasions accelerate over time. We then looked at theoretical extensions that admit the possibility of an invasion that accelerates. Acceleration, it appears, emerges when we allow rare, long-distance dispersals: an outcome that emerges naturally when we allow variation in dispersal ability within a population.

There has been more than seventy years of theoretical development around the ecological dynamics of invasions. We have here skimmed very lightly across its surface. For those who would like to delve deeper, there are several useful resources. The first is a classic: Shigesada and Kawasaki (1997), *Biological Invasions: Theory and Practice*, which remains a beautiful, concise, and relevant exposition of the reaction–diffusion theory and its extensions. Deeper still, and broader than reaction diffusion theory, is Lewis et al. (2016), *The Mathematics Behind Biological Invasions*. We touch back to both of these books from time to time in the coming chapters. Finally, for the ecologist looking for a broad overview of the ecological theory around invasive species and their management, Hui and Richardson (2017), *Invasion Dynamics* is the latest word.

Evolution on invasion fronts

3

Like death and taxes, variation between individuals is inevitable. Even sheep—those paragons of homogeneity—vary. Individual sheep are considered so indistinguishable that the sheer monotony of counting them eventually bores the brain into unconsciousness. But anyone who has ever looked at sheep carefully—anyone who has ever measured a set of sheep—will know that sheep vary. They vary in size, growth rate, personality, wool yield, and so on. Indeed, choose any biological population, look at it carefully enough, and you will find variation.

Often, this variation is heritable. Sheep with high wool yield tend to have offspring with high wool yield; sheep that grow fast will tend to have offspring that grow fast. This is not to say that genes are destiny, of course; the environment also plays a powerful role. Sheep on good pasture will always grow faster than those forced to eat dust and stubble. But underlying all this variation caused by a fickle environment, there is almost always some variation caused by genetic differences between individuals. This heritable variation is the raw material of evolution.

In recent years, it has become apparent that evolution has a powerful role to play in invasion dynamics. Conditions on the invasion front impose strong selection on the very traits responsible for invasion speed. And because heritable variation is nearly ubiquitous in populations, those traits rapidly evolve, changing the speed of the invasion as they do so. In Chapter 2 we saw that, in the simplest case, the speed of an invasion is determined by the rates of population growth, r, and dispersal, D. These innocuous pronumerals, r and D, are just numbers, parameters of a model, but each is also a masterpiece of minimalism. The parameter r compresses the various lives of many complex individuals into a single number; D attempts the same compression with the various journeys of these individuals. This extreme minimalism is the essence of theoretical work; it is a powerful strategy for understanding how a system works, but we need to remember that in the real world, there is variation in the traits determining r and D, and so these parameters also have variation, and they might evolve.

Towards the end of Chapter 2, we came across models that allowed variation in dispersal kernels within the population. But of course, admitting the possibility of such variation without admitting the possibility of evolution is unsatisfying. It's a half-baked cake. Evolution requires two things: heritable variation, and selection. Natural selection occurs whenever variation in a trait is correlated with fitness (i.e., reproduction or survival). If the ewes in our flock of sheep found green coloured rams very attractive, and we painted 10% of our rams green, then we might see these green rams father not 10%, but 90% of next spring's lambs. Colouration would be correlated with reproductive fitness, and natural selection has favoured green rams. But none of our spring lambs will be green.

The Ecology and Evolution of Invasive Populations. Ben Phillips, Oxford University Press. © Ben Phillips (2025).
DOI: 10.1093/9780191924910.003.0003

We generated variation (by painting our sheep) and saw selection—a correlation between a trait and fitness—but the trait under selection was not heritable, so evolution didn't happen. Spread models that admit individual variation in dispersal parameters but that do not treat that variation as heritable do not allow evolution, despite the presence of selection. They are models that paint the sheep green.

You will recall from Section 1.1 that in 2006 I stumbled across a striking example of an accelerating invasion—cane toads in Australia—and, being wholly naive to invasion theory (Chapter 2), I simply assumed that the toads must have evolved increased dispersal. To test this idea, I collected toads from across their range, from where they were first introduced on the east coast to the oncoming invasion front in the Northern Territory. I brought all these toads back to the Northern Territory, attached radio transmitters to them, released them, and painstakingly followed each of them around, recording where they had moved each night. What I found was a very clear shift in movement behaviour. Toads collected from the invasion front had much higher movement rates than those collected close to where they were introduced (Figure 3.1). This is not definitive proof of an evolutionary shift, but it certainly supports the idea! So if the invasion had accelerated, and toads on the front seemed to have evolved greater dispersal rates, the million dollar question was why. What selective force was favouring increased dispersal?

3.1 Spatial sorting on invasion fronts

Let's imagine that we introduce a population to a vast, homogeneous, and favourable new environment. In fact, let's abandon sheep, cats, toads, and plants for a while and ponder bacteria again. Let's introduce a small population of bacteria to the centre of a Petri dish,[1] much as we saw in Figure 2.1. Now, these particular bacteria are motile: they have propellers (flagella, which are little whippy tails) that they spin around to propel themselves through the water. Our bacteria, like all populations, exhibit trait variation: half of our bacteria have one propeller, and half have two. I'm going to call these two bacterial phenotypes B_1 and B_2.[2] You will not be surprised to learn that B_2 types, having two propellers, swim faster than B_1 types.

Anyway, we introduce this population into the centre of our dish. Our bacteria move around for a while (before reproducing), and so the area occupied by our bacteria grows. Let's freeze time right here, before they get a chance to reproduce. Even before reproduction, the population has expanded its range, purely by the movement of individuals. Individuals have been moving around randomly, but some individuals just happen to have swum further into unoccupied space than others. These individuals now make up the new edge of the population. But we know that B_2 bacteria move faster than B_1, so while both types are equally likely to have swum into unoccupied space, B_2s will have swum farther out. The new population edge must be made up almost entirely of fast-moving B_2 bacteria. Indeed it was the dispersal speed of these B_2 bacteria that

[1] This is a vast, homogeneous, and favourable new environment if you're a bacterium.

[2] 'Phenotype' just means a trait variant: so in this case we have a trait—number of flagella—and two phenotypes for that trait: the one- and two-tailed phenotypes.

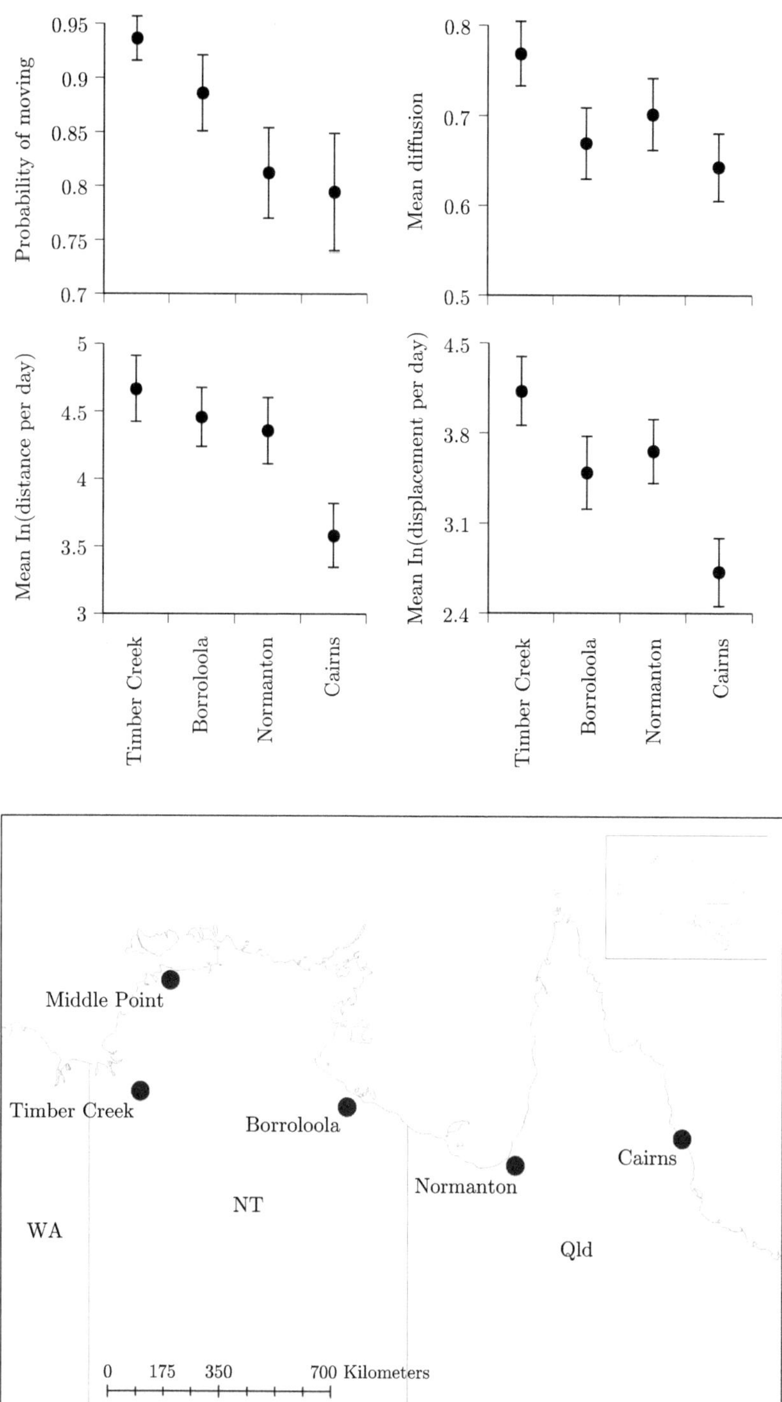

Figure 3.1 *Cane toads from the invasion front (near Timber Creek) show markedly higher rates of movement than toads closer to the introduction point (near Cairns). From Phillips et al. (2008).*

determined how far the edge of the population moved during our short period of dispersal.

We can see that the process of dispersal has sorted our bacterial phenotypes across space. The edge of the population is now dominated by the fast-moving phenotype. If we took a slice through our Petri dish and plotted the frequency of each bacterial type, we would expect to see something like the top panel of Figure 3.2. And when we calculate the proportion of the population that is made up of B_2 phenotypes at each point in space, we see, as expected, that the faster phenotype is dominating the new edge of the population (lower panel of Figure 3.2). The process of dispersal into unoccupied territory has sorted our bacterial types across space.

This spatial sorting of phenotypes is a completely natural consequence of the variation in dispersal rates. As with evolution, invasions, and chemical reactions, it is something that just happens. It happens whenever we have variation in dispersal, and unoccupied space. Now, if this variation in dispersal is heritable (and we know it is in this case: bacteria with two propellers divide into daughter cells, each of which has two propellers of its own), then the offspring of individuals on the edge of the population will have higher mean rates of dispersal than the initial population that we introduced. So if we look only at the edge of the population, it looks very much like there has been an evolutionary shift towards higher dispersal.

If we now repeat this cycle of dispersal and reproduction, it is straightforward to see that there will be a new population edge, and it will tend to be composed of the fastest phenotypes from the old population edge, and so on. We can see that we have unearthed a runaway evolutionary process: if we follow the invasion

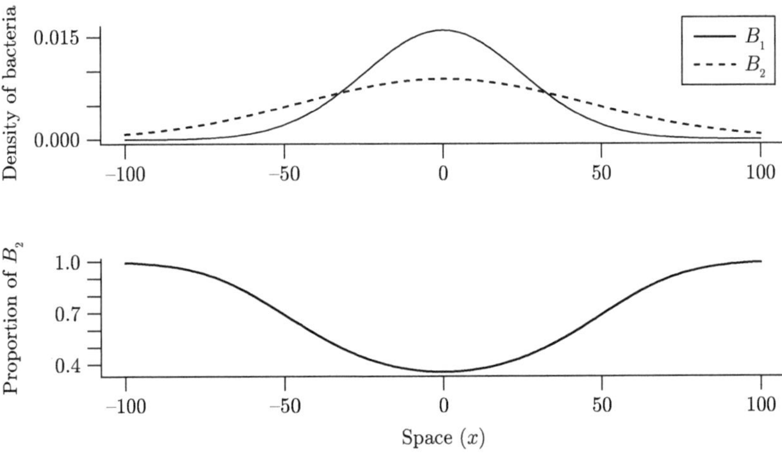

Figure 3.2 *The hypothetical frequency of slow and fast-moving bacterial types* (*B_1 and B_2*) *after a short period of dispersal. The top panel shows the expected density through space assuming each bacterial type has its own diffusion coefficient* (*D_1 and D_2*). *The lower panel shows how the proportion of the B_2 type changes across space.*

front as it expands through space, we will see a steady increase in the proportion of highly dispersive B_2 bacteria in that population.

Indeed that is precisely what happens if, instead of doing thought experiments, we actually do the experiment. In 2013, a team of medical researchers did this experiment (van Ditmarsch et al., 2013). What they saw astonished them. Every time they repeated this experiment starting from an ancestral strain, their bacterial populations on the invasion front evolved additional flagellae. These shifts in dispersal phenotype were due to changes at a single gene. Often different mutations in different strains, but always the same gene affected. Every bacterial strain subject to repeated invasions across a Petri dish evolved into a 'hyperswarmer' on the invasion front (Figure 3.3).

3.1.1 Spatial sorting is a very general mechanism

I've just explained the idea of spatial sorting using bacteria in a Petri dish as an example. But of course, the same mechanism could explain the evolution of dispersal rate in toads. Toads don't move using flagella, they hop, but we just need to imagine that some individuals have aspects of behaviour, physiology, or morphology that lead them to have higher dispersal rates (more propellers) than others. I had been thinking about this idea for a few years and writing a paper about it in 2007—the Olympic games were coming up in China—and so I was naturally enough thinking about this variation as variation in athletic performance. It is the elite athletes of the toad world that end up on the invasion front each generation front and then, the mating game being what it is, they mate with the other athletic toads nearby. Tongue in cheek, and because this mechanism for the evolution of dispersal on invasion fronts didn't seem to have a name yet, I christened it the Olympic village effect (Phillips et al., 2008). It was only some years later, when it became clear that the Olympic village effect was more than just a fun footnote, that it was decided to give it a more serious-sounding name: spatial sorting (Shine et al., 2011).

The reason that spatial sorting needed a more serious name was that it is a general and underappreciated evolutionary force. Spatial sorting is a form of natural selection, but it is not natural selection as we usually envisage it. Classical natural selection is a process whereby some individuals possess traits that cause them to send relatively more offspring into an unoccupied future. Spatial sorting is a process whereby some individuals possess traits that cause them to send relatively more offspring into an unoccupied space. Classical natural selection simply happens when there is trait-based variation in survival and reproduction (the future is naturally unoccupied), and spatial sorting happens when there is trait-based variation in dispersal (and unoccupied space). Spatial sorting is a spatial analogue of classical natural selection, and it is made very apparent on invasion fronts.

Although it lacked a clear moniker in 2007, the idea of spatial sorting had been around for longer than I initially knew. Spatial sorting was first proposed in 1987, in a verbal model by a pair of scientists attempting to explain patterns they were seeing in the seed morphology of lodgepole pines (Cwynar and MacDonald, 1987).[3] Pine seeds are wind-dispersed and have evolved a single torsioned

3 'If biologists, like economists, tended to name ideas after people, we should rightly now be talking about the Macdonald–Cwynar effect, which has quite a nice ring to it, actually.

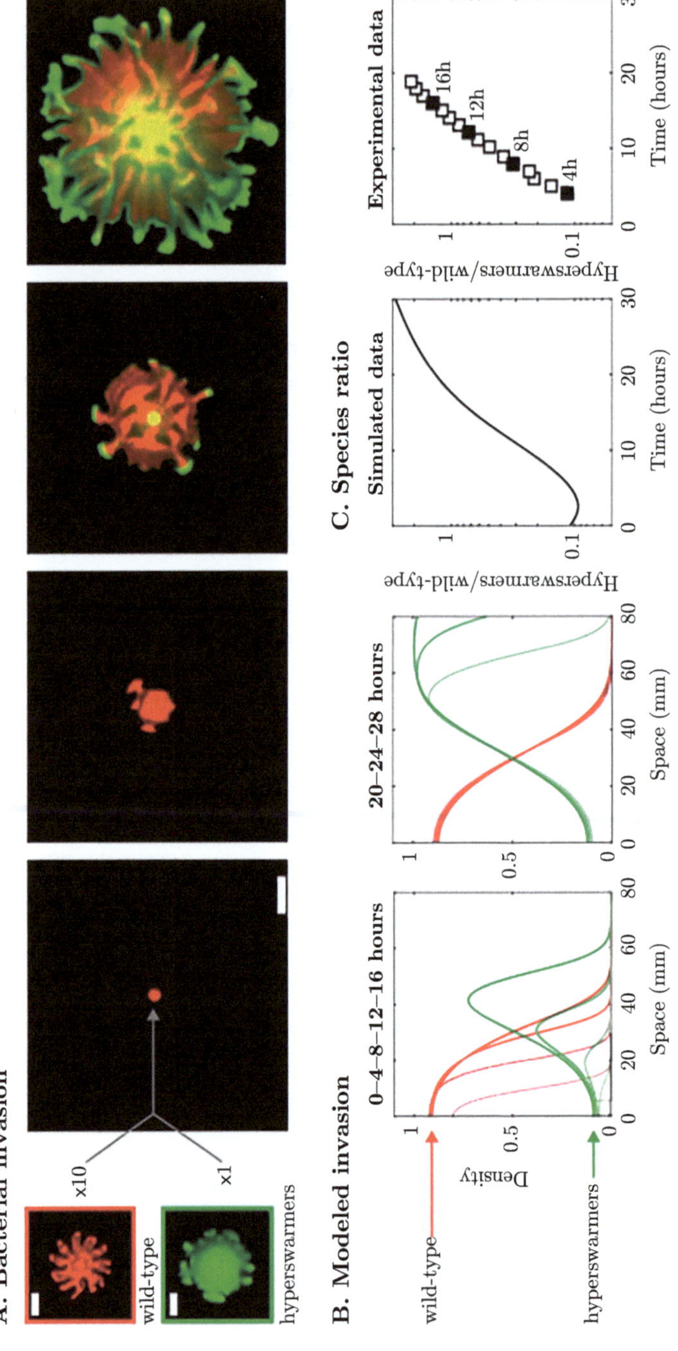

Figure 3.3 *When bacteria are allowed to spread across a Petri dish, we can watch the evolution of dispersal play out. This figure shows results from an experiment in which a mixed population of wild type and hyperswarmers is introduced into a Petri dish. Hyperswarmers (light green) rapidly come to dominate the invasion front. From Deforet, et al. 2019.*

blade—like a single helicopter blade—that causes the seed to spin like a helicopter as it falls. This motion slows the fall of the seed, causing the wind to disperse the seed further from its parent than it would otherwise. Cwynar and Macdonald had noticed that seeds from the northern part of the lodgepole pine's range had larger blades (relative to their mass) than seeds from the southern part of the range. Clearly, this was a shift towards seeds that were more effective dispersers. They also knew that lodgepole pines had been expanding northwards for the last 10,000 years, following the retreat of the glaciers after the last ice age. They very clearly imagined the mechanism we now call spatial sorting:

> We propose that some of the observed geographical variation ... between central and marginal populations, may arise from the process of postglacial migration of a new species across the landscape. ... Each founding requires a dispersal event. The seeds most likely to found a new population at a distance are those possessing morphological features that promote dispersal. We might expect that the repeated long-distance founding of the migration process would result in directional selection for the ability to disperse.

So, the idea was first dreamt up to explain shifts in seed morphology in northward-migrating trees. Ten years later, researchers working on climate-change-driven shifts in insect distribution found that crickets shifting northward under climate change had larger-winged morphs in northern populations (Simmons and Thomas, 2004). The mechanism was dreamt up again to explain accelerating spread rates in toads (Phillips et al., 2006, 2008), and medical researchers pondering the formation of bacterial biofilms found evidence for it again, in 2013 (van Ditmarsch et al., 2013). This is an impressive array of systems: plants, insects, toads, bacteria, and indeed there are now many more documented cases (Phillips et al. (2010a); Chuang and Peterson (2016); see Table C.1 in Appendix C).

An impressive breadth of systems had shown evidence that dispersal evolves on invasion fronts. We had what sounded like a sensible mechanism for explaining that pattern—spatial sorting—but we didn't yet have a clear experimental demonstration that spatial sorting (and not some other mechanism) was driving these patterns. That changed in 2016 when Jennifer Williams published the first experiment involving replicate invasions that attempted to manipulate the evolutionary process (Williams et al., 2016a). Her experiment involved replicate invasions of the fruit fly of the plant world—*Arabidopsis thaliana*—and had treatments in which invasions were allowed to progress either *au naturel* or with spatial sorting foiled. To foil spatial sorting, she simply replaced newly germinating plants with plants drawn at random from the population's initial seed pool. Her experiment, run over six generations, showed clear increases in spread distance in the replicates in which evolution was allowed. It was a neat experiment, and then in 2017 two research groups independently ran nearly identical experiments to test spatial sorting; but this time using beetles rather than plants. Again, these experiments ran treatments in which spatial sorting was naturally allowed to happen, and other treatments where it was undermined; we have already met the system: beetles spreading through arrays of Petri dishes (Ochocki and Miller (2017); Weiss-Lehman et al. (2017) section 1.2.3).

We have already seen data on replicate invasions of these beetles through their arrays of bean-filled Petri dishes (Figure 1.6). These data came from the control treatments, the invasions that were allowed to proceed with no manipulation. Alongside these control treatments, the researchers ran an additional treatment in which beetles were allowed to disperse but were then re-sorted before being allowed to breed. In these manipulated replicates, spatial sorting occurred, naturally, but then was undone by collecting all beetles in the array and randomly assigning beetles back to each patch up to the number of beetles that had been in that patch. The result of one of these experiments can be seen in Figure 3.4.

Even though these experiments proceeded for only 10 generations,[4] the results clearly show that mean spread distance is higher in the invasions where spatial sorting was allowed to proceed. The results of both the beetle and plant experiments also showed effects on the variability of invasion speed; a result we will unpack in detail in Chapter 5. These three experiments cleverly show that

[4] Ten generations is a short time when you are thinking about experimental evolution; it is a very long time when you are running a lab experiment that involves counting thousands of beetles every day for the best part of a year.

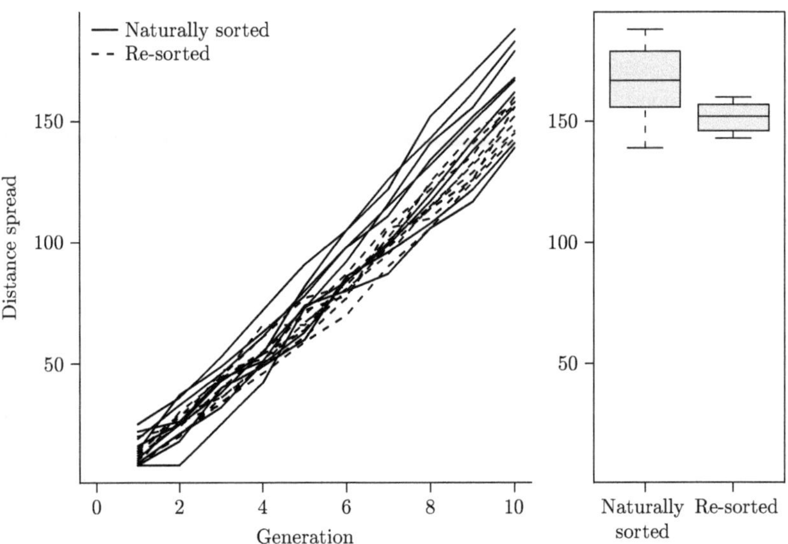

Figure 3.4 *Spread distance over time for replicate invasions of bean beetles. One half of the replicates had spatial sorting disabled, by enforcing random assortments of beetles across patches. The right-hand panel shows variation in distance spread in each treatment after 10 generations. Data from Ochocki and Miller (2017).*

when the invasion front is randomly assorted every generation, it tends to travel slower than when natural assortment is allowed to proceed. These experiments have been joined by several others in a range of organisms (see Table D.1 in Appendix D). While the experiments were directed at testing spatial sorting, it is important to realize that there are other evolutionary processes at play that might also be undermined by this re-assortment manipulation. For now though, it seems very clear that dispersal traits evolve on invasion fronts, and spatial sorting is a natural mechanism that would drive that shift.

3.2   Natural selection on invasion fronts

We have seen that there is a natural process occurring on invasion fronts that can drive the evolution of dispersal. But we also know that invasion speed is controlled not just by the rate of dispersal, but also by the rate of population growth. r matters as much as D. Is it possible that growth rate might evolve also on invasion fronts? The answer is yes, but again, the interesting question is, actually, why?

We have seen that invasion fronts take on a characteristic shape as they propagate through space (see Figure 2.3). Essentially, an invasion front is just a density cline that moves through space. At any moment in time, the leading edge of the population is at a very low density, and as we travel back towards the core of the population, density increases. If we freeze time for a moment and consider this density cline, it doesn't take long to realize that conditions on the leading edge of the invasion are quite different to those in the core of the population. Individuals on the leading edge experience a low density of conspecifics. In these low-density conditions, competition for resources is not really an issue. In such an environment, the main limitation individuals might experience is not their rate of acquiring resources, but the rate at which they can convert abundant resources into offspring. In the core of the range, on the other hand, competition with conspecifics is likely a stronger force. Here it is not resource conversion so much as resource acquisition that might limit the reproductive output of individuals.

This is a cartoon sketch, but it makes the point that very different traits might be favoured in a high-density environment than in a low-density one. It is a cartoon sketch, but it is also a venerable idea. It was developed with particular reference to invasive species by Lewontin, in 1965. Lewontin wrote:

> The absence of population pressures is the mark of a colonizing episode and the population can be assumed to be growing exponentially with unlimited resources, at least for a short time. ... The second aspect of colonization ... is the change in direction of selective forces that may accompany a colonizing episode. A newly formed colony is at first in low density and the individuals are in competition with individuals of

other species, as opposed to the situation in older populations where any competition is primarily with members of the same species.

A 'change in the direction of selective forces' is almost certainly afoot as we move along the density cline of an invasion front. While density regulation may be in play in the core of the range, at the very leading edge the population is effectively growing exponentially (density is low and competition can be considered negligible). And in an exponentially growing population, any trait that increases r will be favoured. Thus, the leading edge of an invasion front is an environment where traits that increase r are favoured. I develop these ideas more formally in Section 3.3.

In 1967, this idea made its way into Macarthur and Wilson's classic, *The Theory of Island Biogeography*, where it became known as r/K selection (MacArthur and Wilson, 1967). Neither Lewontin, nor Macarthur and Wilson were thinking of invasion fronts, necessarily, but they were thinking about colonization events, and the population dynamics following colonization. All argued that we would expect selection to favour different sets of traits depending upon the degree to which the population experiences conspecific competition. Both made the point that at very low density (or in a non-density-regulated population), the only thing that matters to fitness is an individual's reproductive rate; the speed with which it can make new copies of itself. The term r/K selection referenced the logistic population model (the same growth function we saw in the Fisher–KPP model in Section 2.1) and the fact that, under this model, when a population is at or near carrying capacity, variation in r between individuals makes no difference to how many offspring they leave behind (Roff, 1993).[5]

Nowadays, the term r/K selection has largely fallen into disrepute. There are several reasons for this, but primarily it is because in most real systems, complex life histories do not easily boil down to neat, mutually exclusive sets of traits that affect either r or K (Pianka, 1970; Reznick et al., 2002).[6] There was also a tendency among ecologists to label species as either 'r-selected' or 'K-selected' (when most species in a dynamic world would experience both high- and low-density conditions[7]), and also a growing acceptance among ecologists that not all populations are density regulated. Although the term has fallen into disrepute, we should not make the mistake of throwing the baby out with the bathwater; r/K selection remains a useful mental model; it was conceived in direct reference to colonizing populations, and there is very little doubt that selection pressures are different between density-regulated and non-density-regulated environments.

Once we appreciate that there is potentially a difference in selection pressure across the invasion front, we then need to realize that the population on the leading edge of the invasion is where next generation's leading edge population is going to be drawn from (ignoring Chapter 6 for now). The density cline moves through space, and the population closest to the new leading edge is the population of the previous leading edge. When we see things this way, we notice that there is a population on the leading edge that is consistently in a low-density,

---

[5] This can be easily demonstrated by simply setting $N = K$ in the logistic growth equation. When you do this, it makes no difference whether $r = 1$ or $r = 100$—the growth of the population, $\frac{dN}{dt}$, is unchanged. Instead, variation in K between individuals has the greater bearing on fitness.

[6] This was something Lewontin knew.

[7] Something else that Lewontin knew.

exponentially growing state. This population is not just growing exponentially 'for a short time' (as Lewontin put it) but potentially for as long as the invasion is propagating through space.

The above considerations inspired me to measure changes in growth rates along the toads' invasion history. Was there evidence that population growth rates might also have been evolving? It wasn't logistically feasible to measure the full life cycle of so many individuals, but it was feasible to measure juvenile growth rates. All else being equal, individuals who grow to maturity faster will have higher reproductive rates than those who grow slower (Cole, 1954).[8] When I looked at juvenile growth rates in toads, I found a large effect: juvenile toads on the invasion front grew about 30% faster than those from the core of the range (Phillips, 2009).

Another surprising example supporting this idea came from parish records to the east of Québec City. The European colonization of this region began in 1608, and the local parish kept a detailed record of births, deaths, and marriages across the region all the way through to the 1960s. Examining these data, Claudia Moreau and colleagues (2011) were able to reconstruct a clear pattern of spatial spread. By digging into the genealogies, they were also able to show that women on the invasion front had, on average, a 20% larger effective family size than women in the range core. While some of this may be due to simple density release, they also show that family size is heritable on the invasion front, implying that at least some of this shift in reproductive rate could be a result of the r-selection we expect to occur in this setting.

Interestingly, while evidence for the evolution of increased dispersal on invasion fronts has been nearly uniformly supportive, evidence for increased growth rates has been decidedly more mixed (Chuang and Peterson, 2016). I found evidence supporting the idea in toads, but patterns both consistent and inconsistent with the idea have been found since (see a range of examples in Appendices C and D). The reasons for this are, I think, at least fourfold. First, not all populations are density regulated. Many populations are instead regulated by non-density-dependent factors; weather, for example. If there is no density-dependent regulation, then there is no 'change in direction of selective forces' across the invasion front. In such situations, we might not expect growth rates to diverge between front and core. Second, it is very difficult to measure all the individual traits determining r. r is a masterpiece of minimalism and there are numerous measurable traits—juvenile growth, survival, fecundity, maturation time, and so on—that determine an individual's reproductive rate. Some of the studies failing to find an effect may have measured the wrong traits, or not fully described how their particular trait relates to growth rate. I mentioned that I found evidence for increased juvenile growth rate in invasion front toads (Phillips, 2009), but then a few years later we found hugely reduced clutch number in females from the invasion front (Hudson et al., 2015). Invasion front toads might mature faster, but they also make a lot fewer eggs. What is the sum total of these two effects on r? We still don't really know.

[8] The essential reason for this is that reproductive rate is a rate: it is a number of (new) individuals per unit time, *individuals/time*. If you make the denominator, *time*, smaller by maturing earlier, the growth rate increases.

The third and fourth reasons why growth rates might not increase on invasion fronts are more complex, and we will get to these fully in Chapters 6 and 7. In the meantime, it will be useful to step through a simple model of how r and D might evolve on an invasion front. This will consolidate our ideas about the evolutionary forces on invasion fronts, but also give us a prelude into understanding why evolutionary processes might not favour increased growth rates on invasion fronts.

3.3 Spatial and temporal fitnesses

The previous two sections have pointed out that there are evolutionary forces at play on the invasion front that favour increased dispersal and increased growth rate. Dispersal and growth are the same parameters that determine the speed of an invasion (Chapter 2), so invasion front populations appear to be under selection for traits that will cause the invasion to accelerate. Empirically, however, we sometimes see growth rate decline on an invasion front; an effect also seen in some models (e.g., Bartoń et al., 2012). What is going on? In this section, I argue that the evolutionary process on an invasion front can be understood in terms of natural selection, but to do so we need to accept that individuals have both temporal and spatial aspects of fitness. Temporal fitness is classic fitness—the balance of births and deaths. It is the rate at which new individuals are pushed forward in time. Spatial fitness is dispersal—the balance of comings and goings. It is the rate at which new individuals are pushed forward in space. When we see the world in this way, we can re-derive a classic model of natural selection as it applies on invasion fronts. Doing this shows that on an invasion front, evolution is optimizing the product of temporal and spatial fitnesses.

Let us take a moment to imagine another pared-down world: a world in which time and space are discrete. In this world, there is again only one dimension of space, but this space is no longer continuous, it is broken into evenly spaced points—stepping stones—upon which our organism lives. Theoreticians call this a discrete space, as opposed to the continuous spaces we have imagined so far. Again, let's denote locations in space (individual stepping stones) as x. Because space is discrete, x can only take integer values: 0, 1, 2, and so on. We will also treat time as discrete. Time is no longer continuous, but proceeds forward in discrete steps. Again, let's denote particular time points by t. Because time is discrete, t can only take integer values: 0, 1, 2, and so on.

We imagine introducing a large population to $x = 0$. Events in our pared-down world happen in this order: births, deaths, dispersal. Within one time interval, we only allow dispersal between neighbouring patches. Because we are only allowing nearest-neighbour dispersal, our population can only invade space one stepping stone at a time. This is a lot of rules to specify in our imaginary world, but imposing these rules is helpful, because it allows us to know where the invasion front is at any time. If time starts at zero, at any given time, t, the invasion front will be at location $x = t$ (see Figure 3.5). Let's track this invasion front population through space and time, and see how it evolves.

To see how the population evolves, we need to give it some heritable variation. Let's imagine each individual contains only a single gene of interest (a 'locus'), and let the locus have two possible variants ('alleles'). We will denote these two alleles as a and A. At any given time and space, our population has a number of individuals, $n_a(t, x)$, that carry allele a, and a number of individuals, $n_A(t, x)$, that carry allele A. Let's ignore the A individuals for a moment while we work out the dynamics. First, births:

$$n'_a(t, x) = (1 + b_a)n_a(t, x).$$

This says that each a type individual gives birth to b_a offspring. So the total number of individuals (parents plus offspring) is now $(1 + b_a)n_a(t, x)$. We denote this result with a prime, $n'_a(t, x)$ because time hasn't moved forward yet; we are still within an interval of time. Next, deaths:

$$n''_a(t, x) = (1 - d_a)n'_a(t, x) \tag{3.1}$$

$$= (1 + b_a)(1 - d_a)n_a(t, x) \tag{3.2}$$

$$= W_a n_a(t, x). \tag{3.3}$$

A few things have happened here. First, we said that some proportion, d_a, of our individuals die, so we have $1 - d_a$ left of the n'_a individuals we had. We then substitute our earlier value for n'_a, and we find that we have $(1 + b_a)(1 - d_a)n_a(t, x)$ individuals. This term, $(1 + b_a)(1 - d_a)$, is the per-capita balance of births and deaths for the a type. It is the absolute fitness of that type. To simplify our notation, we just denote this absolute fitness of the a type as W_a. We can just remember

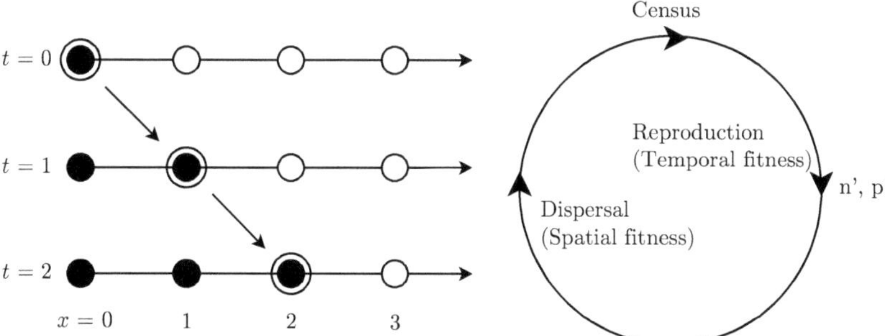

Figure 3.5 *A population on a discrete one-dimensional lattice (left-hand panel). When dispersal is restricted to occurring only between nearest neighbours, the invasion front at time t will be at t = x. The right-hand panel shows the life cycle of our theoretical organism. From Phillips and Perkins (2019).*

that $W_a = (1 + b_a)(1 - d_a)$ move on with our lives, and save ourselves a whole bunch of pronumerals.

At this stage, time has still not moved forward. We are still within an interval of time, but births and deaths have happened. Our last process is dispersal. Here is where we remember that we are only interested in tracking the invasion front population at $x = t$. We now let some proportion, V_a, of our individuals disperse from x to $x + 1$. Because $x + 1$ is currently empty, these are all the individuals that will accrue to that location in this time interval. And with dispersal, we have completed a time interval, so we move from t to $t + 1$:

$$n_a(t + 1, x + 1) = V_a W_a n_a(t, x = t) \tag{3.4}$$

$$n_A(t + 1, x + 1) = V_A W_A n_A(t, x = t). \tag{3.5}$$

I have included the equation for n_A individuals also, because it follows the same logic. Things have been complicated for a few moments, but we now have very simple equations that describe how the number of individuals in the furthermost patch changes over time. Theoreticians call equations like these 'recursion equations', because once you have calculated the value for the first time step, you can feed that result back into the right-hand side of the equation and get the value for the next time step, and so on. In this way, a recursion equation can describe the entire dynamics of a system, one step at a time.

We will take a quick break now from the math and note a few important points. First, this is not a density-regulated population. The balance of births and deaths in no way depends upon the density of individuals. It depends only upon the genotype of the individuals: each a individual will give rise to W_a individuals of that type in the next generation; each A individual will give rise to W_A individuals of that type in the next generation. So this model emulates the notions described in Section 3.2: an exponentially growing population on the leading edge of the invasion.

The next thing to point out is that W_i denotes the absolute fitness of allele i. I didn't choose the pronumeral W at random: this pronumeral is used widely in classic population genetics to denote the balance of births and deaths. In classic population genetics, this balance of births and deaths is often simply called 'fitness', and the mean fitness of a population, $\bar{W}$, is conceptually the same as its intrinsic growth rate, r: if we shrunk our time intervals to be infinitesimally small—that is, we morphed into a continuous-time model—$r = \ln \bar{W}$. Here, it will be useful for us to call W 'temporal fitness', and V 'spatial fitness'. W_i controls the number of individuals of type i that are sent forward in time; V_i controls the number of individuals of type i that are sent forward in space. Simply by looking at these equations, we can see that if, for example, $V_A W_A$ was much larger than $V_a W_a$, then relatively more of the A type would accrue to the leading edge of the invasion. We can probably see that this must be true, but let's show it.

Let's first define a new variable, $p(t, x) = \frac{n_A(t,x)}{n_A(t,x) + n_a(t,x)}$, which is the proportion of the population composed of individuals of the A type. Because there are only two types in our population, if we know p then we also know the proportion of a types: $q(t, x) = 1 - p(t, x)$. So a recursion equation for p will capture the entire evolutionary dynamics of our population. Given we have defined $p(t, x)$, we have also defined $p(t + 1, x + 1) = \frac{n_A(t+1,x+1)}{n_A(t+1,x+1)+n_a(t+1,x+1)}$. We can express this recursion in terms of the numbers of each type at t, x by substituting our definitions of $n(t + 1, x + 1)$ from equations (3.4). When we do this, and divide the top and bottom of the fraction by $n_A(t, x) + n_a(t, x)$, we arrive at a recursion equation for p:

$$p(t + 1, x + 1) = \frac{V_A W_A p(t, x)}{V_A W_A p(t, x) + V_a W_a (1 - p(t, x))}. \tag{3.6}$$

This is the classic model of natural selection (on haploid genotypes: Haldane, 1924; Wright, 1931), but with a new term added that captures spatial fitness. With this equation, starting from any value of p and given values for the spatial and temporal fitness terms, we could use a computer to calculate how p changes over space and time. To do this, we would just use equation (3.6) to calculate $p(t + 1, x + 1)$ and then feed our answer back into the right-hand side of this equation to get $p(t + 2, x + 2)$ and so on. One example can be seen in Figure 3.6, where we see the A allele increasing in frequency over time. Interestingly, this happens despite the A allele having the lower reproductive fitness: in the example in Figure 3.6, $W_A = 1$ and $W_a = 1.4$. The a type leaves 40% more offspring behind each generation and yet, on the invasion front, it is decreasing in frequency. Why? Because it can't keep up; the A type disperses twice as often.

While this example is interesting, it is only a single set of parameter values. We can learn a lot more in this case by doing a little more math. The first question we can ask is whether there are values of p at which the system stops changing. These are called the equilibrium values of p, and they can be found by solving $p(t + 1, x + 1) - p(t, x) = 0$, for p. The answer is that the system no longer changes when $p(t, x) = 0$, or when $p(t, x) = 1$. When you think about it, these must be equilibrium values, because you can't really have change in the proportion of types in the population when one or the other type has gone extinct. The interesting thing here is that there are no intermediate equilibria. This system will always result in the extinction of one of the types. There will always be an ultimate evolutionary winner in this system. Moreover, a little more math can tell you the conditions determining which type wins:[9] A will be the ultimate winner when $V_A W_A > V_a W_a$, and a will be the ultimate winner when $V_A W_A < V_a W_a$. Thus, it is not classical temporal fitness (W) that determines the winner, nor is it the type with the highest spatial fitness—the best disperser. The evolutionary winner on the invasion front is determined by the type with the highest spatiotemporal fitness.

Sometimes, this model tells us, an exponentially growing population might evolve to have lower reproductive rates. And this will happen whenever achieving high spatiotemporal fitness means reducing temporal fitness, as happened in

[9] This little bit more math is a stability analysis; finding the conditions under which each of the equilibria are stable.

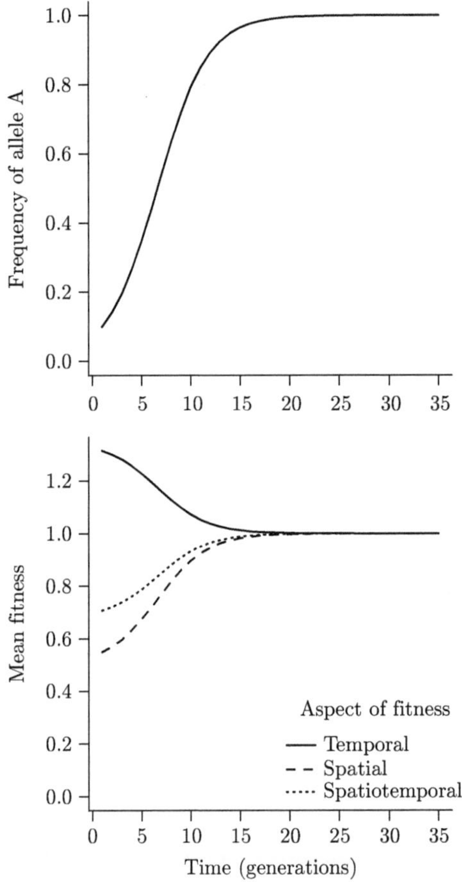

Figure 3.6 *Dynamics of a population starting with an initial frequency of A type of 0.1, and with $V_a = 0.5$; $V_A = 1$; $W_a = 1.35$; $W_A = 1$. The top panel shows the frequency of the A allele over time. The bottom panel shows how population fitness changes over time, for spatial, temporal, and spatiotemporal aspects of fitness.*

Figure 3.6. So one of the reasons that we might not see increases in reproductive rates in invasion front populations (Section 3.2) is that there is some trade-off between spatial and temporal aspects of fitness: the advantages of increasing dispersal have outweighed the cost of reduced investment in reproduction. We delve into this idea in greater depth in Chapter 7.

The other thing to notice about this model is that W in our discrete-time model is analogous to r in a continuous-time model; and V is analogous to D. The implication is that evolution on invasion fronts should maximize something along the lines of rD. We learned in Chapter 2 that the speed of an invasion scales with $\sqrt{rD}$. Thus, the implication of our model (Phillips and Perkins, 2019) is that the evolutionary processes on invasion fronts drive traits in a direction that causes invasions to accelerate. Interestingly, this is precisely the conclusion reached by another team of modellers, who took a pair of competing reaction–diffusion equations as their starting point (Deforet et al., 2019): on an invasion front, the genotype with the highest value of $r_i D_i$ (where i denotes genotype) will dominate the invasion front. This is one of those conclusions that, like natural selection, seems screamingly obvious *after* someone has pointed it out. And if we treat organisms as having both spatial and temporal axes to their fitness, we can see that this obvious conclusion is an obvious outcome of natural selection. While the simplest ecological model of an invasion suggests that it will travel at a fixed speed, the simplest evolutionary model suggests that it will accelerate.

3.4 Spread models incorporating evolution

Models of invasive populations began with Fisher, Kolmogorov, Petrovskii, and Piskunov. These people set out the simplest possible mathematical model of an invasion, and went from there (Section 2.1). Research works since then have elaborated on this work, and investigated more complex scenarios. Eventually, scenarios get sufficiently complex that mathematical analysis breaks down. We then solve equations numerically, and eventually, as things get really complex, we simulate processes on computers. This is how theory usually goes: we start with the simplest possible model, and only after understanding that do we proceed to more complex scenarios. For some reason, the theoretical development of evolving invasive populations went the other way: starting with computer simulations and eventually arriving at some concise and simple mathematical models.

Although verbal models of r/K selection and spatial sorting were advanced in the 1960s and 1980s, the first spatially explicit models of invasion with evolution didn't arrive on the scene until the Travis and Dytham model (2002, see Figure 3.7). These early models were all individual-based (or 'agent-based') simulation models. In these models, a large number of 'individuals' are tracked in a computer's memory. Each individual is given characteristics (its location, its dispersal phenotype, say), and then rules for reproducing, dying, and moving are laid down. At each time step, every individual in the population does what it is allowed to do, and then we progress to the next time step, and so on. While all individuals follow the same rules, they don't all do the same things; stochasticity (randomness) is usually built in. Essentially, these models simulate little biological worlds. They can be put together with basic theory and some basic technical experience in computer coding, and they can be immensely fun to watch play out over time. Such models are a great way to test ideas, and to develop early insight into outcomes and processes, and it was in this spirit that many of the early evolutionary spread models were undertaken.

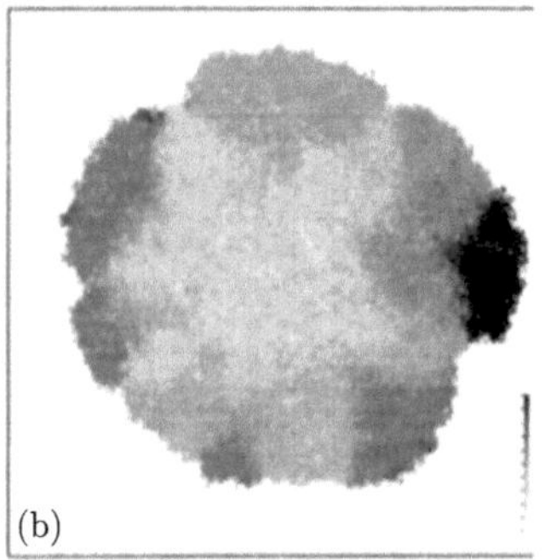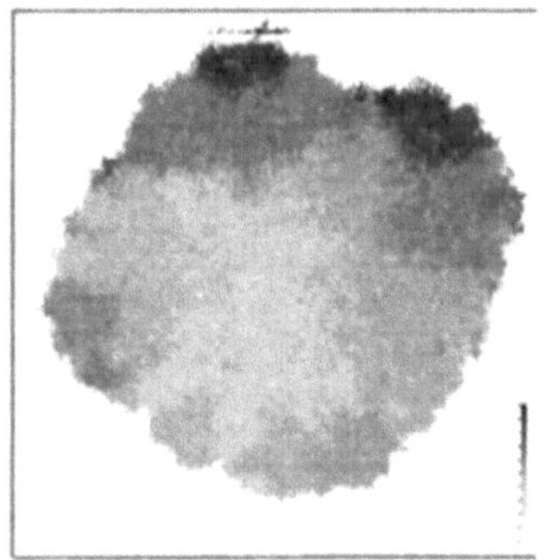

Figure 3.7 *An early individual-based simulation of a population spreading in two dimensions from a central point. Each pixel is a sub-population of many individuals, and darker colours represent higher mean dispersal values in these sub-populations. From Travis and Dytham (2002).*

The first tranche of models, put together by independent people (which is important, given how easy it is to make mistakes in computer code), looked at the evolution of dispersal during invasion (e.g., Travis and Dytham, 2002; Hughes et al., 2007; Phillips et al., 2008). These models, all with slightly different assumptions, showed that dispersal evolved upwards on invasion fronts, causing them to accelerate. Cwynar and Macdonald's (1987) verbal model was looking good. In 2010, an individual-based model was put together that allowed both dispersal and growth rates to evolve under a three-way allocation trade-off between dispersal, reproduction, and competitive ability (Burton et al., 2010). This model was the first to show that both investment in dispersal and reproductive rates might increase on invasion fronts, at a cost to competitive ability. Again, invasions in which evolution was allowed accelerated.

During this time, between 2006 and 2010, I was busily running large field and lab experiments to test these (and other) ideas. While radiotracking death adders, or counting tadpoles, I was also thinking a lot about the 'Olympic village effect' (spatial sorting). Every few months, my postdoctoral supervisor, Rick Shine, would come along to radiotrack death adders or count tadpoles, and we would discuss ideas. Feeling mildly heretical, I would argue that the Olympic village effect was different from, but analogous to, natural selection. Rick would cite classical evolutionary theory and demur, but in an interested and open-minded way. Through these discussions, I eventually realized that we would need a model that separated natural selection from the (then) Olympic village effect. At about the same time, Rick went on a speaking tour of China and, having struggled to answer smart questions from Chinese students about the nature of the Olympic village effect, arrived at the same conclusion I had: the Olympic village effect was different from, but analogous to, natural selection. Rick is a man of confidence, a top-rate thinker, and a very gifted writer: he immediately proposed that we write a paper pointing this out. I was nervous, because I saw this as a claim on some deeply fundamental aspects of evolutionary theory, and we were empiricists really, not theoreticians. Nonetheless, the paper came together very rapidly, and I included an individual-based model in which all individuals had the same number of offspring. In this model, there could be no natural selection, because there was no variation in reproductive (temporal) fitness. Despite this, we still saw dispersal evolve upwards on the invasion front. We made our arguments, and we gave the Olympic village effect a more serious sounding name: spatial sorting (Shine et al., 2011). To my relief, the paper was well received. The model I put together for that paper wasn't particularly excellent (the one in Section 3.3 does a much better job and we can make fitness differences vanish simply by setting $W_a = W_A = 1$), but it made the point that dispersal would evolve on invasion fronts even if natural selection was absent.

One of the excellent outcomes from the 2011 spatial sorting paper was that some hard-core theoreticians noticed. These were people who didn't count tadpoles or radiotrack death adders; they developed theoretical models from the ground up, and used sophisticated mathematics to make their points. They published in journals I had never heard of, and used language I could barely understand. In 2012, two very mathematical papers came out. These papers proposed a generalization of Fisher's equation (equation (2.1)); one in which the dispersal coefficient, D, could evolve (Benichou et al., 2012; Bouin et al., 2012). Finally, we had arrived at where Fisher started: with the simplest possible model that could describe an invasion with evolving dispersal. The model they proposed looks like this:

$$\frac{\partial N(t,x,D)}{\partial t} = D\frac{\partial^2 N}{\partial x^2} + rN(1 - \frac{N}{K}) + \alpha\frac{\partial^2 N}{\partial D^2}. \tag{3.7}$$

You can see that it is nearly the same as Fisher's equation (equation (2.2)), except that it is now tracking the number of individuals of phenotype D at a particular time and place—$N(t,x,D)$—and there is another term added: another one of those vexing second derivatives ($\alpha\frac{\partial^2 N}{\partial D^2}$). This last term, and let's call it the 'dispersal diffusion term', will make most people's heads swim, but it can be built up in the same way as the (now) more familiar spatial diffusion term, $D\frac{\partial^2 N}{\partial x^2}$ (Appendix A). What the dispersal diffusion term does is cause a normal distribution of dispersal phenotypes, D, to exist at each location in space and time. This variation arises through mutation—tiny random shifts in individual D values at each time step. So this model, unlike Fisher's, incorporates individual-level variation in dispersal. Like Fisher's model, however, there is no variation in growth rate, r, so natural selection is excluded. This model is an elegant mathematical statement of what I had been trying to achieve in 2011 with an individual-based model.

As well as boiling the system down to its most basic elements, this team of French mathematicians was able to use everything that had been learned about reaction–diffusion equations and statistical mechanics since Fisher to derive some general results. What they learned, in extremely rigorous fashion, is that spatial sorting does indeed cause dispersal rates to evolve upwards on invasion fronts, and this causes invasion fronts to accelerate. Like Fisher's model, their model is elegant and (just) mathematically tractable. As such, it has generated interest in the theoretical community, and this model is now an active area of theoretical research. Within eight years, there were more than 25 new mathematically heavy papers extending and developing this model, many bringing to life Benichou et al's (2012) understated conclusion, 'applications beyond the field of population dynamics could be expected'.

Some time after the French team laid out spatial sorting in reaction–diffusion terms, various people started weaving it into the lore of integrodifference equations. This started with Ramanantoanina et al. (2014) who developed integrodifference models that admitted the presence of individuals with varying dispersal abilities, and (extending upon the reaction–diffusion framework) also allowed for dispersal kernels that had fatter tails (higher kurtosis). Schreiber and Beckman (2020) extended this to variation in dispersal and reproduction, making an important distinction between heritable and non-heritable variation, and showing that both r and D would tend to increase on the invasion front. At the time of writing, the latest word in the integrodifference model space is Poloni and Lutscher (2023).

But in the meantime, applications in population dynamics were being realized. One nice example was a complex reaction–diffusion– advection[10] model (with evolution) to understand the growth of tumours (Orlando et al., 2013). This model showed that tumours might be experiencing the very same evolutionary processes that were afoot in the cane toad invasion. Spatial sorting had found its way to medical research. But more on that later (Chapter 9).

[10] Diffusion–advection is just diffusion in which there is a bias for individuals to move towards one direction, such that the mean displacement is not zero: the population tends to move in a particular direction.

3.5   Spatial selection and enhanced spatial selection

We have so far been building an argument that there are two evolutionary forces at play on an invasion front: spatial sorting, acting on dispersal; and natural selection, acting on growth rate. The population genetic model developed in Section 3.3 focuses very clearly on those two forces, and it shows that on invasion fronts, evolution is optimizing spatiotemporal fitness. In 2008, when I was fumbling around for language to describe these ideas, I referred to this interaction of forces (acting on dispersal)—natural selection plus spatial sorting—as 'spatial selection'.

In 2013, a very smart colleague of mine, Alex Perkins, put together an integrodifference model in which both growth rate and dispersal could evolve (Perkins et al., 2013). Alex was hoping to use this model to recapitulate the toad invasion (toads on a string, rather than cats, in this case), and he roped me in for assistance with the data. Alex built a really nice model, but we struggled to get it to emulate the toad invasion. It usually fell well short of the spread distance that toads actually achieved. There are a number of reasons why the model may have fallen short, and we will come back to those in due time. But the part of the story that matters now is that in the process of running his model, Alex noticed something very interesting: dispersal rate evolved faster when growth rate was also allowed to evolve.

In Alex's model (as opposed to the model in Section 3.3), growth rate and dispersal could be made to be independent traits. In the spatiotemporal fitness model (Section 3.3), we modelled a single gene that has effects on both dispersal and growth rates, so growth and dispersal are not independent. In Alex's model, he imagined a large number of genes affecting each of dispersal and growth. These sets of genes could be independent, or could be made to overlap to various degrees; it was a fancy model (technically, this was an application of the infinitesimal model of quantitative trait evolution; another of Fisher's inventions). When the traits were set to be independent in this model, they should evolve independently. We wouldn't expect change in the dispersal trait to affect the evolution of growth, or vice versa. So it was intriguing that when growth rate, r, was allowed to evolve (and it always evolved to higher values), this caused dispersal rates to evolve more rapidly. Something about the evolution of higher growth rates caused dispersal to evolve faster. Taking a lead from my earlier term, we called this phenomenon 'enhanced spatial selection'.

As well as pointing out the phenomenon, we provided an explanation for why it might happen. The explanation we provided was quite compelling, but—seven years later—I am fairly sure it was incorrect. I take full responsibility for the compelling but erroneous explanation and, to avoid perpetuating the confusion, will not trot it out again here. The good news is that the correct mechanism explaining enhanced spatial selection is actually quite simple: when growth rate is higher, the evolutionary response to spatial sorting is stronger.

Why would spatial sorting be a stronger force when growth rate is higher? The short answer is that higher growth rates cause the invasion front to take on a steeper shape. When growth rates are small, an invasion front can be a gentle cline between carrying capacity and zero. When growth rates are high, on the other hand, the invasion front can be a cliff (Figure 3.8). In our model of spatiotemporal fitness above (Section 3.3), we modelled a cliff: our dispersing individuals fell off a density cliff and arrived in a completely vacant space. But an extension of that model showed that when our dispersing individuals arrived in an occupied patch, the response to spatial sorting follows a classic selection-migration model (Phillips and Perkins, 2019). Essentially, the effect of spatial sorting is diluted by the other individuals already present in the occupied patch.

If we imagine the other extreme, in which there is no density gradient at all, we would have perfectly balanced immigration and emigration, and we would find spatial sorting completely balanced by gene flow. There would be no evolutionary change, despite spatial sorting happening between every patch. If we consider

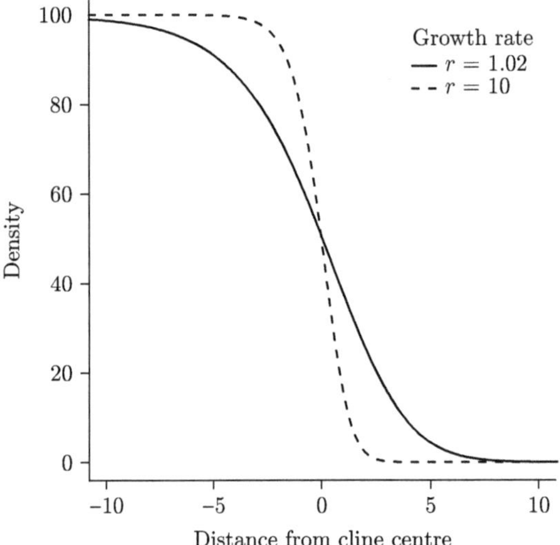

Figure 3.8 *Invasion fronts get steeper when the population growth rate is higher. Here we compare the wave shape in a reaction–diffusion model resulting from* $r = 1.02$ *versus* $r = 10$ *(D = 1.05 in both cases).*

these two extremes, it is straightforward to see that the response to spatial sorting will scale with the steepness of the density gradient. When growth rate is allowed to evolve, growth rates increase; this causes the invasion front to steepen, and that, in turn, increases the evolutionary response of dispersal to spatial sorting.

We can see this effect of growth rate on spatial sorting in other models, where the effect has been noted, but not really discussed. For example, larger base values for growth rate were observed to cause higher dispersal rates to evolve in the model of Burton et al. (2010). And we can see this effect of r on spatial sorting in the analytical model of Benichou et al. (2012); see equation (3.7). In this model, they found that the rate of evolution of dispersal was sensitive to r/α, where α is the mutation rate. For a fixed mutation rate, α, higher r will give faster rates of dispersal evolution on the invasion front.

I promised in the introduction that there was some nomenclatural confusion. Here is a moment when we should eradicate some confusion. We should abandon the use of 'enhanced spatial selection': we should abandon it because what we are actually seeing when r evolves is enhanced spatial sorting. I shall avoid the term 'enhanced spatial selection' henceforth.

3.6 Absolute and relative fitness of dispersal phenotypes

Although we have found reason that growth rates might be under selection on an invasion front, it is striking that the empirical support for increased growth rate is equivocal, and in models that allow both growth and dispersal to evolve under trade-offs (e.g., Burton et al., 2010; Perkins et al., 2013), it seems to be dispersal that usually has the stronger response. While we might find excuses for the apparently patchy empirical response by growth rate (wrong traits being measured, non-density-regulated populations, and so on), it is harder to ignore the simulation results where such issues are absent (e.g., Burton et al., 2010). Why would dispersal have a stronger evolutionary response than growth rate? An obvious possibility is that there might be more than just spatial sorting (enhanced or otherwise) acting on dispersal. It might be that there are several forces conspiring to drive the evolution of dispersal on invasion fronts.

The first obvious contender is, of course, classical natural selection: variation in temporal fitness. At this stage, we have not even considered that classical natural selection—the primary force of evolution—might be favouring increased dispersal. That seems an oversight, so let's bring our old friend natural selection back into play. As it is usually rendered, natural selection arises from variation between individuals in survival and reproduction. On an invasion front, a huge amount of variation in reproduction can be generated simply by where you are placed on the density gradient. To give you a sense of how large this fitness win might be, let's consider toads again for a moment. Female cane toads, famously, can lay more than 30,000 eggs in a sitting. That's a lot of babies. One of the major determinants of the survival of these babies is competition. Baby toads are cut-throat competitors: tadpoles will eat eggs laid in their pond and destroy the majority of them (Crossland and Shine, 2011b); those they don't destroy they inflict with developmental abnormalities using chemical excretions (Crossland and Shine, 2011a); after they metamorphose and become baby toads, they are shameless cannibals, using their toes to lure smaller toads close and then eating them (Pizzato and Shine, 2008). Female toads have a lot of eggs, but in a high-density population, most of those eggs never make it to adulthood; they are killed by other toads. In a population at carrying capacity, of those 30,000 eggs, only one or two will survive to adulthood. But in a low-density population, how many might survive: 10, 500, 10,000? We don't really know, but the point is it could be *a lot* more. So a highly dispersive individual on an invasion front may well find it has much higher absolute fitness compared with a sedentary conspecific.

But here is where we need to be careful. Up until now, we have been very careful to separate spatial from temporal effects. If we compare the number of surviving offspring between our dispersive and non-dispersive individuals, we would conclude that the dispersive individual, by escaping competition, has increased its fitness. But this is a wrong comparison to make, because it is comparing individuals who are now inhabiting different spatial locations. These two individuals no longer compete with each other, but with the individuals they are now nearby to. Our dispersive individual may have more offspring, but *so does every individual at that location*. So the apparently large fitness win made through dispersal may be small or even negative relative to the fitness of the dispersing individual's new competitors.

Thus, when we are considering temporal fitness, we need to define the spatial scale at which it operates. Just as when we consider spatial fitness, we need to define the temporal scale at which it operates. To lose track of these restrictions is an easy trap, and one that leads to confusion. This was, in fact, how we managed to devise the incorrect explanation for why evolving r leads to faster evolution of D in Section 3.5.

So if we exclude the fitness advantage of moving down the density gradient, how else might natural selection favour dispersal? The answer potentially lies in selection operating below the level of the individual, at the level of the gene: kin selection, and inclusive fitness.

3.7 Kin competition

Our starting point from Chapter 1 has been understanding invasions. We have established that invasions often accelerate and that at least some of this acceleration is due to dispersal rates evolving on invasion fronts. In uncovering the importance of the evolution of dispersal, however, we are opening the door on a long-running evolutionary puzzle. At least since the early 1970s, theoreticians have wondered why organisms disperse: there are clearly risks to be faced in moving beyond known territory—you don't know where you are going; you could get hit by a car—but dispersal is, nonetheless, ubiquitous. The evolution of dispersal is an important and long-standing topic in evolutionary biology, and has developed quite independently from our understanding of invasions. We will come back to this larger history at various times in the book, but for now we need to talk about kin competition.

One of the earliest models of dispersal evolution made the point that if individuals don't disperse, they are competing for space with their very close relatives (Van Valen, 1971). As a consequence, it is in mom's best interest to have at least some of the kids leave home. Forcing some offspring to disperse increases her inclusive fitness. To put it another way, when there is a competition for space, and excess offspring, sending some offspring out into the world is effectively like buying additional lottery tickets. Thus, avoiding kin competition is seen as an important driver of dispersal evolution.

Early models that investigate kin competition imagine that available habitat is always full. Essentially, the population is always at carrying capacity: offspring are produced, and they compete for space in the next generation. There can only ever be K individuals in the population. Even under these conditions in which spare space is very limited, dispersal evolves (Hamilton and May, 1977; Comins et al., 1980). These saturated spaces mean that, for every ticket (i.e., offspring), there is only a small chance of winning the lottery. But still, dispersal is worthwhile. We have seen that in real invasions, however, space is not saturated. Even after a location is colonized, it can take a while for that location to reach carrying capacity. During this period of time—between colonization and hitting K—the per-ticket chances of winning the lottery are substantially higher. Thus, we might expect kin competition, or at least intergenerational fitness effects, to be amplified somewhat on invasion fronts (Travis et al., 2009). It is worth pointing out, however, that this argument could also be cast purely in terms of spatial sorting: there is vacant space, and it tends to be colonized by more dispersive individuals.

There is another, and perhaps more important, reason we would expect the effects of kin competition to be amplified on invasion fronts. After a while, everyone on an invasion front is kin. I've taken some liberties with the notion of kinship here, but relatedness should increase on invasion fronts as an invasion progresses. The reason for this is, again, assortment through space. We have already seen that only the most dispersive individuals make it to the invasion front, and these are a subset of the initial population. These individuals take a small, approximately random sample of the rest of the gene pool with them. The fact that there is selection on some parts of the genome and a small sample of the rest means that the invasion front population only represents a subset of the initial population's genetic diversity (see Chapter 5). And this process repeats as the invasion propagates through space. After a while, we have a subset of a subset of a subset of the original genetic diversity. Genetic diversity drops as the invasion proceeds, and this means that pairs of individuals on the invasion front will increasingly tend to have the same alleles at any given location in the genome. The population becomes increasingly related.

So the argument goes that avoiding competition with your kin is a good thing, because it buys your genes additional tickets in the lottery of life. Invasion front tickets have an edge in the lottery, and kinship increases as invasions progress. For these reasons, we might expect kin competition to be a driver of dispersal evolution on invasion fronts. These are clever and intriguing arguments, and they are complex. To test this idea, a group of European modellers did an interesting simulation experiment (Kubisch et al., 2013). They built an individual-based simulation model in which individuals carried genes affecting their disperal probability. They imagined a world very similar to that of the spatiotemporal fitness model of Section 3.3 in which space and time are discrete. Instead of a one-dimensional lattice, however, their space was a two-dimensional lattice (with locations given by $\{x, y\}$), wrapped around a cylinder. This arrangement made a world that was much longer (in x) than it was wide (in y), and in which it was impossible to hit an edge if you travelled in the y-direction. They allowed only nearest-neighbour dispersal, and they started their invasion by sprinkling individuals across the lattice at $x = 0$, and tracking the invasion as it moved across increasing x values.

One of the fun things about individual-based models is that they have emergent properties. Things happen in the model that were not explicitly written in to them—things like spatial sorting and kin competition. Kubisch et al. (2013) argued that earlier simulation models had conflated kin competition and spatial sorting: they are different mechanisms and we need to work out how much each contributes to disperal evolution on invasion fronts. These were good points at the time, and the challenge these modellers set themselves was to separate these two effects. But because these effects are not explicitly coded into the model, you

[11] Technically, they are only preserving the mean of the spatial sorting effect, and that technicality matters in this case.

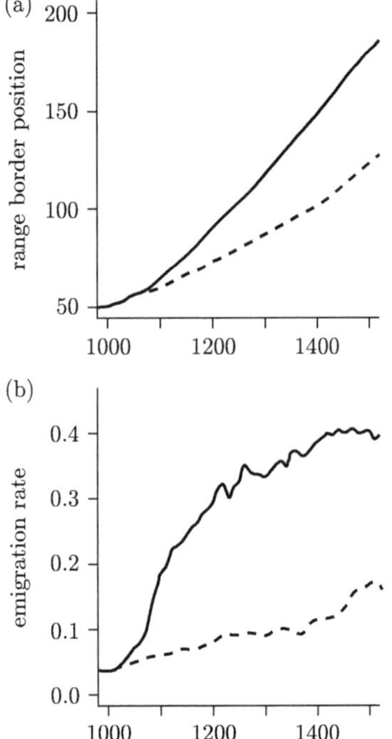

Figure 3.9 *Shuffling individuals across the invasion front (dashed line) results in slower evolution of dispersal (emigration rate) and slower acceleration of an invasion, compared with an unshuffled invasion. This effect was attributed to the removal of kin competition in the shuffled treatment. From Kubisch et al. (2013).*

can't just turn one of them off. But you can manipulate the simulated world in a way that disables one of the mechanisms. Kubisch et al. (2013)attempted to disable kin competition. How? By shuffling individuals in the y-direction each generation. In this way, they destroy any genetic substructure that emerges in the y-direction, but preserve substructure that develops across the invasion, in the x-direction. This has the effect of dispersing kin for free, as it were. Kin competition has been obviated, because kin have been artificially moved across the lattice before natural dispersal occurs and so are no longer competing with each other for space, even if they don't disperse on their own. But they have been moved in a way that preserves local population sizes and spatial sorting.[11] This was a sneaky trick.

Their model showed that shuffling individuals across the invasion front really did cause the evolution of dispersal to slow dramatically, with obvious consequences for the rate of spread (Figure 3.9). This was not a small effect, and the authors felt that this demonstrated that kin competition was a major force driving dispersal on invasion fronts. There is no doubt that this was an ingenious *in silico* experiment to remove kin competition, but theoretical developments happening parallel to this work would show that their manipulation does more than just remove kin competition; it also hobbles spatial sorting (see Chapter 5 and Section 7.1.4). So while kin competition may play a role on an invasion front, the jury remains out on precisely how important it is. It remains very challenging to separate spatial effects from kin selection effects in this context.

In the meantime, there are also some concerns, I think, with the basic arguments for kin competition during invasions. These are the arguments I sketched earlier in this section. First, the argument that there is particular advantage to avoiding kin competition in the interval of time between colonization and K. This is just spatial sorting in another guise: available space tending to be colonized by the more dispersive individuals. The second argument—that relatedness increases as invasions progress—ultimately leads to large regions of space with very little genetic variation (see Chapter 5). In this case, it is impossible to avoid related individuals by dispersing, because the population in the neighbouring patch is very closely related to the population in an individual's home patch. So the inclusive fitness benefit from dispersing may actually be reduced rather than increased as an invasion progresses. There is still work to be done in this space.

3.8 Inbreeding depression

Another classic explanation for the evolution of dispersal is the avoidance of inbreeding (e.g., Gandon, 1999). We all know that breeding with close relatives is a bad idea. A great way to avoid inbreeding, of course, is to disperse. If you are living a long way from your relatives, you are unlikely to mate with them. Thus, one of the classic arguments for the evolution of dispersal rests on the idea that it is possible to avoid the costs of inbreeding by dispersing. Perhaps avoidance of inbreeding depression might be a driver of increased dispersal on range edges? Well, probably not. To see why, we need to dig a little deeper into inbreeding.

The basic issue with inbreeding is that it causes deleterious alleles to be expressed, so offspring of a close mating will often have lower fitness. Let us consider a single genetic locus, and we imagine that both parents are heterozygous at that locus, with both parents carrying a functional and a (rare) non-functional copy of a key developmental gene. That one functional copy that each parent carries is enough to allow normal development, so each of the parents has been able to grow happily to reproductive age despite the fact that they carry a rare deleterious copy of the gene. If we mate these two individuals, however, 25% of their offspring will carry two copies of the non-functional gene. These offspring will not develop normally, and are likely to die well before they reach reproductive age. Thus, in this simple example, we can see that the potential fitness of this pair has been reduced by 25% because it is a close mating. If, instead of mating with a close relative, each of these individuals had chosen an unrelated individual (one less likely to carry the deleterious allele), there would have been no loss of offspring. As a consequence, we can see that there will often be strong selection to avoid mating with close relatives. This is why we all know that breeding with close relatives is a bad idea.

Now, let us imagine a scenario where the non-functional allele is not just rare, it is completely absent. Now (at least from this one gene view of things) it makes no difference who you mate with; relative or not, it makes no difference. This is a toy example, but the point is that inbreeding depression requires genetic variation in the population. If there is no genetic variation, then inbreeding depression is not a thing. Genetic diversity tends to decline (sometimes very dramatically) on invasion fronts (Chapter 5), so we can see that invasion fronts might very often reduce selection for inbreeding avoidance (Pujol et al., 2009). If inbreeding is less costly, then why do we need to disperse in order to avoid it? It seems very unlikely, then, that inbreeding avoidance is a strong additional driver of dispersal evolution on invasion fronts.

3.9 Local adaptation

As populations spread through space, they are very likely to encounter novel environments; conditions different from those they are adapted to. This is probably more true for an invasive species than it is for, say, an invading pathogen, or a tumour. Nonetheless, space is often not homogeneous, and natural selection will work to optimize populations towards their local conditions. In this chapter, we have focused almost entirely on the evolutionary processes that are playing out on invasion fronts. These processes optimize the phenotype for invasion; they favour increased rates of dispersal and reproduction. It is important to remember, of course, that local adaptation—adaptation to local environmental conditions—is also occurring (Perkins et al., 2016; Szűcs et al., 2017).

Often, we might expect adaptations to local conditions to also increase the growth rate (Perkins, 2012). Plants adapted to industrial contaminants, for example, will spread faster across contaminated land than will their maladapted conspecifics (Macnair, 1987), and a nice experiment on flour beetles clearly shows that adaptation to novel environments causes much higher growth rate in the novel environment and so can contribute substantially to invasion speed (Szűcs et al., 2017). Thus, local adaptation may often drive change that is also favoured on the invasion front. So we would expect selection on standing variation to drive local adaptation on the invasion front; a result that plays out in models that allow such behaviour (e.g., Perkins, 2012). But how long before genetic variation on the invasion front is exhausted? Chapter 5 shows that we might often see genetic variation on the front drop rapidly. So, a relevant question is whether we would expect novel genetic variation to arise on, or reach the invasion front. Invasion front populations are, compared to populations behind the invasion front, relatively small. This means that fewer new mutations can be expected to occur on invasion fronts than will occur behind the invasion front.

We know from Fisher's original model that advantageous alleles (i.e., new adaptations) spread in a travelling wave through a population—waves of adaptation happening within a wave of invasion—so if an adaptation emerges behind the invasion front, would we expect its wave of advance to overtake the invasion front? Can an invasion accelerate because it gains local adaptations that have arisen behind the invasion front? With the theory we have developed so far, such acceleration seems unlikely, for two reasons. First, dispersal rates on the invasion front (subject to strong spatial sorting) can evolve beyond what is optimal behind the invasion front (where spatial sorting is absent or greatly diminished; Perkins et al. (2016)). As a consequence, dispersal rates behind the invasion front are expected, and have been observed, to drop dramatically after an invasion has passed (e.g., Lindstrom et al., 2013). This means that the diffusion rate behind the invasion front D^- is often slower than diffusion on the invasion front D^+. Second, the growth rate driving the population forward is different from the growth rate driving the advantageous allele forward. The population is driven forward by the growth rate of the population, λ, whereas the allele is driven forward by the relative fitness of the favoured allele. If the favourable allele is denoted A and the wildtype a, then this relative fitness is approximately $w_A = \frac{\lambda_A}{\lambda_a}$, where the λ's in this case are the growth rates at equilibrium density (i.e., when $N \approx K$). So for the growth rate of the allele to be greater than the growth rate of the population requires the advantageous allele to have a fitness advantage such that $\frac{\lambda_A}{\lambda_a} > \lambda_a$ or $\lambda_A > \lambda_a^2$. Unless the new allele gives a major improvement in fitness, it is unlikely to have a growth rate larger than that of the population. Taken together, then, these arguments suggest that in most cases $\lambda_a D^+$ is likely to be larger than $w_A D^-$. (You will recall that invasion speed in a simple reaction–diffusion model is given by $2\sqrt{rD}$ and that growth at high density, λ, is typically much lower than growth at low density, r, so if $\lambda_a D^+ > w_A D^-$, then $r_a D^+ >> w_A D^-$; so the population invasion front is likely to move faster than the adaptation invasion front.) Although I have made qualitative arguments, it seems that any adaptation occurring behind the invasion front will find it hard to ever catch the constantly moving invasion front. Certainly, the one experimental study that transplanted advantageous mutants into invading populations (Deforet et al., 2019) found that mutants were unlikely to establish upon the front unless they were transplanted on or near the invasion front.

These qualitative arguments paint a picture in which adaptations to local conditions occur, but these adaptations largely happen after the invasion front has passed. This leaves the invasion front as a transient and strangely isolated process, rolling across the landscape under its own steam and under its own rules. It would be merely a quirky phenomenon were it not for the fact that all the colonizers in an invasive population spring from that process. All the colonizers in an invasive population are founded from a population that is being optimized for dispersal and growth, and the founding population adapts towards local conditions from that starting point.

We take this basic model of the isolated invasion front as true for now, but there are invasions where the front is not an isolated entity, and we explore these in Chapter 6; there are also cases where the growth rate on the invasion front could be expected to be lower than the fitness advantage of a new allele, which we explore in Chapter 5.

3.10   The wrap

In this chapter, we have introduced the idea that, within a population, there may be heritable variation for the traits determining the rates of population growth and dispersal, r and D. When we admit this variation in invasive populations, it turns out that we expect r and D to evolve on the invasion front in a way that maximizes the product rD and that, as a consequence, causes invasions to accelerate. While the simplest ecological model suggests invasions should propagate forward at a constant speed, the simplest evolutionary model suggests invasions should accelerate.

Dispersal evolves on invasion fronts due to an underappreciated evolutionary force—spatial sorting—which is a spatial analogue of classical natural selection. Population growth is also expected to evolve, in response to classical natural selection favouring individuals with high values of r on the (low density) invasion front.

While empirical work strongly supports the prediction of increased dispersal on invasion fronts, evidence for the evolution of an increased population growth rate is more ambiguous. We discuss four reasons for this discrepancy. We also consider the roles of kin competition, inbreeding avoidance, and local adaptation on invasion fronts.

While this book focuses on invasive *populations* and how evolution might play out within the invaded range as a consequence of the invasion process, the idea of evolution in invasive *species* more broadly has been kicked around for quite a while. Rather than focusing on the invasion process itself, this broader work on invasive species has mostly examined how evolution plays out on independent populations (in the native versus invaded range, for example) and is not closely tied to the ecological theory of invasions. This broader question of evolution and invasive species really got started with a conference in Asilomar, California reported in the now classic edited volume by Baker and Stebbins (1965), *The Genetics of Colonizing Species*. There are a bunch of fascinating papers in that volume, with a who's who list of authors and recordings of some great repartee between the participants. If you can find a copy, it is worth a read (Lewontin's paper in particular, in my humble opinion). In this vein, another two edited volumes worth perusing are Sax et al.'s (2005) *Species Invasions: Insights into Ecology, Evolution, and Biogeography*, and more recently Barrett et al.'s (2016) *Invasion Genetics: The Baker and Stebbins Legacy*. Edited volumes are, by their nature, somewhat idiosyncratic in what they cover (many cooks and all that), so a couple of single-author books that are also valuable, mostly for the range of ideas they cover and a rich set of empirical examples. First, Cox's (2013) *Alien Species and Evolution*. And more recently, Le Roux's (2022) *The Evolutionary Ecology of Invasive Species*.

Stochasticity and invasion fronts

Uncertainty is cooked into most games that people enjoy. If horse races always ended in horses placed in alphabetical order, there wouldn't be much point in watching them. If the person who chose the red marker always wins a game of snakes and ladders, no one would play it. Uncertainty keeps things interesting.

Uncertainty is also a fact of life. No one knows if they will be alive tomorrow, or not. Ten-year-old children usually do not know if (and who) they will marry, how many children they will have, or who their best friend will be when they are sixty. No one can tell you what the temperature will be outside your house at noon two weeks from now. And what is true for humans in this regard is true generally, for all organisms: the future is uncertain.

The Scottish poet Robert Burns had the truth of it when he penned 'The best laid plans of mice and men oft go awry / And leave us nought but grief and pain for promised joy.' He wrote this as part of his apology to a mouse whose nest he accidentally dug up. This mouse, snug in its nest, nearly weaned six young; instead, nothing. Its fitness was severely and unexpectedly curtailed, whereas the mouse 10 metres away that did not accidentally fall to the shovel of a famous poet may have gone on to successfully wean its six offspring. These mice were, of course, part of a population of mice, and each mouse in this population was subject to the fickle hand of chance; only one got written about.

These chance events, happening every day to every individual in a population, affect the ecology and evolution of populations. In fact, these chance events, in aggregate, can matter to populations (and so invasions) in profound ways. Our early understanding of invasions was built entirely from observing invasions in nature: Elton's book was a compendium of such examples (Elton, 1958). For well-documented systems (like muskrats, cane toads), we often notice that some parts of the invasion travel faster than others, but we mostly ascribed that variation to something about the environment. In recent years, workers around the world have started running replicate invasions under laboratory conditions. These replicate invasions—run in near-constant environments—have revealed considerable variation in invasion speeds (e.g., Figure 4.1). This variation is clearly arising not from the environment, but from within the population, from the chance events that make life interesting. This chapter and the next look at how chance can be woven into our understanding of invasions, and why that understanding is important.

The Ecology and Evolution of Invasive Populations. Ben Phillips, Oxford University Press. © Ben Phillips (2025).
DOI: 10.1093/9780191924910.003.0004

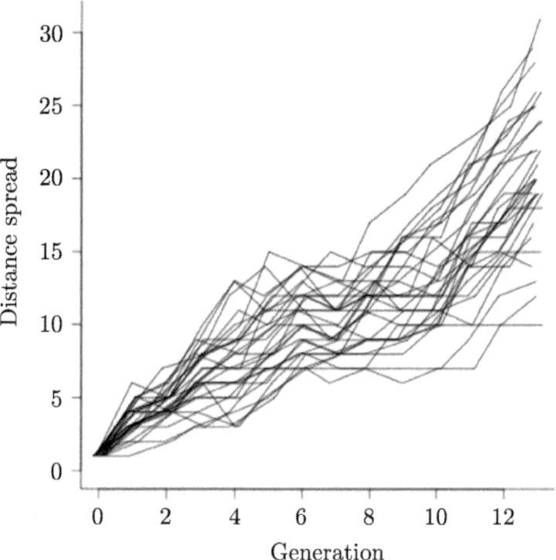

Figure 4.1 *Replicated invasions of flour beetles across habitat arrays in a laboratory setting. There is a nearly threefold variation in spread rate across replicates over 12 generations. Data from Melbourne and Hastings (2009)*

We have already learned (Chapter 2) that invasions move forward as a consequence of dispersal and population growth, so this chapter looks at how chance events in dispersal and population growth affect invasions. It turns out that we expect the basic pattern shown in Figure 4.1: replicate invasions will tend to show divergent trajectories over time. But we also discover that the divergence we see in Figure 4.1 is more than can be explained by chance events in dispersal and population growth. Something else is going on. Evolution, of course. In this chapter, we will also set up some basic ideas that will help us in Chapter 5 as we see how chance can also play a major role in the evolution of invasive populations.

4.1 Stochastic processes

It was one of the great (re)discoveries of the renaissance that, while an individual outcome might be uncertain, the distribution of outcomes across a population might follow quite striking patterns.[1] Figure 4.2, for example, shows the distribution of litter sizes in a population of pigs. It is very clear that there is variation in litter size, but that there is also a pattern to this variation: most sows have about 10–12 piglets, with diminishing numbers of sows having greater or fewer than this amount.

If a farmer pointed to a random sow in this herd, you couldn't say precisely how many piglets she will have, but 'between 5 and 17' would be a very good guess. Why would it be a good guess? Because there is a high probability that you would be correct: most litters fall within this range. When we make such guesses, we are thinking probabilistically. In fact, we can estimate that this guess has a probability of about 0.94 of being correct.

[1] I refer here to 'population' in the statistical sense—the complete collection of some thing—as opposed to the biological sense—the number of individuals of a particular organism in a particular time and place. You can go back to the biological sense now.

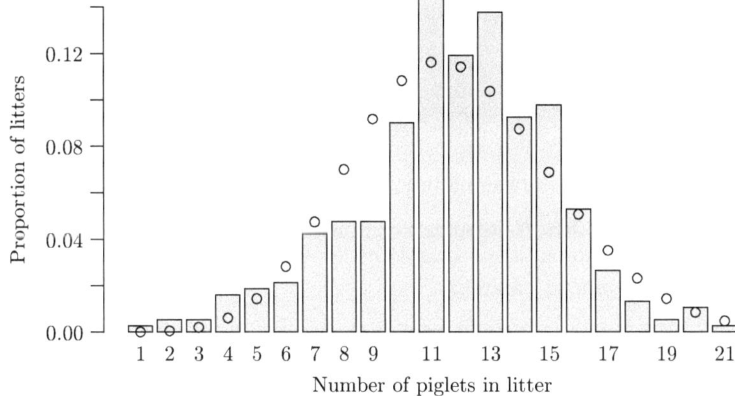

Figure 4.2 *Distribution of litter sizes of pigs over 20 years on a farm in England. Data from Braude (1954). Points represent the probability of each litter size from a best fit Poisson distribution.*

More generally, were we to model the growth dynamics of this pig population, we could assign a probability to each litter size, S. Litter sizes of 10 are much more probable than litter sizes of 20: $P(S = 10) \approx 0.09$; and $P(S = 20) \approx 0.011$. In placing probabilities on litter sizes, we would be making reproduction a probabilistic, or 'stochastic', process. 'Stochasticity' refers to the fact that many processes have a random element to them, and we can assign probabilities to the outcomes of these processes. When you roll a dice, you cannot really say in advance what the outcome will be; you might roll a three or a five (or any other number between one and six), and each of these outcomes would have an equal probability of 1/6 of occurring. So a dice roll might be referred to as a stochastic process (and this stochastic process is why the winner of snakes and ladders is uncertain). Similarly, the speed at which a horse will run varies from day to day, so race speed could be treated as a stochastic process (and this is why the winner of a horse race is uncertain). Stochastic processes are what lead to the uncertain outcomes, in games and in life.

4.2 Stochastic processes in invasive populations

The beetle invasions in Figure 4.1 were all run in a laboratory, and started with the same number of founders all drawn from the same population. Despite the best efforts of the researchers, an astonishing variety of invasion speeds emerged from that experiment (Melbourne and Hastings, 2009). Were the experiment to be run again, it would be impossible to say in advance how fast any particular invasion would progress. The best we could do would be to provide a probabilistic answer. It is clear that stochastic processes are at play here.

In the context of invasions, we can recognize four broad classes of stochastic processes:

- demographic stochasticity—random variation in births, deaths, sex ratio, survival;
- dispersal stochasticity—random variation in dispersal and settlement decisions, distances moved;
- evolutionary stochasticity—random variation in the production or transmission of genetic variation;
- environmental stochasticity—random variation in the quality of the environment through time or space.

All of these stochastic processes occur in a natural invasion, and they build upon and interact with each other. We can see that the first three sources are endogenous sources of variation; they come from processes internal to the population. The last source—environmental stochasticity—is an exogenous source of variation; it comes from processes external to the population.

In our beetle example, the researchers did their level best to make environmental stochasticity go away, so nearly all of the variation they uncovered must have been from the first three endogenous sources: demographic, dispersal, and evolutionary stochasticity (Melbourne and Hastings, 2009). When we look at invasions in nature (e.g., Sections 1.2.2 and 1.2.1), we see variation in the speed of different parts of the invasion front. We might simply interpret this as due to environmental variation, but the fact that replicate laboratory invasions (in which environmental variation is negligible) can produce so much variation suggests that much of the variation in natural invasions may also be caused by endogenous (demographic, dispersal, and evolutionary) rather than exogenous (environmental) sources of stochasticity. One thing is certain, however. In the beetle experiment, the clean lines of the deterministic models of spread and evolution (Chapters 2 and 3) have given way to much messier outcomes.

In this chapter, we will dive into the non-heritable stochastic processes: demographic, dispersal, and environmental stochasticity. In Chapter 5, we will examine evolutionary stochasticity. We will see that although demographic, dispersal, and environmental stochasticity can make invasion outcomes very unpredictable, they quite often don't have a major impact on the mean invasion speed. Evolutionary stochasticity, on the other hand, often causes invasions to progress much slower than they otherwise would.

4.3 Population size matters

Although the beetle experiment shows highly variable invasions, not all laboratory invasions give such variable results. In recent years, several groups have begun to run replicate laboratory invasions using ciliates or motile bacteria as a study organism. These experimental setups with microorganisms are really fantastic.

With microorganisms, it is possible to run an invasion spanning many genera-tions and involving hundreds of thousands of individuals. Each of these replicate invasions can be done in the space equivalent of a toothpick. These are great systems for running replicate invasions, and it is notable that when replicates are run they also tend to give much less variable results than what we see in the bee-tle experiments (e.g., Giometto et al., 2014; Fronhofer and Altermatt, 2015). In these systems, the deterministic models of Chapter 2 tend to do a pretty good job of describing outcomes.

So, what is going on? Why do replicate invasions in one system give wildly variable results, while those in another seem much more consistent? At least part of the answer is population size. Intrinsic stochastic effects (demography, dis-persal, and evolution) become much more important when population sizes are small: the beetle invasions involved tens to hundreds of individuals per patch; the ciliate invasions involved tens of thousands of individuals.

Why do endogenous stochastic effects matter more in small populations? Essentially, it comes down to the law of large numbers (LLN): a large sample will give you an estimate for the mean that is very close to the true value; a small sam-ple will give a much less certain estimate. To put it another way, if our herd of pigs consisted of only five sows, the mean number of offspring—the per-capita birth rate—in any year would be the average of five numbers drawn at random from the distribution shown in Figure 4.2. Because this is a small sample, our result can be easily biased by one or two unusual litter sizes. By contrast, if our herd con-sists of 100 sows, our mean number of offspring will be the average of this much larger sample and so will, each year, tend to be much closer to the true mean. The occasional low or high litter number is averaged out when the population is large. Because of this, there will be less year-to-year variation in the per-capita birth rate when our population is large (Figure 4.3).

Thus, stochasticity matters more in small populations and less in large pop-ulations. In fact, as population size approaches infinity, stochastic models often converge to the deterministic model. Thus, we can see deterministic models as very good descriptions of the world when population size is very large (as it is in the experimental systems that use ciliates and bacteria). By contrast, when population sizes are small (such as in the beetle experiments), stochastic models become necessary and useful.

In this context, it is important to remember that, on invasion fronts, there is always a small population on the leading edge. Because of this, endogenous stochastic processes might very often prove to be important on natural invasion fronts. Until recently, these endogenous sources of stochasticity did not get a great deal of attention in invasion biology. It is almost as if Skellam's concentric circles were taken too seriously. Where effort had gone into understanding stochasticity, the assumption was that the environment was the primary source of variance in spread rate. Most of the effort had gone into imagining environmental effects on invasion speed, and how a stochastically changing environment might affect invasion speed (Lewis et al., 2016).

In population and evolutionary biology more broadly, it was well understood that endogenous stochasticity was important, particularly for small populations.

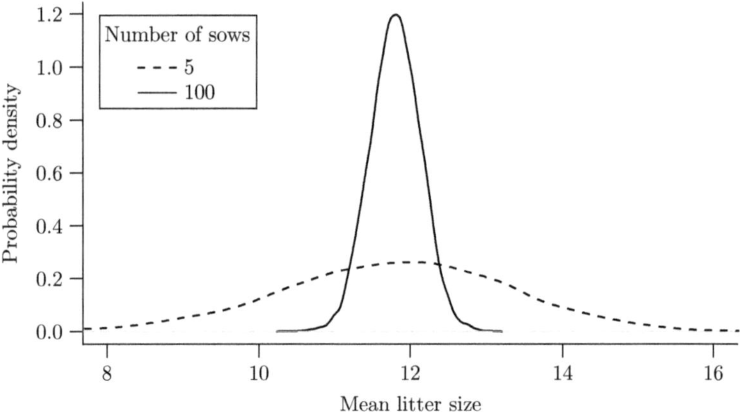

Figure 4.3 *The law of large numbers causes stochasticity to be a milder force in large populations. An example is the annual per-capita birth rates resulting from sampling the distribution of litter sizes in Figure 4.2. Here we estimate (by resampling) the distribution of mean litter sizes resulting from samples of either 5 or 100 sows.*

Here we had long-standing ideas about stochastic extinction (the probability of going extinct increases as population size gets small), and evolutionary ideas around genetic drift (drift—a stochastic process—can overwhelm selection when population size gets small). But these ideas had not really been applied to invasions.

This started to change in the early 2000s, and the results have been remarkable. Demographic, dispersal, and evolutionary stochasticity playing out in the small population of the invasion front can cause a surprising amount of variation in outcomes. And all of that variation can happen in a completely stable and homogeneous environment. By 2009, researchers were talking about fundamental limits to our ability to predict invasions, and this is purely from considering demographic stochasticity. When evolution joined the party a few years later, things got even messier.

4.4 Sampling distributions, and the law of large numbers

Before we step through each source of stochasticity, it is worth generalizing our observation about small population sizes (and if you are rusty on probability distributions, you might want to read through Appendix B first). Doing so will help us understand quite a range of upcoming phenomena.

We have made a verbal argument that small populations are more strongly influenced by the occasional weird individual. We have talked about litter size in pigs as a random variable, S. A sow that has a litter size of 1 will have a big impact on the mean litter size of her cohort if she is one of only three individuals in the cohort. Her weight with regard to calculating the mean is 1/3. If she is one of 1,000, on the other hand, her weight with regard to the mean is only 1/1,000 and her effect on the mean will be tiny.

To simplify the world, we often try to average out random outcomes. In probability models, we refer to the mean value of a random variable, X, as the 'expected value of X', $E(X)$. In the real world, we refer to the mean value of a sample of observations (a sample of x's) as $\bar{x}$. $E(X)$ and $\bar{x}$ are not the same thing. The uppercase X denotes a theoretical possibility, an unevaluated value; the lowercase x denotes a real (evaluated) observation. $E(X)$ denotes the true mean value for the population; $\bar{x}$ denotes an *estimate* of the true mean, taken from data.

You may not have noticed, but to make Figure 4.3, I made a quiet little assumption. I took the observed distribution of litter sizes in Figure 4.2 to be the true distribution of S, and I simulated what would happen if we took lots of samples from that true distribution. By making that assumption, I was claiming to know that $E(S) = 11.8$, and that the distribution of litter sizes shown in Figure 4.2 is the true distribution of S. I then took a random sample of $n = 5$ litter sizes $\{s_1, s_2, s_3, \ldots, s_5\}$ from that distribution, found the sample mean, $\bar{s}_{n=5}$, recorded that value, took another sample of size $n = 5$, and repeated that process a thousand times. The result is that I got a distribution of values for $\bar{s}_{n=5}$. I repeated this process with a larger sample size ($n = 100$) to get a distribution of values for $\bar{s}_{n=100}$.

When we plot these two distributions (Figure 4.3), it is very clear that they differ. What we have plotted is the *sampling distribution* for the mean, $\bar{s}$, estimated for two different sample sizes. In both cases, you can see that the distribution of $\bar{s}$ is perfectly centred on $E(S)$, but results vary from sample to sample. The distribution of $\bar{s}_{n=5}$ is much wider (more variable) than the distribution of $\bar{s}_{n=100}$. In short, larger sample sizes tend to give estimates of the mean, $\bar{s}$, that are close to the true value, $E(S)$; smaller sample sizes give much more variable estimates.

This idea of the sampling distribution can be applied to any parameter in a probability model. For example, the normal distribution has a parameter for the true mean, $E(X) = \mu$, and a parameter for the true variance, $\text{Var}(X) = \sigma^2$. Each of these parameters can be estimated from a sample, and each estimator will have its own sampling distribution. Statisticians have usually worked out the theoretical sampling distribution for important estimators. For example, the mean is estimated with the sample mean, $\bar{x}$, and this is distributed according to a normal distribution: $\bar{x} \sim \text{Normal}(\mu, \frac{\sigma^2}{n})$. The variance is estimated with the sample variance s^2 and this also has a sampling distribution, but it is a scaled chi-squared distribution rather than a normal distribution (if you are a glutton for punishment: $(n-1)\frac{s^2}{\sigma^2} \sim \chi^2(n-1)$). In both cases, the variance in the sampling distribution declines as sample size increases.

This tendency for the variance of the sampling distribution to diminish as sample size (n) increases is a very general result. It is so general that it has been

[2] This 'law' does have limits. Statisticians have managed to find distributions/estimators for which the LLN does not hold. 'Evil distributions' as a colleague once described them.

elevated to the status of a statistical law: the law of large numbers.[2] The LLN is a nice shorthand for understanding why small populations suffer relatively greater stochasticity. In small populations, we are averaging over a small number of individuals: n is small, and so the LLN tells us we can expect a lot of sampling error.

While we average over individuals to find mean demographic parameter values at particular places and times, we might also average over other variables, like time or space. An invasion that has only been progressing for three generations, for example, will give us a much messier estimate of invasion speed than an invasion that has been progressing for 100 generations. In much of what follows in the next few chapters, we often end up with an argument stemming from the LLN.

4.5 Demographic stochasticity

What are the processes underlying the endogenous demographics of a population? Births and deaths, obviously. But there are also less obvious processes. In a population with two sexes, for example, sex allocation is a demographic process. We might also imagine that there is (non-heritable) heterogeneity in demographic rates caused by, for example, variation in somatic growth, age at maturity, and so on.

Each of these processes—sex allocation, births, deaths, demographic heterogeneity—has a stochastic element to it. Each of these processes works together in a multiplicative way to determine the number of individuals in the next generation. To bring this back to the dice analogy, it is like we flip a coin to determine sex; if we have flipped a female, we then have a (many-sided) di that determines her expected fitness; a (many-sided) di that, given expected fitness, determines the actual offspring number for that female; another di that, given the offspring number, determines the number of surviving offspring for that female. We flip this coin and roll all of these dices to work out the number of surviving offspring that recruit to our population from that individual. We then repeat this process for each individual in the population. Finally, we add up these results across all individuals to get the total number of offspring recruiting to the population in this round of the game.

That's a lot of dice rolling. And when we look at things this way, it is not surprising that populations rarely follow nice smooth growth trajectories. We also start to appreciate another reason why invasions in ciliates (that reproduce by binary fission) might be more predictable than invasions in beetles (that reproduce sexually): beetles have a more complex life history, and so generate more avenues for stochastic effects to compound upon each other.

These stochastic growth trajectories matter because stochastic variation in population size, all on its own, can cause populations to go extinct. In deterministic population models, by contrast, extinction never happens. In deterministic models, it is possible to have fractions of an individual.[3] In deterministic models, this fraction can just become arbitrarily small, forever, without actually hitting zero.

So how do we add demographic stochasticity into population models? Well, there are two pathways. The first (and intuitively simplest) is simulation: we can draw realized values at each time step from a distribution of possible values, and iterate our population through time. The second (and more powerful) pathway is a formal analysis of transition probabilities. We will sketch these two pathways now, imagining only a single source of stochasticity. We will start with simulation, and then give a brief sketch of how these relate to transition probabilities.

4.5.1   Simulating demographic stochasticity

Let's start with a very simple model of geometric growth. Let's imagine a population of hermaphrodites in which all individuals lay eggs each spring and then die. These hermaphrodites live in some halcyon environment that is perfectly homogeneous in space and time. With these assumptions, we get rid of most sources of stochasticity, except births. There is no overlap of generations, and the population grows in discrete intervals of time, so a deterministic model would say that

$$N(t + 1) = bN(t). \tag{4.1}$$

Here, $N(t)$ is the number of individuals at time t, and b is the mean number of surviving offspring produced by an individual. As stated, this is a deterministic model. If we specify a starting value for N and a value for b, we will get the same growth trajectory every time we iterate this equation.

One way to make this model stochastic would be to draw the number of individuals in the next generation from a probability distribution where the mean of that distribution is given by the deterministic expectation. We introduce stochastic variation around a deterministic skeleton. So equation (4.1) says we should have $bN(t)$ individuals; we can make this the mean of some distribution and then take a random sample from that distribution (see Appendix B). A convenient and not so arbitrary choice of distributions[4] is to draw a sample from the Poisson distribution with a mean $\lambda = bN(t)$.[5]

$$N(t + 1) \sim \text{Poisson}(bN(t)). \tag{4.2}$$

This notation should be read as $N(t + 1)$ is 'drawn from' a Poisson distribution with a mean of $bN(t)$. There is no 'equals' sign any more, because we are trying to denote that $N(t + 1)$ is no longer precisely determined; it is a stochastic outcome.

If we start at 10 individuals (*i.e.*, $N(0) = 10$), and iterate over these two equations, we can see the very obvious stochastic effect at play (Figure 4.4). Compared to our deterministic model, some of the stochastic realizations grow a lot faster, some grow slower, and others go extinct. Generally, the variation

[3] This fractionalizing of individuals is often justified by redefining N as a density of individuals, rather than a count. So it becomes, say, 0.1 individuals per hectare, instead of 0.1 individuals.

[4] As we will see later in Section 4.5.2.

[5] The Poisson has only a single parameter, usually denoted λ, that is both the mean and the variance.

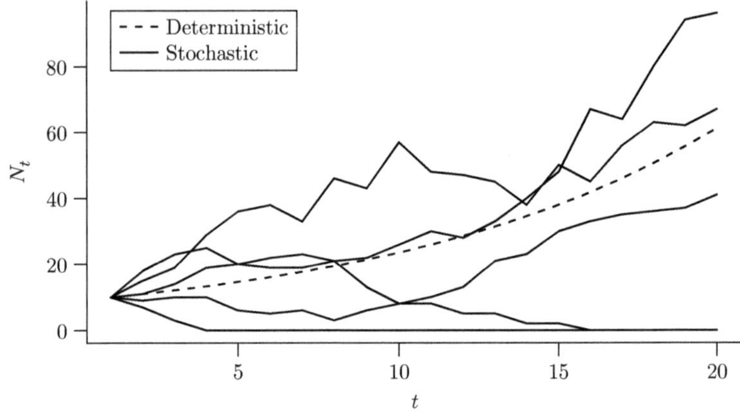

Figure 4.4 *Comparison of deterministic and stochastic models of geometric population growth. Here, b = 1.1, and we compare five stochastic realizations of the model in equation (4.2) versus the deterministic model in equation (4.1).*

between realizations steadily increases as time goes on. There is a sensitive dependence of these longer term trajectories on the initial few time periods. A lucky break early on can lead to hugely larger population sizes 10 generations later. Conversely, an unlucky break can be hard to recover from. In fact, these non-linearities mean that the expected growth rate for the stochastic model is less than the growth rate of the expected number of individuals (Lewontin and Cohen, 1969): if we ignore stochasticity, we might tend to overestimate the growth rate of the population.

So here we have a simple population model that can give very different outcomes, depending on chance. And while the deterministic model yields endless, predictable, smooth growth, the stochastic model allows populations to fluctuate over time, and even to go extinct. The stochastic model also indicates that the growth a population achieves will often be less than what we might naively expect from a deterministic model. This last result is a bit intriguing, and so worth unpacking a little; we will do so in Section 4.5.3.

4.5.1.1 *Logistic growth*

The previous model had a constant growth rate, determined by births, b. Often, however, we are interested in populations that self-regulate, through negative density dependence. A simple model here is the classic logistic growth model, in which

$$N(t+1) = N(t)\left(1 + b\left(1 - \frac{N(t)}{K}\right)\right). \tag{4.3}$$

Here we have the same assumptions as the earlier model (equation (4.1)), but now we are assuming that the number of births decreases linearly as density increases. As before, we can make this a stochastic model by the simple expedient of drawing the realized number of individuals at $t+1$ from a Poisson distribution with the expected value of $\lambda(t) = (1+b)N(t)(1 - \frac{N(t)}{K})$. Or,

$$N(t+1) \sim \text{Poisson}\left((1+b)N(t)\left(1 - \frac{N(t)}{K}\right)\right). \qquad (4.4)$$

As before, we can compare the outcomes from iterating deterministic and stochastic versions of this model, and the results can be seen in Figure 4.5.

We can see that the logistic model seems to settle into jiggling around the carrying capacity with what looks like a constant variance.

4.5.2   Formal description of demographic stochasticity, and transition probabilities

We will now look at a more formal way to approach this same problem: one in which we start at the individual and aggregate up to the population. To make this model stochastic, we might instead say that each individual has the same expected number of offspring, $\bar{b}$, but the actual number of offspring accruing to individual i is a stochastic process. That is,

$$b_i = \bar{b} + \epsilon_i, \qquad (4.5)$$

where ϵ_i is some random deviation from $\bar{b}$ that is experienced by individual i; these deviations will be in the negative or positive direction, and each individual gets its own deviation (independent of every other individual). The values of

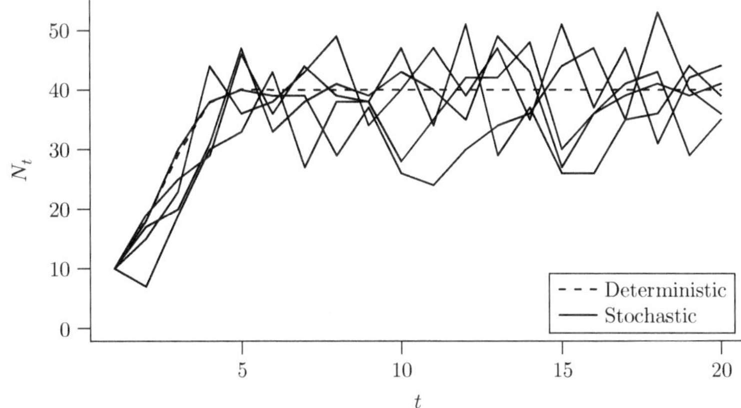

Figure 4.5 *Comparison of deterministic and stochastic models of logistic population growth. Here, b = 1.1, and we compare five stochastic realizations of the model in equation (4.4) versus the deterministic model in equation (4.3).*

epsilon are drawn from some probability distribution with an expected value, $E(\epsilon) = 0$. The population-level mean birth rate is now time-dependent (because every generation will have a different set of ϵ_i): $b(t) = \frac{1}{N(t)} \sum b_i = \bar{b} + \frac{1}{N(t)} \sum \epsilon_i$. That is, the mean birth rate at any given time is the overall mean birth rate, plus the mean of the individual random deviations in that time interval. Given the LLN, as N becomes very large, we can expect the mean of our random deviations, $\frac{1}{N(t)} \sum \epsilon_i$, to approach $E(\epsilon) = 0$. Our deterministic model has now become stochastic, because the birth rate now changes (stochastically) each time step:

$$N(t + 1) = b(t)N(t). \tag{4.6}$$

The only question remaining, really, is how we generate each value of $b(t)$. So far, all we have said is that each $b(t)$ represents some random deviation from the mean. Here's where probabilities come into play; we need a probability model for $b(t)$.

The Poisson is a probability model (see Appendix B) that describes the number of events that might happen in an interval of time. Given some mean rate (events per time), the model tells us the probability that we might see 0, 1, 2, ..., ∞ events in any given interval of time. Applied to reproduction, we might claim that sows produce mature egg cells at a given rate, and in the interval of time that fertilization is possible, we might thus see a Poisson distribution of offspring across individual sows. It is a reasonable argument, and the Poisson also often provides a reasonable fit to birth data (e.g., Figure 4.2). The Poisson is a discrete probability distribution—it only describes integer values—so numbers drawn at random from a Poisson can only be integers, which is handy given we are trying to describe numbers of individuals and fractional individuals don't really make sense. The Poisson probability density function says that, given a rate at which events happen, λ, the probability of seeing a given number of events, k, is: $P(X = k \mid \lambda) = e^{-\lambda}\frac{\lambda^k}{k!}$. The Poisson has the very handy property that λ also happens to be both the mean and the variance of the distribution.

So now, instead of saying that $b_i = \bar{b} + \epsilon_i$, we can specify the probability of a given number of offspring ($B = b_i$) accruing to an individual:

$$P(B = b_i) = e^{\bar{b}}\frac{\bar{b}^{b_i}}{b_i!}. \tag{4.7}$$

To bring this back to the growth of the population, we need to sum offspring across all our N individuals in the population. We will skip a little math here, but it turns out that this sum is also distributed as a Poisson distribution, with a mean given by $\bar{b}N(t)$. So now we have:

$$P(N(t + 1) = k \mid N(t)) = e^{-\bar{b}N(t)}\frac{(\bar{b}N(t))^k}{k!}. \tag{4.8}$$

This is the formal probability model describing the probability of transitioning from $N(t)$ to any given value of $N(t + 1)$. It is, like the deterministic model, a recursion equation. As such, it is amenable to analysis in which we ask what the equilibrium is: that is, an analysis in which the transition probability no longer changes over time. Such analyses can unearth what is known as the 'stable distribution'. In our case, the stable distribution would be a distribution of population sizes that the population samples over time. You can more or less see the stable distribution of population sizes emerge in Figure 4.5 as the simulated populations wiggle around the carrying capacity, sampling their stable distribution. By the same token, you can see that a stable distribution in Figure 4.4 is likely to be a function of time, because the mean population size endlessly grows, and the variance around it also appears to increase over time.

4.5.3   Divergent expectations

At this point, we are in a position to show how it is that the growth rate of this stochastic model is less than that of the deterministic precursor; another little insight from Lewontin (Lewontin and Cohen, 1969). If we are modelling geometric growth with a fixed value of b, the general (deterministic) solution is

$$N(t) = b^t N(0). \tag{4.9}$$

But when b varies over time, we have to multiply out across all of the $b(t)$s:

$$N(t) = N(0) \prod_t b(t). \tag{4.10}$$

We might naively think that we could calculate the mean of $b(t)$ over time, $E(b) = \frac{1}{t} \sum_t b(t)$, and use this to work out the number of individuals at time t, by simply replacing the b in equation (4.9) with $E(b)$. But this is incorrect, because we can see from equation (4.10) that we would be saying that $E(b)^t = \prod_t b(t)$, which is clearly untrue (except when b is a constant). But surely the expected growth to time t could be explained by the mean growth rate? Well, kind of. Let's take equation (4.10), divide both sides by $N(0)$, take the log of both sides, and then divide both sides by t:

$$\frac{1}{t}(\ln N(t) - \ln N(0)) = \frac{1}{t} \sum_t^t \ln b(t). \tag{4.11}$$

Now we have an arithmetic mean on the right-hand side, but it is the mean of the log of the growth rates, $E(\ln b)$. And on the left-hand side, we have the mean change in population size (on the log scale) per unit time. So for geometric growth, we have to calculate the mean growth rate on the log scale, not the original scale. Well, that's fine; we can now put that mean back onto the original scale: $e^{E(\ln b)}$. So now we have a quantity that we can stick into the deterministic equation (equation (4.9)) in the place of b:

$$N(t) = e^{E(\ln b)t}N(0). \tag{4.12}$$

So far, so good. But there is a hitch: it turns out that if you take the mean of a set of varying numbers on the log scale and then push the result back to the original scale, you will *always* get a result that is less than the mean on the original scale.[6] In our case, this manifests as $e^{E(\ln b)} < E(b)$ whenever there is variation in b over time. When the growth rate is identical every generation (as in the deterministic model), we don't notice this issue because we aren't averaging over any variation, and Jensen's inequality has slipped out the side door. But when there is variation, as in a stochastic model, then the growth achieved by the population is less than what we would expect from the mean of the growth rates over time.

Hopefully, you followed that argument. It took me a while to get it the first time. And it seems like a very important result: stochasticity slows growth (and so, presumably, invasions). It seems important, but there is something a little unnerving about this argument: it vanishes as soon as we go from discrete to continuous time. The reason for this is that in continuous time (the exponential model), the growth rate is already sitting up above Euler's number:

$$N(t) = e^{rt}N(0). \tag{4.13}$$

In this case, were we to go through the same argument, we would find that $E(r)$ is the correct value for calculating population size at any given time.

So this argument vanishes when there is no variation in birth rate, and it vanishes when there is variation and growth in continuous time. It only matters when we consider discrete-time models, and so it seems unclear to me how important this effect is. Does it really happen in nature, or is it just an idiosyncrasy of discrete-time models? I think the jury is out on that one, but it is worth noting that there are many populations where discrete time seems like the best description. Many species reproduce seasonally, for example, and are probably best described with a discrete-time model. And of course continuous-time models are very closely approximated by discrete time if we make the time intervals sufficiently small. On balance, then, it seems likely that this effect may be quite common in nature.

4.6 More complex models of demographic stochasticity

When presented with a wiggly population trajectory such as that in Figure 4.5, most ecologists would suggest that this random bouncing around is most likely caused by the environment. We all know that there are good years and bad years. Sometimes it rains, sometimes it doesn't, and sometimes it floods. Sometimes rogue poets are roaming the fields, and sometimes they are not. It is very easy to see these effects and so, naturally, we assume that they cause most of the variation

[6] This is just one of those true things; it's called Jensen's inequality. Jensen's inequality refers to any curving function, not just the log function.

in population size. But we know (having just made those population trajectories) that all the stochasticity is coming from random variation in the number of births. In our case, the environment has nothing to do with it.

So what are the relative roles of environment versus demography in generating variation in population size? This question matters, because stochasticity can cause extinction, and it would be nice to know why populations go extinct. And here is where models matter. Typically, to work out how much variation is caused by the environment, we would first account for the endogenous sources of variation. Whatever is left over—the residual variation—must be from the environment. If we don't account for all endogenous sources of variation, we will overestimate the environmental effect.

In 2008, Brett Melbourne and Alan Hastings set out to carefully enumerate and model all sources of demographic variation in a population (Melbourne and Hastings, 2008). As well as working out the math, they wanted to test their models against an experimental system: flour beetles grown in a constant environment. (This is the same beetle system that generated the variable invasions in Figure 4.1, but in this case they were just looking at population growth in discrete jars, rather than invasion speed across arrays.) Given that they have provided a constant environment, could they build a stochastic model of population growth that recapitulates all the variation they saw in their beetle populations?

As mentioned earlier (Section 4.5), births are just one of several processes in an individual's life that have a stochastic element to them. We could add additional processes to the simple models above to capture these additional sources of stochasticity; we could add more dice to our game. Doing so would add more variation to any distribution that emerged. The growth trajectory of the population would become more uncertain with every additional source of variation that we included. We could add sex determination as a draw from a binomial distribution, survival as an exponential process, and so on. When Melbourne and Hastings did this, they found that formally incorporating all these stochastic effects into the population model (as we did in Section 4.5.2, but with more processes) could have a huge impact on the variance in population size (Figure 4.6).

Importantly, when they fitted these models to their data from beetle populations, they found that the best fit model was the most complex one ('NBBg' in Figure 4.6); the model accounting for all sources of demographic stochasticity. This model did a very good job of accounting for their data. By contrast, the simplest model, accounting only for stochasticity in the birth process (as we did above), did an abysmal job.

Their work argues strongly that endogenous demographic stochasticity can be a big source of variation in population size (and so a big source of extinction risk), particularly at small population sizes. Their work also implies that we might often have underestimated the role of (endogenous) demographic stochasticity, because unless we account for it very carefully there will be a lot of leftover variation, and for want of a better plan we will assign that residual variation to (exogenous) environmental effects.

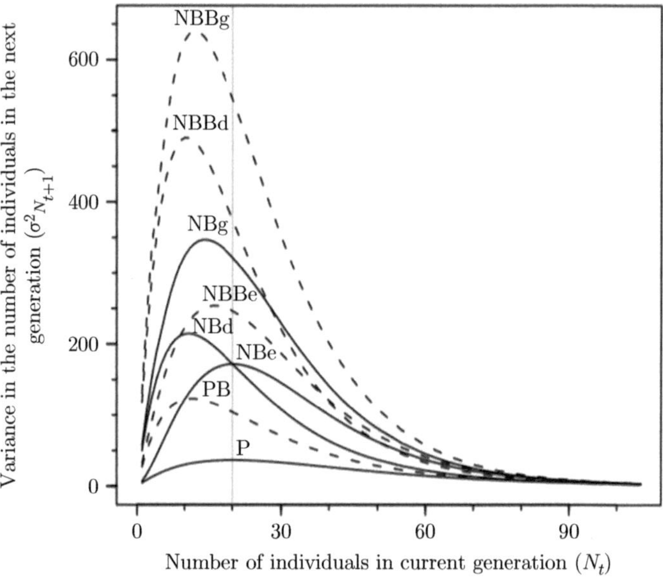

Figure 4.6 *Model assumptions make a huge difference to variance in population size. Variance in the number of individuals in the next generation as a function of the number of individuals in the current generation is shown for eight models developed by Melbourne and Hastings (2008). Model 'P' is the simple Poisson model, similar to that developed in equation (4.4). Models get more complex from there, eventually incorporating all forms of demographic stochasticity in model 'NBBg'.*

All of this has bearing on invasions, because population growth is one of the two fundamental processes that cause populations to spread. Make that population growth stochastic, and spread rates will become stochastic. Make that population growth stochastic for demographic reasons, and spread rates will be stochastic even in a homogeneous environment.

4.6.1 Hierarchical stochastic models

For those interested in how such a complex model might be put together, a quick sketch. The analogy I made earlier about flipping coins and rolling dice is, in fact, a hierarchical probability model. We can gang together lots of stochastic effects into arbitrarily complicated models using this basic strategy in which we let random outcomes determine the settings for the next random outcome. For example, we might say that

$$N(t + 1) \sim \text{Poisson}(bN(t))$$

but then make b also a random variable. We might say that b varies because of environmental effects such that

$$b \sim \mathrm{Normal}(0, \sigma_e).$$

And then we might make σ change over time as a random variable, drawn from, say, a Gamma distribution. And so on.

Clearly, we would need to put very careful thought into precisely how and why we are incorporating these various levels of stochasticity. But the point here is that by making the parameters of our probability models themselves subject to random effects, we can build up models that describe stochastic processes operating across levels of organization, from individuals through populations and the environment. These models are very flexible, and are brilliant for incorporating processes into analytic and inferential models alike (e.g., Royle and Dorazio (2008)). They can, however, become very complicated, and as a consequence will often defy formal analysis and require us to make conclusions from simulation.

4.7 Stochastic dispersal

Population growth is only one-half of the dynamic duo that drives invasions. Dispersal is the other half. By contrast to many demographic processes, dispersal has nearly always been treated as a probabilistic process (e.g., Neubert et al., 1995). Diffusion, for example, is a model arising from random movements of particles (Appendix A), and in simple discrete space models individuals are often treated as having a fixed probability of dispersing. In the same way that stochastic demographic models imagine processes at the individual level, and aggregate up to form probability distributions describing the entire population, so too with dispersal. Dispersal kernels can often be built from process models describing probabilistic movement rules at the 'micro' (*i.e.*, individual in a small time interval) level, and then aggregating over time and individuals to generate a dispersal kernel—a distribution of dispersal distances.

While dispersal models nearly always talk in terms of probabilistic dispersal, they are often made deterministic by making the simplifying assumption that the population size is infinite. When we do this, the population-level randomness vanishes: for example, precisely $x\%$ of individuals at A move to B, every time, or the mean dispersal distance in the population becomes a constant. This large population assumption is just another manifestation of the LLN. This is how the reaction–diffusion model (Section 2.1) can give deterministic spread rates even though diffusion itself arises from random processes at the individual level; population size is considered infinite.[7] Deterministic integrodifference models also use a dispersal kernel that is effectively assuming an infinite population size. A stochastic model would instead have a finite population size and so only *roughly* $x\%$ of individuals at A moving to B each time: the mean dispersal distance in the population varies randomly each generation.

To make this distinction concrete, let's compare how dispersal distances might be distributed as our population size increases. Let's take the same dispersal kernel

[7] Here we reveal a slight internal tension in deterministic invasion models: population size is infinite for the purposes of dispersal, but finite in that the population has a fixed carrying capacity.

(and let's use the normal distribution again; a kernel consistent with diffusion), but we compare three different-sized populations: $N = 20$; 100; $1,000$ (Figure 4.7). It is very clear that as our population size increases, the 'realized' dispersal kernel starts to look like the 'expected' kernel. When population size is small, however, things are a little messy. The LLN again.

We can also make this concrete with some simple statistical arguments around sampling distributions (Section 4.4). Imagine the simplest dispersal model: nearest-neighbour dispersal. Under this model, space is discrete and individuals can only disperse to the neighbouring patch. This is the dispersal model used in our model of spatial and temporal fitness (Section 3.3). Under nearest-neighbour dispersal, a fixed proportion of individuals, m, migrates to the next patch. So if we have N individuals in a patch, $N_m = mN$ individuals disperse. Note that this is a precise number; a deterministic model. We can make this stochastic by simply redefining m as a probability rather than a proportion, and saying that individuals disperse with a fixed probability, m. Now, instead of a precise number, we have a distribution of outcomes. Instead of the number of dispersers, N_m, being identical for a given N and m, it varies according to a binomial distribution:

$$N_m \sim \text{Binom}(m, N).$$

Now, each time we go to work out how many of the N individuals will disperse, we don't know what the answer will be in advance: it will be a number sampled from

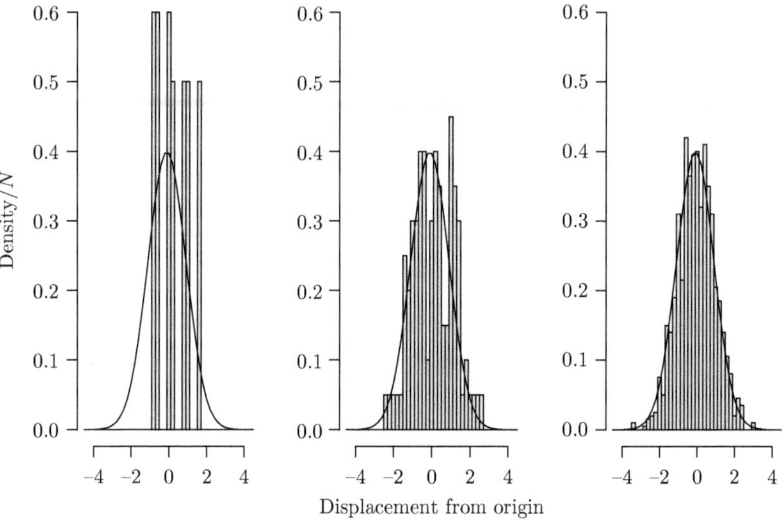

Figure 4.7 *Comparison of distribution of dispersal distances arising from population sizes of 20, 100, and 1,000, respectively. All populations assume that dispersal follows a normal distribution with a mean of 0 and a standard deviation of 1. This expected distribution is plotted over the realized distribution.*

this binomial distribution.[8] If $N = 10$ and $m = 0.5$, for example, the actual number dispersing might be anywhere between 0 and 10, though 5 (i.e., mN) would be the most likely number (see Figure B.1). The variance in the *proportion* of individuals dispersing comes from the sampling distribution of the proportions and so will be $\frac{m(1-m)}{N}$. We can see the LLN again here: as N increases, stochasticity in dispersal decreases.

In a continuous space, the most common dispersal model is diffusion. Diffusion arises from the sum of many random back and forth movements, and gives rise to a normal distribution of displacements with a mean $\mu = 0$ and variance $\sigma^2 = 2Dt$ (Appendix A). So were we to simulate diffusion, we could skip the millions of random back and forth movements of individual particles, and just draw individual displacements in one unit of time, d_i, from a normal distribution according to

$$d_i \sim \mathrm{Normal}(0, 2D),$$

where $\mathrm{Normal}(\mu, \sigma^2)$ is the normal distribution. This is, in fact, what we just did to produce Figure 4.7. These individual samples, in aggregate, would be a sample of size N from the above distribution. And of course a finite sample of size N from this distribution will rarely yield a mean displacement of zero. Instead, the mean displacement in any given generation will be a sample from another normal distribution with $\bar{x} = 0$ and variance $s^2 = \frac{\sigma^2}{N}$, this being the sampling distribution of the normal distribution. The mean squared displacement (the square of the mean distance moved from a starting position) will also vary randomly according to a χ^2-distribution with $N - 1$ degrees of freedom (Section 4.4), this being the sampling distribution of the variance of a normal distribution. Given these considerations, it is once again easy to see why a small population size will give more stochastic variance in dispersal than a large population size.

The above are informal arguments about how we might formally model stochastic dispersal. Making dispersal stochastic often feels like less of a leap than making demography stochastic because dispersal is nearly always described with a population-level distribution of dispersal probabilities. Demography, by contrast, is more commonly treated with point values for growth rate and carrying capacity, so it takes an adjustment to treat these as probabilistic. As with demographic stochasticity, however, we can expect stochasticity in dispersal to be a much stronger force when population size is small.

4.8   Population size on invasion fronts

One of the points I have been making *ad nauseum* is that demographic and dispersal stochasticity matter more in small populations than in large ones. I have also pointed out that the leading edge of an invasion is, by definition, a much smaller population than that in the core. And all follows from the leading edge,

8 For details on the binomial distribution and sampling, see Appendix B.

so we have a sense that stochasticity might be quite a big deal in invasions. But we have also seen that some invasions seem to exhibit a great deal of stochasticity in spread rate, whereas others are much more consistent across replicates. Presumably, this reflects variation in population size on the invasion front, but how do we measure population size on the leading edge of an invasion?

In a 'well-mixed' population, population size is simply a count of all the individuals. To measure population size in a spatially explicit population, however (one that is not well-mixed), requires us to define a spatial domain across which we count individuals. How do we define this spatial domain in a useful way on a moving invasion front? It is not a trivial question to resolve.

If we focus on the processes in question, we might see that there are actually two relevant spatial scales and so 'population size' might vary depending on the process we are considering. For demographic stochasticity, the relevant scale is the scale at which the demographic processes of birth, death, and intraspecific competition play out. For dispersal, the relevant scale is the scale at which dispersal plays out. For many organisms, birth, death, and competition are intensely local phenomena: they are often defined by interactions between individuals in the population. This fact of demography—that it involves interactions with conspecifics—makes demography strongly localized. Dispersal on the other hand does not usually require interactions and so is free to act across a much larger spatial scale. As a consequence, it may often be the case that the population size for demographic processes will be much smaller than the population size for dispersal processes. As a consequence, we might expect demographic stochasticity to be a stronger force than dispersal stochasticity.

Regardless of these considerations, when we look at an invasion front, we can see that the shape of the wave front has a very large bearing on population size on the front. The population size in a spatially continuous population will equal the area under the density curve. If we consider the area under a steep invasion front, versus the area under a shallow invasion front, it is clear that the shallow invasion front contains a larger area, and so a larger population size (Figure 4.8). We learned in Section 3.5 that, for a given dispersal kernel, increasing the population growth rate will cause the wave front to steepen (Figure 3.8). In fact, the shape of the wave is a function of the growth rate and the diffusion coefficient (Fisher, 1937): the wave gets steeper as the growth rate increases or as the diffusion coefficient decreases.

The above considerations lead us to the intuition that the correct spatial scale for assessing population size might actually be set by the parameters r and D. In fact, it is. Under the Fisher–KPP model, the length of the invasion front is proportional to $L = \sqrt{D/r}$. This value, L, is the spatial scale of the invasion front. The units of L are units of space (metres, kilometres, and so on). Invasions with a large value of L will have wider invasion fronts and larger population sizes than invasions with a small value of L. We still do not have a precise measure for the population size on the invasion front, but we have a powerful method for ordering different invasions by population size: invasions with a high ratio of $D{:}r$ will experience much less stochasticity than those with a low ratio of $D{:}r$.

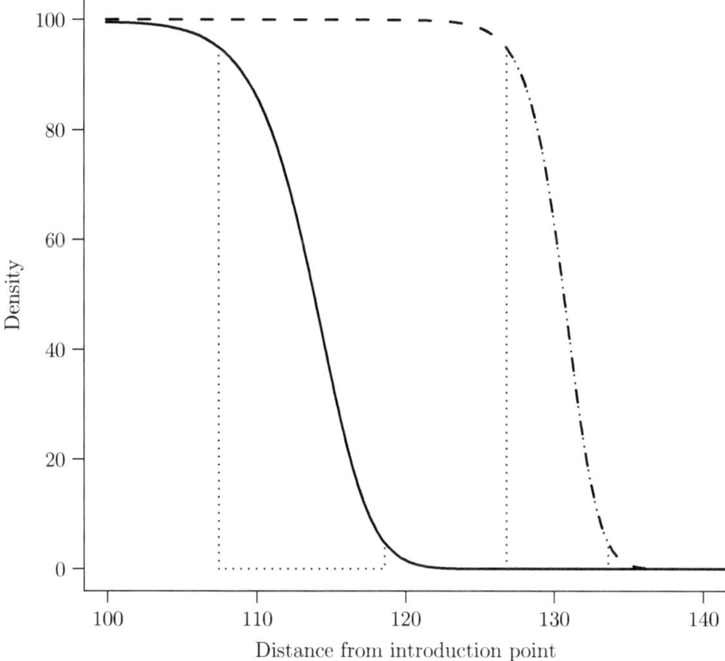

Figure 4.8 *Population size depends on wave shape. Here we compare the leading edge of two reaction–diffusion invasions that differ only in growth rate. The dashed line shows the invasion with the higher growth rate. For each invasion, population size is the area under the curve. If we define the region running between a density of 5% and 95% as the invasion front, it is clear that the length of the two invasion fronts is different and so the shaded areas under the two invasion fronts are different, with the shallower (low growth rate) invasion front on the left-hand side containing a larger population size.*

4.9 Environmental stochasticity

In the same way that we can introduce stochasticity to the processes affecting individual dispersal and reproduction, we can also introduce environmental stochasticity. Environmental stochasticity sits at the population, rather than individual, level. While demographic and dispersal processes apply stochastic effects to each individual, environmental stochasticity applies stochastic effects to the entire population in each time period (and/or in each part of space). Because environmental stochasticity applies to all individuals equally, its effect on population is not really mitigated by large population size. The variation across years is constant, and not intrinsically tied to population size as it is with demographic stochasticity.

In a model with both demographic and environmental stochasticity, each individual gets stochastic effects associated with the individual, and with the time (or space). To equation (4.5), for example, we might add a temporal environmental effect:

$$b_{i,t} = \bar{b} + \epsilon_i + \epsilon_t, \tag{4.14}$$

where ϵ_t is a temporal stochastic effect accruing to all individuals in the population caused by the environment at that time. In the dice analogy, it is like we just went through that long process of flipping coins and rolling a family of dice to work out how many offspring each individual has (Section 4.5), we add up all those results across all individuals, and then, at the end, we roll another dice that adjusts all individuals by the same small (and random) amount.

Where environmental stochasticity is sometimes made more complex is in making ϵ_t correlated over time or space. We certainly expect environmental effects to be correlated across both space and time. With regard to space, this spatial autocorrelation is often stated as the first law of geography: 'everything is related to everything else, but near things are more related than distant things' (Tobler, 1970). And the same basic premise applies to time also.

4.10 Stochastic invasions

A formal description of stochasticity in both demography and dispersal and how this affects spread is a formidable mathematical undertaking. Certainly beyond what I am able to explain. It is also a mathematical frontier: a place where the mathematics 'gets interesting', which I am fairly sure is how applied mathematicians say that they don't really know what is going on yet, either. So for this section, we will try to wrap our heads around it using a combination of analogy and intuition, rather than formal analytic models.

But before we embark into theory again, let's revisit those flour beetles that Melbourne and Hastings were messing around with. You will recall that they went to considerable effort to build a mathematical model that accounted for all the demographic stochasticity in their beetle populations. They found that it was only by accounting for all sources of stochasticity that they could get their models to fit their data. But their models did, ultimately, fit very well (Melbourne and Hastings, 2008). Perhaps feeling encouraged by this initial success in explaining population growth, they embarked on a slightly more ambitious undertaking: to run replicate invasions of their beetles, and use a fully stochastic model to explain all the variation they observed in invasion speed (Melbourne and Hastings (2009); see Figure 4.1). Despite working really hard to account for stochasticity in both demography and dispersal, their full model was unable to account for the full range of variation in invasion speeds thrown up by those beetles.

It was Melbourne and Hastings (2009) who suggested that there might be fundamental limits to our capacity to predict invasions. I suspect they are correct. But there was also a hint in their data that there may be additional stochastic processes afoot; one they had not accounted for. Evolutionary processes are also affected by chance, and there is a growing realization that they might also play a large role in the outcome of invasions. More on that in Chapter 5.

But first, a quick review of what we know about how demographic and environmental stochasticity can affect invasion speeds. This is still very much a part of our understanding that is under construction. It is still difficult to draw clear generalities, and broad results often vary depending upon the modelling strategy used, and how, precisely, stochasticity is introduced. Because experimental work on this topic requires numerous replicate invasions to be observed under controlled conditions, empirical studies are still relatively rare.

4.10.1   Divergent fronts, convergent speeds

One result that does seem clear (although slightly boggling) is that for stochastic invasions, uncertainty in the position of the invasion front tends to increase over time, even though the speed of the invasion converges on the equilibrium speed (Figure 4.9; Neubert et al. (2000)). This increasing variation over time in the position of the invasion front is precisely what we see occurring in replicated laboratory invasions (e.g., Figures 1.6 and 4.1), but it seems strange that we can have such divergent positions, but a convergent speed. We can understand this outcome by once again using the LLN. This time, by thinking of the mean speed of the wave as an average of speeds over time: $\bar{C} = \frac{1}{t} \sum C_t$, where C is a random variable denoting the speed of the wave. The sampling distribution for $\bar{C}$ has a variance $\sigma^2 = \frac{\text{Var}(C)}{t}$ (this is just the sampling distribution of a mean). So we can see that as time increases, the variance in speed decreases according to $1/t$. As time increases, we have a larger and larger sample of speeds, and so our mean speed converges on the true equilibrium speed.

So far, so good. The distance an invasion has moved to time t, however, $X_t = \bar{C}t$. This is just distance = speed × time. But this means that the variance in distance moved, $\text{Var}(X_t) = t^2\sigma^2 = t\text{Var}(C)$. So we can see that the variation in the position of the invasion front actually *increases* linearly with time.

Taken together, this implies two things. First, that invasion speeds measured early in an invasion are likely to give you only a vague sense of the equilibrium speed. You might recall that deterministic invasions also take a little while to reach equilibrium speed (Section 2.4). So between this, and the effectively small sample size (over time), early stage invasions can only give an approximate sense of what the final mean invasion speed will be.

The second implication of this result is that forecasting where an invasion will be at a given time gets harder as we look further into the future. This is a perfectly normal and reasonable problem to have. Increasing uncertainty over time is as true for forecasting the weather as it is for the stock market or a biological invasion. It probably should be called the first law of temporal dynamics.

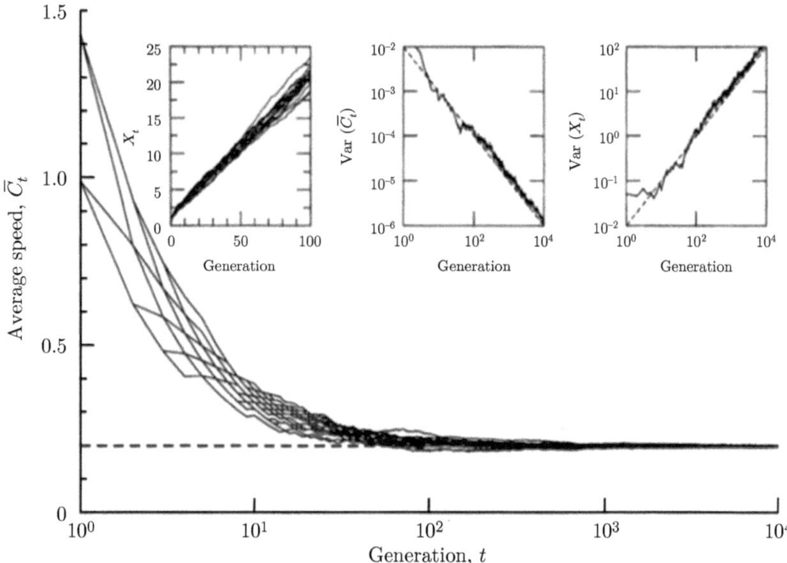

Figure 4.9 *The speed of stochastic invasions can converge on an equilibrium even as the position of the wavefront becomes increasingly uncertain. Twenty realizations of a stochastic invasion model are shown. The main panel shows the speed of each invasion (calculated from starting time to t). Inset 1 shows the position of the invasion front at time t. Inset 2 shows the variation across realizations in the mean speed of the invasion over time. Inset 3 shows the variation across realizations in the position of the invasion front over time. Reproduced from Neubert et al. (2000).*

4.10.2   Stochastic brakes

Another interesting result from stochastic invasion models is that stochasticity may tend to slow invasions down. All that chance acts as a kind of brake on invasion speed. This is a very clear phenomenon of discrete-time models, and arises from the basic argument first set out by Lewontin and Cohen (1969) about divergent expectations in growth rates (Section 4.5.3): stochasticity reduces the mean growth rate of a population and so can be expected to slow an invasion. The argument has been elaborated to invasion speeds by several authors (e.g., Neubert et al., 2000; Schreiber and Ryan, 2011). Although it is a clear effect in discrete-time models, the effect disappears in continuous-time models. Thus, although this is a very clear mathematical argument, its importance depends upon how much your population of interest follows a temporally discrete pattern. A more prosaic reason that stochastic populations may spread more slowly than deterministic models is that deterministic models—tracking density rather than individuals—allow growth and dispersal of even infinitesimally small population densities. Stochastic models, on the other hand, at low population densities go from one individual to none. End of story. So in many stochastic models (particularly simulations), invasions are not pulled forward by the growth and dispersal of fractions of individuals, and so will tend to move more slowly as a consequence (Bridle et al., 2010; Lewis et al., 2016).

4.10.3 Covariation between growth and dispersal

Perhaps the most interesting result from stochastic spread models to date is that covariation between stochastic effects matters, a lot. For example, we might imagine stochastic environmental effects that apply to both reproduction and dispersal. Some years are better than others, and this variation might affect both reproduction and dispersal: when times are good, there is more free energy available for both activities. A nice analogy from plants (Lewis et al., 2016) is that in a good year, the grass stalks might grow taller, and produce more seeds per stalk. The extra height facilitates higher dispersal rates, and the extra seeds facilitate higher growth rates. These positive correlations have been shown in a few models to cause invasions to move faster than they would otherwise (e.g., Schreiber and Beckman, 2020). This is interesting, because while good years might yield a larger than average value of rD, bad years will produce a lower than average value of rD.[9] so it is not immediately apparent that covariation should matter. So why the change in expected speed?

One way to understand this result is that we can specify that

$$r_t D_t = (\bar{r} + \epsilon_{rt})(\bar{D} + \epsilon_{Dt}) \tag{4.15}$$

$$= \bar{r}\bar{D} + \epsilon_{rt}D + \epsilon_{Dt}r + \epsilon_{rt}\epsilon_{Dt}, \tag{4.16}$$

where those ϵs represent small random deviations from the mean of each parameter. When there is a positive correlation between these two random effects, positive values of ϵ_{rt} tend to be associated with positive values of ϵ_{Dt}, and likewise for negative values (see Figure 4.10 as an example). As a consequence, when there is a positive correlation, $\epsilon_{rt}\epsilon_{Dt}$ will tend to be a positive number, and so the mean over time, $\overline{rD} > \bar{r}\bar{D}$. The same argument applies to negative correlations, except that in that case, $\epsilon_{rt}\epsilon_{Dt}$ tends to be negative and so $\overline{rD} < \bar{r}\bar{D}$. Thus, positive covariation between r and D speeds up an invasion, whereas negative covariation slows it down.

4.10.4 Covariation over time and space

Covariation is also important when we consider the nature of stochasticity imposed upon the system. It is easy to see that environmental stochasticity might also include autocorrelation. That is, the random deviations at a given time or space depend on the deviation in the previous time or space. If we consider time, positive autocorrelation will tend to give long runs of good times, and long runs of bad times; negative autocorrelation will tend to give strong swings between good and bad times. No autocorrelation will just give a good or bad time, independent of what happened previously. And the same holds for spatial autocorrelation.

[9] Remember that invasion speed in a reaction–diffusion system is given by $2\sqrt{rD}$.

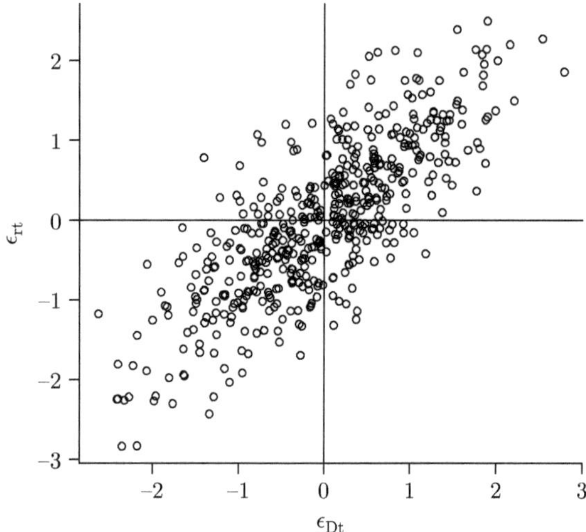

Figure 4.10 *Positive covariation between stochastic effects. Here we imagine that random deviations in r and D covary across time. Deviations in r are denoted ϵ_{rt}; deviations in D are denoted ϵ_{Dt}. As a consequence of this positive correlation, deviations in r and D tend to have the same sign and so mean invasion speeds will be greater than that predicted by the mean of r and D.*

Intriguingly, when theoreticians examined spatial autocorrelation in the stochastic environmental effects on growth rate, they found that positive autocorrelation caused a higher effective value of r and so caused waves to travel faster than expected from mean values of r and D (Méndez et al., 2011). This is a difficult result to explain intuitively, although it is possible to imagine the above argument about covariation (Section 4.10.3) applying in this case also.

How such environmental variation affects populations (and so invasion speeds) depends on many details of the population. Many complex life histories have evolved directly in response to environmental variation (Cole, 1954), so it is a little surprise that life history matters when we try to understand the effects of environmental variation. Populations with long life histories (e.g., many tree species) can yield very complex responses, depending on details of how environmental variation affects each age class in the population (Schreiber and Ryan, 2011).

Autocorrelation is often referred to as a kind of 'memory' built into the system, and it is intriguing to see that environmental autocorrelation can affect invasion dynamics in non-trivial ways. Of course, there is another ubiquitous sort of memory built into biological systems: genes. We have already seen how the memory provided by genes can affect invasions in important deterministic ways. In the next chapter, we will turn to how stochastic evolutionary processes can affect invasions.

4.10.5   The evolution of demographic stochasticity

But before we do that, one last note on demographic stochasticity. It may well have a genetic basis, and so also be able to evolve. Yu et al. (2021), in a fascinating series of experiments, measured the demographic noise that emerged on invasion fronts. They did this in expanding colonies of bacteria and yeast. This meant that they could do lots of replicates, but also that they could choose strains with various gene deletions, run invasions with each of these strains, and so examine the effect that each mutation had on the emergent demographic stochasticity on the invasion front. They found that a single gene deletion can substantially alter the strength of demographic stochasticity on the invasion front.

They looked at 191 genes. The average effect of a gene deletion was an 8% reduction in demographic stochasticity, but one of the genes yielded an 81% reduction in stochasticity! This is a large effect from a single mutation. It is an astonishing result and, at the very least, demonstrates that there is a genetic basis to demographic stochasticity. As such, demographic stochasticity itself can evolve; a possibility that has barely been explored.

4.11   The wrap

In this chapter, we looked at how chance events are ubiquitous and how they change our expectations. We point out that the processes that cause invasions to propagate forward—population growth and dispersal—are subject to both endogenous (individual-level) and exogenous (population-level) stochasticity.

We develop the idea of the LLN. This law means that endogenous stochastic processes carry greater weight in small populations, such as those on the leading edge of an invasion. As a consequence, invasions themselves can be strongly stochastic: outcomes can vary substantially across replicate invasions, even in a homogeneous environment. This can make invasions difficult to predict. Even though the location of the invasion front becomes more uncertain over time, our capacity to estimate the mean invasion speed rapidly improves over time. Forecasting remains difficult, but estimation gets easier as time progresses.

We also develop the idea that stochasticity, by reducing the mean growth rate, can act as a kind of brake on invasion speed; at least where invasions are best described as being discrete in time.

We also build the idea that stochastic processes can interact in non-trivial ways. Not only do multiple stochastic processes tend to increase our uncertainty, but they can also interact in ways that speed, or slow invasions. Positive correlation between exogenous stochastic processes influencing r and D, for example, may cause an increase in invasion speed.

This chapter has attempted a light touch on mathematical ideas that can be tricky to grasp. Appendix B attempts to shed light on a few key concepts. More generally, stochastic processes are actually just tricky things both to grasp and to deal with mathematically. A great introduction can be found in Otto and Day (2007) (already recommended in Chapter 1), but for those interested in a deeper dive, Karlin and Taylor's (1998) *An Introduction to Stochastic Modelling* is essential reading. For applications of stochastic modelling to invasions, Lewis et al. (2016) provide both a great overview, and the ominous quote about how the math 'gets interesting'.

Stochastic evolutionary processes on invasion fronts

5

5.1 Fixation in sector 5!

When you were a child, you were no doubt taught the magic of mixing colours. Mix yellow and blue together, and you get green. It is a very satisfying transformation, and I well remember trying various other combinations to see what might happen (mostly yielding brown, I recall). In all this experimenting, it never occurred to me to take green and somehow reverse the process to get blue and yellow. I presume that some of the basic physical principles of the world had already embedded themselves in my understanding. Unmixing something is usually hard to do.

Instead of paints, let's now take two strains of bacteria: one that fluoresces red, and the other that fluoresces green. Let's mix together these two populations in a test tube full of water. No doubt, the result will be some shade of brown. Now, let's take a drop of this brown mixture and drop it into the centre of a Petri dish filled with growth medium. We have just set off an invasion, and what happens next is surprising: red and green appear to unmix themselves, and dominate different sectors of the invasion (Figure 5.1).

A very reasonable response to this result is astonishment. What is going on here? It appears as though the bacteria have somehow self-organized into green and red camps, and then established state boundaries. We normally do not ascribe such organizational ability to bacteria, so there must be something more basic and fundamental going on. One of those fundamental things is stochasticity.

This chapter examines evolutionary stochasticity in invasive populations. Just as stochasticity can affect the number of individuals through space and time (Chapter 4), so it can also affect the genetic composition of those individuals. Evolutionary stochasticity can have profound effects on invasion dynamics and (as we have just seen) also on the patterns of genetic diversity across an invasion history. We are starting to realize that in invasive populations, evolutionary stochasticity matters, a lot.

Another theme that emerges in this chapter is the idea that invasion fronts reveal spatial analogues of temporal processes. Just as spatial sorting can be seen as a spatial analogue of natural selection (Chapter 3), in this chapter we discover that a founder event is a spatial analogue of a genetic bottleneck. A series of

The Ecology and Evolution of Invasive Populations. Ben Phillips, Oxford University Press. © Ben Phillips (2025).
DOI: 10.1093/9780191924910.003.0005

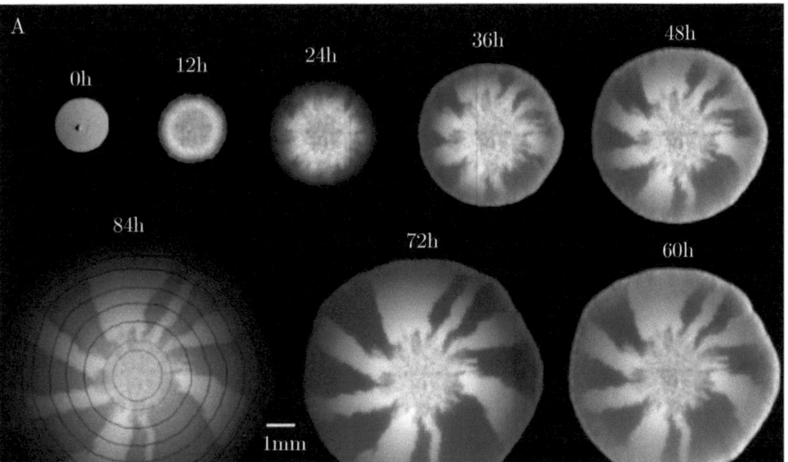

Figure 5.1 *When an invasion is established with two bacterial strains differing in a gene for colour, sectors emerge in which one or the other colour dominates. From Hallatschek and Nelson (2007).*

founder events can be usefully thought of as a spatial analogue of genetic drift. We see how these ideas can explain these unmixing bacteria, and we then start to develop the logical consequence of these ideas for how invasions might be affected by evolutionary stochasticity. In doing so, we bump into new ideas, such as mutation surfing and expansion load.

5.2 Serial founder events

In 1816, the United Kingdom unilaterally claimed ownership of a tiny but strategically located island in the middle of the Atlantic Ocean, Tristan da Cunha. This was something that the United Kingdom did back then. This kind of claim of course required this lonely rock to be defended, and so a garrison was built and around 15 people founded a colony on the island. Through sheer dumb chance, a few of these people carried a set of rare mutations that, when homozygous, cause a disease called retinitis pigmentosa, resulting in loss of vision. They didn't know they carried these alleles; no one did.

Two hundred years later and the population on Tristan da Cunha sits at around 250 people. The incidence of vision impairment in this community is about 70 times higher than for other human populations. By chance, the founders of this population had a higher than average frequency of alleles causing retinitis pigmentosa; all of the island's population descends from those people, and so, even many generations later, the frequency of the disease remains higher than average in this population.

This is, of course, an example of the founder effect. Whenever a new population is founded from a small number of individuals sampled from a larger population, the founder effect comes into play. That small population of founders is a small sample of the larger population and, as we have learned, small samples can show substantial deviations from the mean of the population. This is as true for a sample of litter sizes (Section 4.4) as it is for a sample of genes.

We can see this very simply when we consider the sampling distribution for a proportion. If the proportion of allele A in a population is p, then we can expect the sample proportion $\hat{p}$ in a sample of size N to be approximately normally distributed around p with a variance given by $\frac{p(1-p)}{N}$. As the number of founders, N, decreases, the variance of this sampling distribution increases and we can expect to see values of $\hat{p}$ that are quite different from the proportion in the source population, p. As with demographic stochasticity (Chapter 4), the founder effect gets stronger as the population size (number of founders in this case) gets smaller.

In this case, however, we are talking about alleles that generate phenotypes: so as the founder size gets smaller, newly founded populations get weirder. This link between founder effects and the weirdness of the resulting population has long been known to evolutionary biologists, and was even once proposed as a major mechanism of speciation (Mayr, 1999).[1]

The invasion of Tristan da Cunha is a much simpler invasion than most of the invasions considered in this book. It involved a single colonizing event and then a 'well-mixed' (i.e., aspatial) population thereafter. When a population establishes and then spreads through space—when an invasion front forms—we can easily see that this process is one of repeated, serial, founder events. A small sample of the population establishes the invasion front at $t = 1$, a small sample of that population establishes the invasion front at $t = 2$, and so on. Invading populations are typically not characterized by a single founder event, but by many founder events repeatedly subsampling the population. This serial foundering has long been considered to be a process that reduces genetic diversity on invasion fronts. Each sample has a chance of not capturing all the diversity in the population, each sampled population is a sample of the previous population, and so on.

5.2.1 Simulating serial foundering

We can capture this idea in a very simple genetic simulation. Let us imagine a one-dimensional array of habitat patches, with dispersal between nearest neighbours. This is the same landscape we saw in Section 3.3. We will introduce to the left-hand side of our array a clonal species, much like the bacteria in Figure 5.1: 500 individuals of the green type G and 500 of the red type R. The fitness of the two types is identical (so there is no natural selection here) and, as soon as a patch is colonized, it grows to 1,000 individuals.

As in Section 3.3, we will denote location along our array as x and time as t. We introduce our 1,000 individuals at $t = 0$, $x = 0$, and let us track the proportion of

---

[1] A notion that was largely put to bed by Barton and Charlesworth (1984).

G type in the leading edge of our invasion, at $x = t$. For the invasion to propagate, of course, we need some dispersal, so let's say that exactly 5% of the population (i.e., 20 individuals of 1,000) disperses into the leading patch each generation. Dispersers are chosen completely at random from the population (so there is no spatial sorting here).

We have now specified a system in which we start with equal frequencies of both types, there is no natural selection or spatial sorting, and the only real process is that we have a random sample of the population at $x = t$ going on to make up the population at $t + 1, x + 1$. Let's run this simulation 100 times, each time starting from 1,000 individuals at $x = 0$ and running our invasion for 100 generations each time. The result is shown in Figure 5.2.

This figure looks more or less the same as simulations of genetic drift that you can find in any introductory population genetics textbook. You can immediately see that, although the initial proportion of green genotypes always starts off at 0.5, things rapidly get messy. This is the random sampling imposed by serial founder events at play. For any given realization of this invasion, it is impossible to say in advance what the frequency will be at a given time in the future. But you can also see that, in the long run, the invasion front is either completely dominated by green types, or completely dominated by red types. If we run this simulation for long enough, we will eventually have a situation in which there is no genetic diversity left on the invasion front. One or the other of the genotypes will have reached fixation. Again, you can see that this has happened for most of our simulated invasions within 100 generations.

Now we are in a position to see how the sectoring in Figure 5.1 might have happened. The two-dimensional world of the Petri dish is more complex than

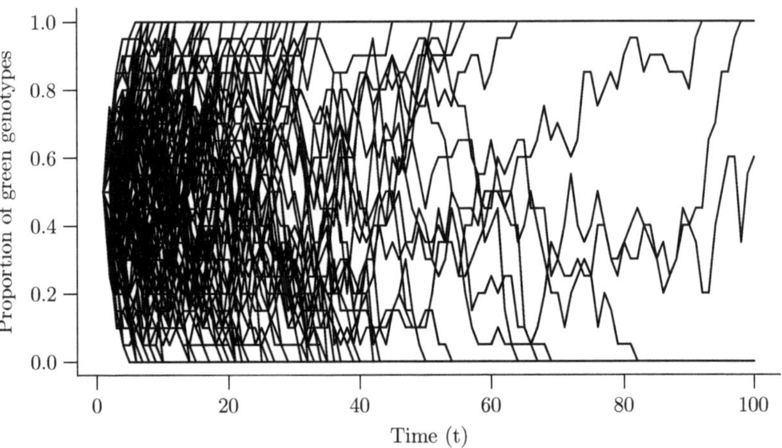

Figure 5.2 *The proportion of type 'G' on the leading edge of 100 replicate invasions over time, each invasion run over 100 generations of spread.*

our one-dimensional model, but there is still an invasion front undergoing serial founder events each generation. We can now see how these serial founder events might eventually cause either the red or the green type to be the only type on the invasion front. As soon as that happens, all of the populations founded as the invasion front passes will be of that type. On some parts of the two-dimensional invasion front, we might see the red type fixate; on others, we might see the green type fixate. The result is the beautiful sectoring that spontaneously emerges in these mixed-cohort laboratory invasions.

5.3 Drifting in space

This process of serial foundering should remind us of another fundamental evolutionary process: genetic drift. The previous simulation showed us what can happen on an invasion front when a small random sample of the population is chosen to found the population in the next vacant space. Genetic drift is what happens when a small sample of individuals is chosen to found the population in the next time. Just as spatial sorting is a spatial analogue of natural selection, serial foundering is a spatial analogue of the much better known process of genetic drift (Slatkin and Excoffier, 2012).

In the previous simulation, we had the furthest-forward population founded by 20 individuals from the next door, these 20 individuals immediately grew to 1,000, then we sampled from that 1,000 to find the 20 founders for the next space, and so on. We could imagine a very different scenario in a single 'well-mixed' population. We have a population of size 20, this population produces 1,000 offspring, but only 20 survive to the next generation, those 20 produce a 1,000 of which only 20 survive, and so on. Our first scenario is a description of serial foundering; our second scenario is a description of drift. The two processes are mathematically identical.

So founder events are just drift playing out through space; drift is just founder events playing out through time. That's pretty interesting.

With drift, the population we see before us is the population at the end of a long line of history. We can't immediately see that history because those historical states of the population are no longer in existence. The nice thing about serial foundering is that the history of serial foundering is written out in space: at any given time, that history is right there in front of us, if only we look.

5.3.1 Clines in diversity, and in allele frequency

Let's take the previous simulation and make it a little more complicated. Instead of only tracking the allele frequency in the furthest forward population, let's track the allele frequency in every habitat patch over time. This just requires a little more accounting: we need to allow 20 random individuals to move left (to $x - 1$) and

[2] And individuals moving to $x = -1$ will promptly be returned to $x = 0$.

20 to move right (to $x + 1$) from every colonized patch in every time step.[2] This is the kind of accounting that would cause a human headache, but computers are good at this kind of thing. Figure 5.3 shows us results from 20 such simulations.

It is apparent from this simulation that the process of serial foundering has caused clines in allele frequency to emerge through space. We started with the frequency of the A allele set to 0.5, but after 100 generations of spread, the frequency on the invasion front has usually gone to either 0 or 1. Over the same time, the allele frequency in the founding population hasn't really changed that much. As a consequence, over 100 generations, we are left with a cline in allele frequency going from about 0.5 to either 0 or 1.

These clines also represent a cline in genetic diversity. In our simulation, genetic diversity was maximal at founding, with an equal frequency of the two genotypes. After 100 generations of spread, we usually only have one of these original two alleles present: genetic diversity has declined across space as the invasion progressed.

Our simulation system is highly contrived, but we can see the basic processes at play. We have made natural selection and spatial sorting disappear, and we have the most rudimentary of growth functions (instantaneous growth to

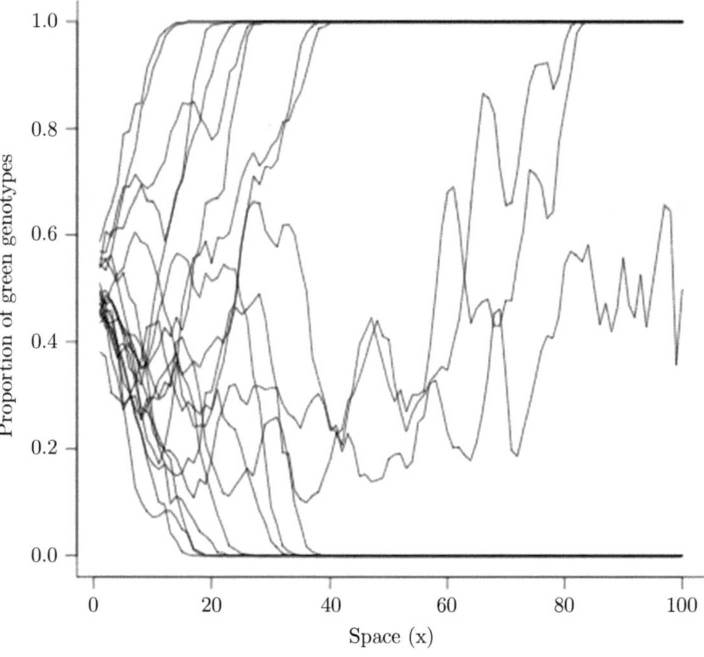

Figure 5.3 *The proportion of type 'G' across 100 habitat patches following an invasion from left to right across these habitat patches. Results are shown for 20 replicate invasions.*

carrying capacity). We have introduced no stochasticity except that due to randomly sampling a set of individuals to disperse each generation. We have found that this single stochastic process can cause a cline towards lower genetic diversity on the invasion front. This single stochastic process—serial foundering—can also explain the emergence of the beautiful sectoring we saw in Hallatschek and Nelson's (2007) experiment on bacteria in a two-dimensional space (Figure 5.1). While this last claim is not particularly strong from our simulation, a more sophisticated simulation of invasion with serial foundering in two-dimensional space shows very similar patterns to those in Hallatschek and Nelson's (2007) experiment (Figure 5.4).

Population geneticists have long been aware that founder events should cause reductions in average genetic diversity (e.g., Nei et al., 1975). It is a short step from acknowledging that founder events may cause weird jumps in allele frequencies to seeing that continuous invasions are a serialization of that same process (Austerlitz et al., 1997). And while a single founder event just causes a random jump in allele frequency, serial founder events start to cause more systematic effects: systematic loss of genetic diversity, and the emergence of clines in allele frequencies. While these effects are quite simple, Hallatschek and Nelson's experiment hints that quite striking things might emerge from this basic process. It is worth burrowing a little deeper into possibilities.

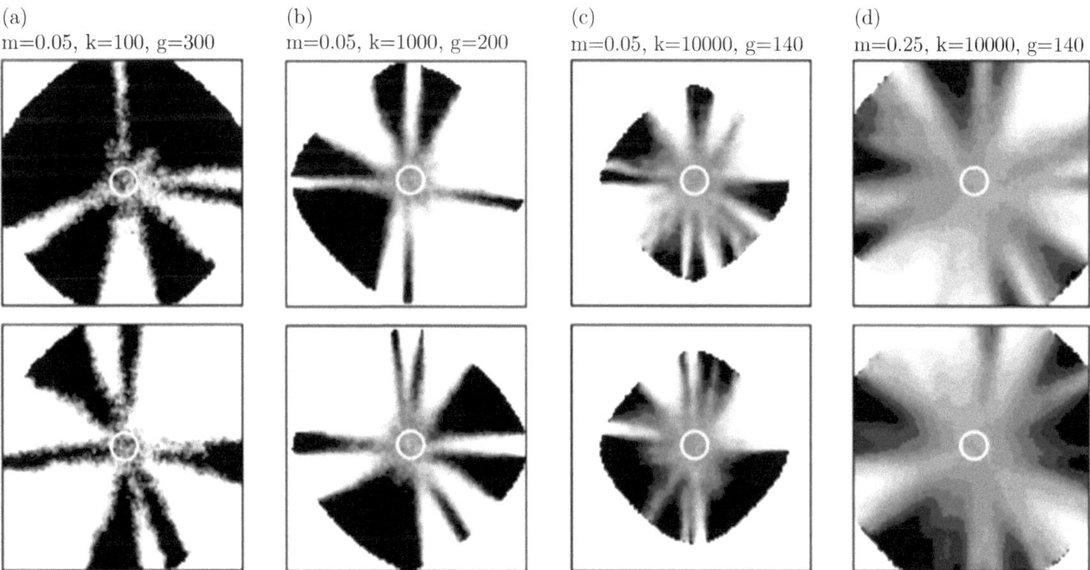

Figure 5.4 *Simulations demonstrating the emergence of sectoring in populations allowed to invade a two-dimensional space. Figure reproduced from Excoffier and Ray (2008).*

5.4 Surfing mutants

We have just seen that a simulated invasion can lead to fixation of one or another allele through purely random processes. In about half of our simulations in Figure 5.2, the 'green' genotype completely dominated the invasion front population—not through natural selection, but through random chance. In our simulations (and in the real experiment), we started with some standing variation, and we saw this winnowed down on the invasion front to (usually) just one of the original two alleles. But what if we start with no genetic variation in the population and allow variation to emerge through mutation? If we did that, we would be asking, 'what happens when new genetic variation appears in an invading population?' The answer is, it depends.

It depends first upon when and where the mutant arises. If the mutation event occurs well behind the invasion front, nothing particularly remarkable occurs. The mutant's fate is more or less the same as we would expect in a well-mixed population. The mutant has some chance of going extinct, and this chance depends upon the selection acting upon the mutant: is it a favourable new allele, is it neutral, or does it reduce fitness? The other determinant of a mutant's fate is the population size: population size determines stochasticity, and so the strength of drift. So the fate of a mutant arising behind the invasion front is more or less as we would expect from classical theory on the balance between drift and selection (Crow and Kimura, 1970).

We have repeatedly mentioned that the population size at the forefront of an invasion is relatively small, so we can expect stochastic effects (including drift, and its spatial analogue, serial foundering) to be more pronounced on an invasion front. So if a mutant arises on the forefront of an invasion (rather than well behind the invasion front), surprising things might be expected to occur. In 2004, a group of researchers from Stanford University—motivated by understanding human evolution—described what that surprise might look like (Edmonds et al., 2004). They used a simulation of a population spreading across a discrete two-dimensional space (a grid of habitat patches in this case). They tracked a single gene in this population and, after some generations of spread, introduced a mutation occurring in a single individual in the furthest-forward population. The mutation was neutral—it had no effect on fitness.

Most of these mutations simply went extinct. It was a single mutation in a single individual, after all. There is always a good chance that a neutral mutation such as this will just wink back out of existence. But about 18% of their mutants did not go extinct. Instead, they drifted to some reasonably high frequency and were still present in the population at the end of the simulation. So far, no real surprises. The surprise was not that some mutants survived; it was in where the mutations were at the end of the simulations. Most of the surviving mutants did what you would expect: at the end of the simulation, they were most commonly found around where the mutation happened (at $x = 20$). But a pretty high proportion of the surviving mutants did something else entirely; they ended up being smeared across space, with high frequencies from where the mutant first occurred, all the way to where the invasion ended. Figure 5.5 shows some clear examples of this phenomenon.

If you think about it, this outcome makes perfect sense against the previous ideas about serial foundering. If by chance a mutant goes from very low frequency to very high frequency over a few generations of serial foundering, it can become more or less the only allele present on the invasion front. And as soon as that happens, all of the subsequent populations founded as the invasion propagates forward will be fixed for the mutant allele. In this way, the mutant can, if it is lucky, catch the invasion wave and surf across space (and we need to thank Klopfstein et al. (2006) for this memorable surfing metaphor), achieving a surprisingly large geographic range for no particularly good reason other than sheer dumb luck.

This is exactly the same process as we saw in our earlier simple simulations (Figure 5.3). It is evolutionary stochasticity at play: serial founder events; drift through space. The only thing that makes this process different from drift through time is that the history of that drift process remains evident across space. With temporal drift, that history is, well. . ., history. With serial foundering, that history is stored as allele frequency changes across space.

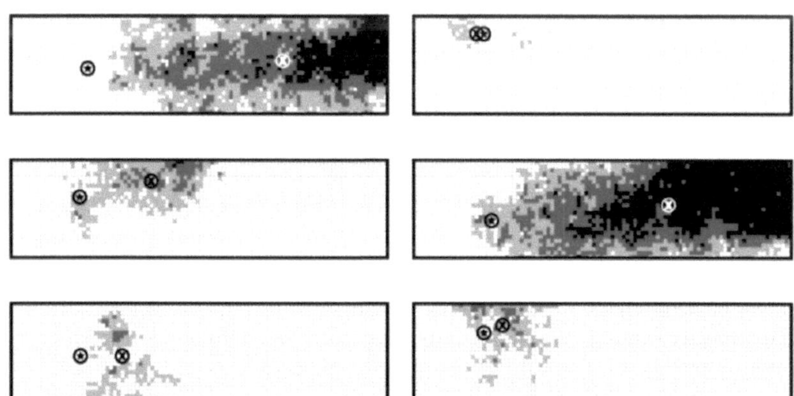

Figure 5.5 *The fate of neutral mutations occurring on an invasion front. Each panel is a different invasion. Invasions have progressed from left to right and a mutation has occurred in a single individual on the invasion front at $x = 20$. Darker areas are regions with a high frequency of mutant types at the end of the simulation. The star shows the location where the mutation first arose, and the cross shows the centroid of the mutants at the end of the simulation. Reproduced from Edmonds et al. (2004).*

5.4.1 Adaptation, or surfing mutants?

Allele frequency clines emerge from serial foundering. This happens if we start with some standing variation or if a mutation occurs near the invasion front. Clines in allele frequency are often interpreted as evidence of local adaptation: an allele is favoured in one area, a different allele is favoured in a different area, and so a cline in allele frequency emerges as populations adapt to these differing local conditions (Reznick and Travis, 1996). We have now seen that this local adaptation model is not necessary for clines to emerge. Clines might often, in fact, be echoes of past invasion rather than evidence of current selection (e.g., Streicher et al., 2016).

About 50,000 years ago, modern humans swept out of Africa and into Europe. This was a continental invasion of precisely the sort this book is interested in. It just happened a long time ago, and happens to involve our species. One striking pattern unearthed by researchers raking over the human genome is a mutation at the human microcephalin gene that occurs in high frequency in non-African populations. The microcephalin gene affects brain size and so, when this pattern was first noticed, arguments were mounted that possibly there had been strong natural selection operating on brain size during the Paleolithic invasion of Europe.

Well, maybe. Science has a humbling way of delivering bad news to our egos. We know the arrival of humans in Europe was a biological invasion, so there is every chance that this pattern in microcephalin diversity could well be driven not by natural selection, but by the genetic echoes of a 50,000-year-old invasion (Edmonds et al., 2004; Klopfstein et al., 2006). This frequency cline is quite possibly just a thing that happened. Indeed, the relatively high frequency of deleterious mutations in non-African populations suggests this kind of thing might have happened quite a lot (Subramanian, 2012; Peischl et al., 2013).

5.4.2 Catching the wave

One clear result from work on surfing mutations is that mutations need to happen in the right place and time in order to have a chance of catching the wave. This is not too surprising, really: anyone who has ever spent time surfing knows that it is impossible to catch a wave if you start paddling after the wave has passed. Several early simulation studies allowed mutations to occur in different places relative to the invasion front and found, unsurprisingly, that mutations occurring right on the leading edge of an invasion had a higher chance of fixing (and surfing) than those occurring further back (Klopfstein et al., 2006; Travis et al., 2007).

This need for mutations to be well placed in the low-density part of the invasion front does, however, raise an interesting question. If mutations happen randomly at a given rate in a population, then we will see more mutations (in an absolute sense) occurring in large populations than in small populations. Take this logic to an invasion front in a snapshot of time and we will see that most

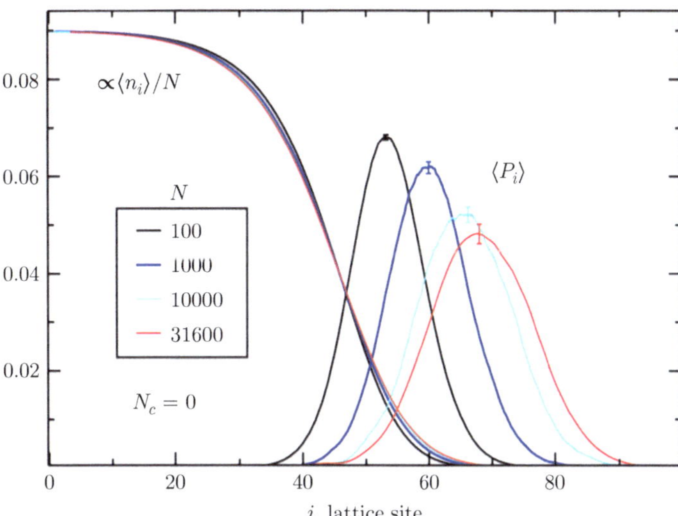

Figure 5.6 *Mutations that successfully go on to surf the invasion front tend to arise towards the tip of the invasion rather than deeper into the wavefront. This despite the fact that more mutation events occur deeper into the wavefront. The figure shows bell-shaped distributions of origin locations against various wavefronts. Reproduced from Hallatschek and Nelson (2008).*

mutations actually happen deep in the invasion front (or in the long-colonized populations) rather than on the very leading edge of the invasion. So, of the mutations that successfully surf, it is not clear where on the wave most of them will have started.

This is the kind of question that keeps physicists up at night, and so it is not altogether surprising that it was both posed and answered by a pair of talented physicists at Harvard University. The answer is fascinating. Most of the surfing mutants actually arise from mutations that happened in the front half of the invasion front (Figure 5.6). Even though mutations here are relatively rare, they make up the bulk of the surfers. The stochastic effects of drift and serial foundering are absolutely critical for a mutation to surf, and these stochastic effects are much stronger at the tip of the invasion. If you are trying to catch a wave, it pays to start paddling early.

5.5 Non-neutral mutations

So far, we have been considering alleles—genetic variants—that don't actually change anything. The vast majority of mutations that can be found in a population are probably of this nature; little slips in the genome that don't make any material difference to the individual that carries them (Kimura, 1985). This is not to say that most mutation events are neutral. If a mutation does occur and it has some effect, nearly all such effects are likely to be bad (Fisher, 1930). If you take an incredibly complex machine like a living cell, and you randomly change some aspect of it, you are much more likely to break the machine than to make it run better. Most functional mutations are deleterious.

Non-neutral mutations, because they change phenotypes, are subject to natural selection. Beneficial mutants are selected towards higher frequencies, and deleterious mutations are selected towards lower frequencies. But drift adds a big dose of stochasticity, so even deleterious mutants might actually increase in frequency and become fixed in populations subject to strong drift. This balance of processes, one deterministic (selection) and the other stochastic (drift), was a major source of disagreement between the two giants of modern evolutionary synthesis, Fisher and Wright. The result of this dispute was a lot of very careful and clever thinking about such matters; thinking that resulted in the Wright–Fisher model of drift–selection balance.

We have learned that serial foundering is very closely related to genetic drift, so using the Wright–Fisher model, we can probably make some guesses about what might happen to non-neutral mutations on an invasion front. We would expect them to behave the same as a non-neutral mutation in a small well-mixed population subject to drift and selection. In such a population, the probability that a new mutant is lost scales with $N_e s$, where N_e is the effective population size, and s is the strength of selection operating against that mutant. This is drift–selection balance. When N_e or s gets small, there is less chance that the mutant is lost, and more chance that it actually becomes fixed. When N_e is small, that fixation can happen even though the mutant might be deleterious (i.e., s is large: see Crow and Kimura (1970)).

That this basic result of drift–selection balance also occurs on invasion fronts is now supported by several theoretical models examining serial foundering during range expansion (Hallatschek and Nelson, 2008; Slatkin and Excoffier, 2012). If we focus just on the furthest-forward population (as we did in the simulations of Section 5.2.1, and the model of spatial sorting in Section 3.3), then it turns out that we can replace the effective population size of classical drift theory with the effective number of colonists, k_e, and achieve a near identical result. The probability that a deleterious allele is lost on the invasion front increases with $k_e s$. But as the effective number of colonists drops, the probability increases that a deleterious allele will drift to fixation, and so catch the wave and surf the invasion front.

Thus, mutation surfing is not a phenomenon restricted just to neutral alleles; deleterious alleles can also surf. This is, again, the result of serial foundering; drift in space. As well as being an analytical expectation, the result has also been manifested in a diversity of simulation models (Travis et al., 2007; Hallatschek and Nelson, 2008; Slatkin and Excoffier, 2012). Although deleterious mutations might also surf, we would expect them to have a lower chance of surfing compared with neutral mutations.

As well as a lower chance of surfing, we would also expect the frequency of deleterious alleles to drop over time after an invasion passes. After the front passes, advantageous alleles that have been left behind will gradually reinvade and be lifted by natural selection. We will see a wave within a wave: Fisher's 'wave of advance of an advantageous allele' happening well behind the invasion front (e.g., Travis et al. (2007)).

5.5.1   Expansion load

I spend a bit of time doing conservation biology. A lot of my colleagues in this field are not particularly interested in evolution: it is peripheral to their concerns; it happens on timescales they are not concerned with (or so they claim). Despite only marginal interest in evolution, however, all of these people know about genetic drift. They know about drift, because, when populations start to get small, drift becomes a stronger player. Deleterious alleles can drift to high frequencies, impact population fitness (i.e., population growth rate), and so cause 'mutational meltdown' of a population (Lynch et al., 1995). In conservation, drift is one of the horsemen of the apocalypse; it conspires with other small population processes to create an 'extinction vortex' that can drive populations extinct. In conservation biology, small populations are the norm, and drift is part of the vernacular.

In evolutionary biology, drift is only one of the processes causing populations to be maladapted. We refer to the reduction in fitness caused by maladapted alleles in a population as 'genetic load'. Genetic load can have several causes. We can have 'mutation load' (that arising from deleterious mutants that are yet to be purged by selection), we can have 'migrant load' (caused by gene flow from populations not adapted to local conditions), we can have inbreeding and outbreeding loads, and so on. We now need to add a new horse to this stable: expansion load.

We now know that invasion fronts experience serial foundering and so have an increased chance of seeing deleterious alleles rise to fixation. As a consequence, we might expect deleterious mutations to accumulate on invasion fronts in the same way that they do in the small populations that my conservation biologist friends are concerned with. In 2013, Peischl et al. (2013) set out to examine this possibility. They imagined a model in which mutations were allowed to occur in a large number of loci. In their model, most of the mutations were deleterious, and the relative fitness of individuals in the population was determined by the aggregate effect of all the mutations carried by an individual. They found dramatic declines in fitness associated with invasion: the invasion front accumulated mutations, and most of this accumulation was due to mutations that first appeared on the invasion front (Figure 5.7). Not only did deleterious mutations accumulate, but they remained evident in the population for thousands of generations after the invasion front had passed.

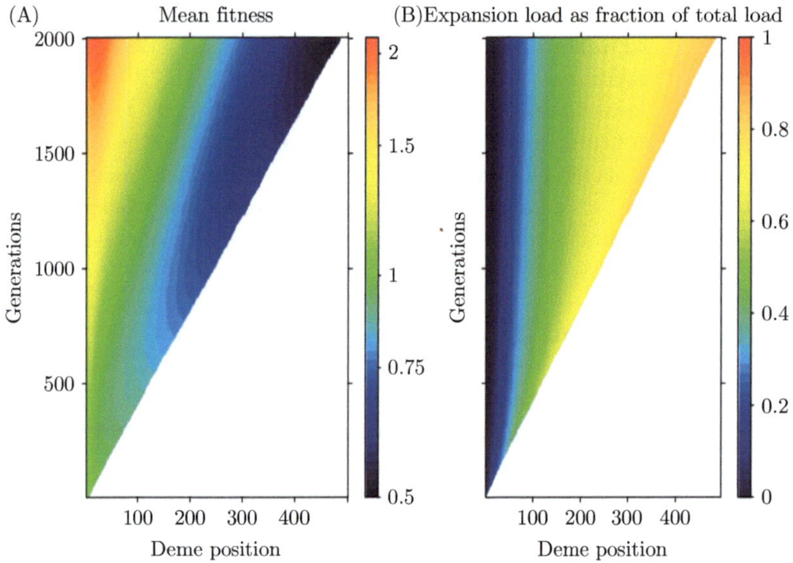

Figure 5.7 *Deleterious mutations arising on an invasion front can become fixed through drift and serial foundering. The result is 'expansion load'; a reduction in fitness caused by the accumulation of deleterious mutations on an invasion front. These figures show the mean results of 50 invasions across a one-dimensional space. Each row of pixels is a snapshot of the mean state of the invasion at the specified time (generation). Reproduced from Peischl et al. (2013).*

This theoretical work was followed up some years later by experiments on bacteria invading agar plates (Bosshard et al., 2017). The experiment involved two bacterial lines—a high mutation rate (HMR), and a low mutation rate (LMR) line—and compared outcomes between situations in which these lines were either allowed to invade an agar plate (effectively a two-dimensional space), or were kept in a well-mixed solution. This experiment confirmed the theory nicely: HMR bacteria from the invasion front had accumulated around six times more mutations than the same strain kept in a well-mixed solution. These mutations had an aggregate negative effect on fitness, with these invasion front HMR bacteria being less fit than their well-mixed brethren as well as the invasion front LMR bacteria. In this experiment, there is very little doubt that deleterious mutations accumulated on the invasion front (over about 1,200 generations), and had a big impact on fitness.

While empirical confirmation is a nice step, it is important to note that this first theoretical model of genome-wide expansion load did not actually link the fitness effects of mutation to population dynamics. Fitness, in the model, was the relative fitness of competing genotypes within a deme, with population dynamics unaffected by fitness values. The population grew in the same way regardless of its mean fitness; a situation referred to as 'soft selection'. As a consequence, the invasion progressed at a constant speed (see the straight line boundary between occupied and unoccupied space in Figure 5.7) despite the accumulation of fitness effects.

Of course, the accumulation of deleterious mutations may often cause lowered absolute fitness, and so have consequences for population dynamics. This is 'hard selection' and it is the type of fitness consequence that my conservation-minded friends worry about. It could well be happening on invasion fronts, and if it does, we would expect it to have consequences for invasion speed, because (as we have often repeated) invasion speed emerges from rates of population growth and dispersal.

5.6   Expansion load and invasion speed

If drift and serial foundering cause deleterious mutations to accumulate on an invasion front, we may expect the invasion to slow down. If these accumulating mutations impact survival and reproduction, then they affect the intrinsic growth rate of the population, r. As we know, r is one of the primary determinants of invasion speed. This is an intriguing thought, and one that Stephan Peischl and his colleagues developed with characteristic rigour.

To examine this possibility, they extended their previous model, allowing hard selection. Now, the aggregate fitness effects of mutation did affect population dynamics. Lower fitness caused lower values of both r and K. And, sure enough, when this was implemented, expansion load caused invasions to slow down. In their simulations, invasions steadily decelerated, and then appeared to settle into a new quasi-equilibrium: a new, roughly constant, speed that was a lot slower than the initial speed the invasion had achieved (Figure 5.8). Interestingly, the experimental work of Bosshard et al. (2017) also lends some support to this modelled phenomenon: their invasion front HMR strains, highly burdened with expansion load, had slower initial invasion speeds when initiating a new invasion than did the control lines. This suggests these loaded strains may have been slowing down the invasion front from which they were sampled.

This slowing down opens up an additional question: if the front slows down, does this give a chance for well-adapted alleles to catch up (see Section 3.9)? The answer appears to be yes, and the new (slower) equilibrium speed that emerges reflects a balance between the rate at which expansion load accumulates and the rate at which it is eroded by the arrival of well-adapted alleles (these well-adapted alleles can 'arrive' either through migration or mutation). In a simpler analytic version of their model, they show that, when the effect of mutations is small, the expected equilibrium speed of the invasion front is about the same as that at which a beneficial allele would spread behind the invasion front (Peischl et al., 2015). That is, the invasion front speed is limited to the speed at which beneficial alleles can arrive. That's a pretty cool result.

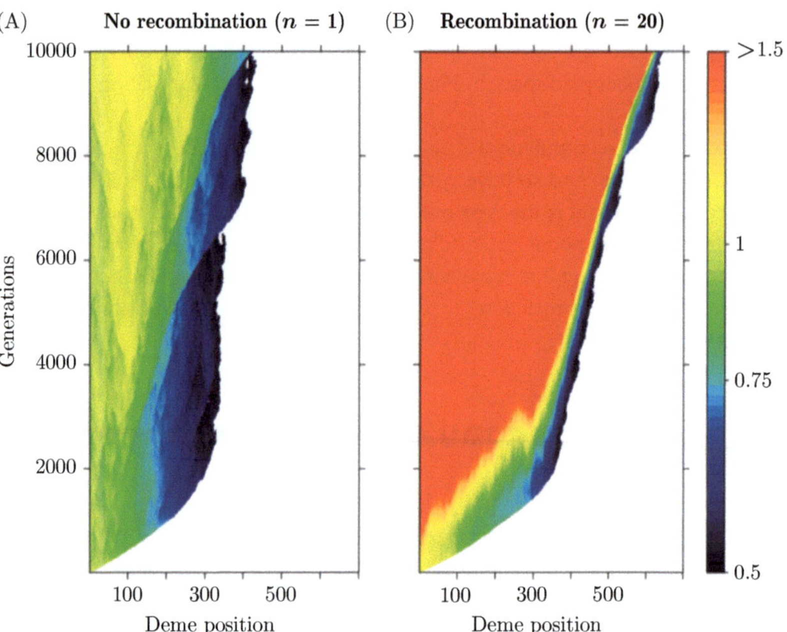

Figure 5.8 *Expansion load causes fitness on the invasion front to drop and can cause invasions to slow dramatically. Each panel shows results for a single simulated invasion across a one-dimensional space, and contrasts an asexual species (no recombination) with a sexual species (recombination). Each row of pixels is a snapshot of the state of the invasion at the specified time (generation). Reproduced from Peischl et al. (2015).*

In their simulation model, however, spread rate varied within and between invasions. Interestingly, the spread dynamics depend upon whether our species is asexual (reproduces without recombination) or sexual (with recombination). In the sexual case (i.e., high recombination between loci), there is a reasonably smooth transition to the equilibrium speed: expansion load accumulates, but it is diluted by the arrival of, and recombination with, fitter genotypes. A rough equilibrium is reached and the invasion progresses at a roughly constant speed. Constant, but much lower than the speed that would occur in the absence of expansion load.

In the asexual case, however, we see only a few lineages dominating the invasion front. These eventually accumulate sufficient expansion load that the invasion largely stops: the frontal population has reached mutational meltdown. Eventually, this failing frontal lineage is overtaken by a fitter lineage from behind the invasion front, and the invasion takes off again, before eventually melting down, and so on. The expansion dynamics that emerges in the absence of recombination is episodic; periods of decelerating expansion, stasis, and rapid acceleration. It is a highly stochastic invasion speed; with this stochasticity caused almost entirely by drift and serial foundering.

In Chapter 4, we saw how demographic and environmental stochasticity can cause invasions to be unpredictable. Now, we have the first inkling that evolutionary stochasticity (drift and serial foundering) may actually act as a brake, slowing an invasion down below the value it would be expected to take otherwise. We also see that drift and foundering might also contribute to stochastic invasion speeds. In this model from Peischl et al. (2015), this stochasticity in speed was most apparent in the absence of recombination, and when mutations had relatively large fitness effects.

5.7   Drift–selection balance, and invasion speed

We have just learned that stochastic evolutionary processes (drift and serial foundering), by generating expansion load, can cause invasions to decelerate. On the other hand, we have deterministic evolutionary processes (selection + sorting, Chapter 3) that can cause invasions to accelerate. How might these two processes balance? Which wins?

In 2019, Kimberley Gilbert and Stephan Peischl gave us a first look at that problem (Peischl and Gilbert, 2019). They extended the model of Peischl et al. (2013) by allowing dispersal to evolve under spatial sorting by plugging in the model of spatial and temporal fitness we developed in Chapter 3 (Section 3.3). They found that the stochastic process was mitigated when dispersal was allowed to evolve. Essentially, as dispersal evolves, a larger number of individuals disperse to make up the new furthest-forward population.[3] We have a larger sample, and so less stochasticity as a consequence. Thus, spatial sorting can act to slow down the loss of fitness caused by expansion load.

In fact, their results imply that spatial sorting might sometimes even completely rescue populations from expansion load (Figure 5.9). In this case, when mean dispersal starts above a given threshold, we see dispersal rates evolve upward (sorting outweighs serial foundering), and every increase in dispersal further shifts the balance towards the deterministic processes of sorting and selection. As a consequence, populations starting above this threshold value of dispersal will tend to see both dispersal and fitness increase over time on the invasion front, despite being subject to the fitness-eroding effects of drift and serial foundering.

It is a really fascinating result, and the first to analytically incorporate all four processes we have unearthed: natural selection, spatial sorting, drift, and serial foundering. I suspect this will prove to be a landmark paper. To incorporate these four moving parts in a simple way, the model of course made quite a few assumptions. Space was discrete and one-dimensional, with only nearest-neighbour dispersal; dispersal only happens after carrying capacity has been reached; and fitness is relative, rather than absolute (so it doesn't affect invasion speed). These assumptions allowed a tractable model to be built that aligns

[3] Recall from section 4.8 that higher dispersal rates equate to a higher effective population size on invasion fronts.

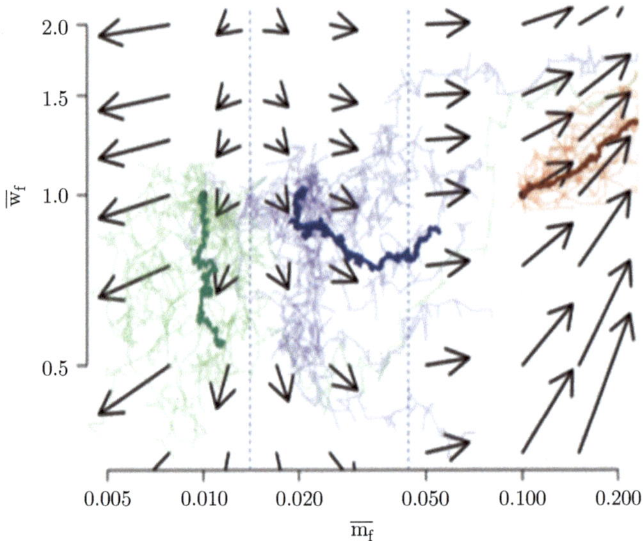

Figure 5.9 *The evolution of mean fitness ($\overline{w}_f$) and mean dispersal ($\overline{m}_f$) on an invasion front. This model from Peischl and Gilbert (2019) incorporates deterministic (natural selection + spatial sorting) and stochastic (drift + serial foundering) evolutionary processes. Outcomes depend on the initial starting value of a population, with expected change given by vectors, and coloured traces representing replicate simulations starting at three different locations (given by colours). From Peischl and Gilbert (2019).*

very nicely with classic theory developed around well-mixed populations in discrete time. They have allowed us to crystalize the fascinating insight that the drift–selection balance of classical evolutionary theory can be generalized to drift–selection happening in both space and time. It remains a formidable challenge, however, to extend these ideas into continuous space and time, two dimensions, and to fitness effects involving hard selection.

In a clever experiment, Lars Bosshard et al. (2019) tested the possibility that spatial sorting and natural selection might overwhelm the fitness loss caused by expansion load. You will recall our expectation that selection and sorting together favour genotypes that yield the fastest invasion speed (see section 3.3). When genotypes with a higher invasion speed emerge on a two-dimensional invasion front, the front gets a bulge in it as these genotypes get further ahead, and this bulge goes on to colonize the rest of the invasion front from the side (Hallatschek and Nelson, 2010). Bosshard and colleagues ran an experiment in which they let a bacterial colony invade an agar plate for three days and then sampled from the invasion front of that plate to found a new invasion on a fresh agar plate. In this way, they could effectively keep an invasion progressing across an agar plate for 39 days. The interesting part of the experiment is that for half of their lines, they sampled the invasion front at a random location (control lines), and in the other half, they always sampled from the part of the invasion that had progressed the furthest (selection lines). In this way, they sampled either a random sample of the genotypes on the invasion front, or the genotypes that would inevitably come to dominate the front. Their control lines decelerated over time, as expansion load came to bear without the full mitigating force of spatial sorting. Their selected lines, however, accelerated over time, as spatial sorting eventually made itself heard through the serial foundering and drift that causes expansion load.

5.8 Evolutionarily stochastic invasion speeds

In early 2013, an invitation arrived in my inbox to attend a workshop in France, 'Biological invasions: stochastic and deterministic models'. The invitation came from Vincent Calvez, who had recently conceived the generalization of Fisher's equation (Section 3.4). I figured it would be good to go: I'd get to meet people interested in the same stuff as I was, and I was bound to learn something. The only problem was that I needed to give a talk, but what about?

So far, all my thinking has been about the deterministic evolutionary processes described in Chapter 3: processes driving increases in dispersal and growth rates on invasion fronts. I had noticed the papers about mutation surfing and also the work of Melbourne and Hastings (Chapter 4) about the fundamental limits to our ability to predict invasions. It struck me that the serial foundering that drives mutation surfing must put quite a stochastic spin on the deterministic evolutionary processes I had been working with. We expected dispersal and growth rates to evolve, but the extent to which this happens in any given invasion must be stochastic.

To put this another way, we know that the invasion front is a repeated sample of the population. In this chapter, we have been focusing on serial foundering as a random sample of the invasion front each generation. We saw in Chapter 3, however, that this repeated sampling is not entirely random. Individuals with higher rates of dispersal and reproduction are more likely to be represented in this sample. The sample is biased towards certain phenotypes, and this is why we see directional evolutionary shift on invasion fronts. But it is important to remember that while these evolutionary biases are occurring (natural selection + spatial sorting), the invasion front is still a sample, and so the outcome is stochastic. More to the point, the invasion front is often a *small* sample, so the processes of natural selection and spatial sorting are operating in a world of substantial stochasticity. How predictable is the evolutionary trajectory of dispersal and growth rate?

To get at this idea, I built an individual-based simulation model in one-dimensional space. Each individual carried two traits: one that determined its expected fitness at low density, and the other that determined its expected dispersal. The traits evolved according to a simple quantitative genetic model. I imagined a situation that was similar to many invasions I was familiar with, the kind of invasion detailed in Elton's (1958) book: a small number of founders is sampled from a large population; they are introduced into a nice new world; they establish a population that begins to spread. I was interested in short periods of time—30 generations or so—and so my system did not consider mutation, but focused completely on sorting of the standing variation that was introduced with those initial founders.

Once I had this system, I could run the kinds of experiments that were impossible in nature. I could make lots of replicate invasions, and I could record the exact distance each one spread in left and right directions. By running these replicate invasions, I could quantify the variation in spread rates, and I could separate out how much of this contribution was due to demographic stochasticity (evolution turned off); initial founder events (variation between replicate invasions), and the serial founder effects and drift that occur during invasion (variation between speeds within invasions, i.e., right versus left).

It turned out that, in the scenarios I tested, most of the variation in invasion speed actually happened within invasions. The left-hand side of one of my invasions would often travel at a very different speed to the right-hand side. Sometimes dispersal would rapidly evolve to high levels on the invasion front; sometimes it would have a more muted response or even decrease. The same variation in outcomes occurred with growth rate. In my system, the random chance of which individuals dispersed furthest in the first few generations made a very big difference to the subsequent dynamics.

The results were, of course, specific to the system I had in hand. But they made the point, I think, that serial founder events and drift, interacting with spatial sorting and natural selection, could generate substantial variation in invasion speeds between replicate invasions (Phillips, 2015). Invasions *were* hard to predict, and not just because of demographic stochasticity.

Again, this was an idea whose time had come. When I happened to meet Stephan Peischl at a conference in Asilomar in 2014 (see Barrett et al. (2016) for a complete account of that excellent conference), I was delighted to discover that we had independently been working on essentially the same idea. Given our different backgrounds, we were coming at the idea from quite different directions, with completely different timescales in mind, but both finding substantial stochasticity in invasion speed driven by serial foundering and drift. While I was trying to get the idea across like a ham-fisted field biologist recently given his first computer, Stephan had developed an elegant simulation model and some nice analytic results that were well-connected with earlier theory (Peischl et al., 2013, 2015).

But by hook and crook the point had been made, and in 2017, two groups independently ran experiments on bean beetles that were designed not only to test the notion of spatial sorting, but also to assess whether evolving invasions were more variable. Both experiments came back with a resounding 'yes'. In both cases, the invasions that had been allowed to evolve showed notably more variation in invasion speed than those that had not (see Figure 3.4; Ochocki and Miller (2017); Weiss-Lehman et al. (2017)). It seems that we had found an explanation for the missing variance in invasion speeds that had been unearthed by Melbourne and Hastings (Sections 4.6, 4.10).

Since then, of course, more empirical studies have emerged, and the answer has been a resounding 'maybe'. At least two other empirical studies, using replicate invasions, have yielded evolving populations in which variance in spread rate is either the same (Szűcs et al., 2017; Van Petegem et al., 2018) or lower (Williams et al., 2016a) in the evolving treatment compared with the non-evolving treatment. It is an open question as to why, precisely, some evolving systems seem to show lower variability compared with non-evolving counterparts.

Williams et al. (2019) named this issue 'expansion variability' and spent some time thinking about the evolutionary factors affecting this variance between different realizations of an invasion. As with demographic stochasticity (Section 4.3), population size on the invasion front is clearly a big player here. Invasion fronts with smaller effective population sizes will experience stronger drift and serial foundering (as well as greater demographic stochasticity). But it is hard to see how evolutionary processes will cause reduced expansion variability. Williams *et al.* point out that deterministic evolutionary processes (selection, spatial sorting) act to reduce phenotypic variance in r and D, and so, if these processes dominate, we might see all realizations of an invasion converge on some optimum value of rD (Deforet et al., 2019; Phillips and Perkins, 2019). This might be true, but while we are converging on this optimum, we are also drifting, and so our total expansion variability will be greater than if we had started at this optimum value. It also remains possible that we never achieve the optimum, because less-than-optimal combinations of alleles fix on our invasion front (Phillips, 2015; Peischl et al., 2015).

Another possibility is that the observation of lower variability in some evolving treatments may actually be an experimental artefact. There are two possibilities here. The first is that the experimental systems, constrained by logistics, are simply in very early phases of the invasion and so realized invasion speeds are still highly variable (see Section 4.10.1); the empirical results are subject to pretty severe sampling error and so we could simply be seeing that sampling error give the occasional strange result. The second (and more likely) possibility is due to a subtle nuance: the systems that have seen higher variance in the non-evolving treatments have halted evolution by replacing, each generation, all individuals in the invasion with individuals drawn from a large reference population. This is a perfectly sensible way to halt evolution, by essentially making traits non-heritable. What this manipulation also does, however, is it takes heritable variation in demographic traits, and it transfers that variance into non-heritable demographic variation (Section 4.5). Thus, in the evolving treatment, we have a source of variance (heritable variation in demographic traits) that probably declines over time, and in the non-evolving treatment we are adding heritable variation onto non-heritable demographic variation, and as the population grows we are taking a larger and larger sample of that variance. So, yes, in these experiments, we can see reduced expansion variability in the evolving treatment, but this may well occur because we are actually artificially increasing the demographic variation of the non-evolving treatment.

Of course, it also remains possible that my suspicions here are incorrect, and that there is some evolutionary mechanism that delivers lower expansion variance. One possibility is a large difference in the standing genetic variation available around r and D. Given stochasticity is driven by the ratio of $D:r$ (Sections 4.8), with low values being associated with high stochasticity, it is possible that in some settings D increases dramatically while r stays unchanged (or is reduced slightly, see Sections 7.2.2 and 7.2.4). Such a scenario may admit the situation in which there is lower expansion variance under evolution. Another possibility, now that we know that demographic stochasticity can have a genetic basis, is that stochasticity itself evolves on invasion fronts (Yu et al., 2021). Perhaps under some conditions, we might see the evolution of reduced stochasticity. What those conditions might be, we currently don't know; there is work to be done in this space.

5.9 The wrap

In this chapter, we point out that, in addition to demographic stochasticity, there is stochasticity in the evolutionary processes playing out on invasion fronts. We develop the idea that invasion fronts are subject to serial founder events, and that this serial foundering amounts to a process that is a spatial analogue of genetic drift. While classic genetic drift plays out through time, the spatial drift we see on invasion fronts plays out through space. While the history of classic drift is, well, history, the history of spatial drift, to some extent, remains evident: preserved in space.

Spatial drift can cause several surprising phenomena to arise. First, genetic variants that become fixed on the invasion front get propagated across space as the invasion front progresses. Frontal variants can 'surf' the invasion wave, and we see situations emerge where even deleterious variants spread to occupy large areas of space.

The second surprising situation is that, under mutation, we expect deleterious variants to accumulate on invasion fronts. The accumulation of this 'expansion load' can cause invasions to dramatically slow or even stop altogether (at least temporarily).

The third surprise is that the strong spatial drift on invasion fronts causes the evolutionary trajectories of r and D on the invasion front to be stochastic. Natural selection and spatial sorting are still at work, but they are working through a strong stochastic filter. Thus, the emergent speed of the invasion is also affected: evolutionary stochasticity can contribute, quite strongly, to making invasions difficult to predict.

For those interested in a deeper dive here, the primary literature is still probably the only place to do that. For a deeper dive into the concepts of drift–selection balance, and the Wright–Fisher model, many introductory population genetics texts could give you that, though Crow and Kimura (1970) is a classic.

<table><tr><td>

6

</td><td>

Pushed and pulled waves

</td></tr></table>

In September 1894, the barque *Cambus Wallace* was shipwrecked on Minjerribah, a large sand island on the east coast of Australia. The *Cambus Wallace* was fully laden at the time, and its cargo, a large quantity of explosives, was considered too dangerous to salvage. Too dangerous to leave also, so some enterprising sailors figured they should just blow the whole thing up.

The explosion must have been quite a spectacle. It made short work of the salvage job: the barque vanished, and the fwoomp completely destroyed the foredune. It probably left those nearby partially deaf. The sailors staggered off, but they had inadvertently set off a process that would play out over the coming years, with dramatic effect. Sand, blown by the wind accumulated in the gap in the foredune, eventually formed a tall sand dune completely devoid of vegetation. The wind continued to blow, and grains were picked from the peak of the dune to cascade down the leeward side. These wind-driven sand grains were legion, and they were relentless. They swamped the vegetation behind the dune, and over a number of years the dune steadily advanced inland. It steadily advanced across the entire island, in fact, killing all the plants that were in its way and eventually cleaving the island in two. Minjerribah is now called the North and South Stradbroke Islands.

The cleaving of Minjerribah was effected by a wave of advance that was driven entirely by dispersal (sand grains don't generally reproduce themselves). It was those sand grains cascading down from the top of the dune that pushed this wave forward. This book is concerned with biological invasions (*i.e.*, grains that do reproduce themselves), and so we have been dealing with invasions where both reproduction and dispersal have been at play. To keep things simple, however, we have been focused up until now on invasions that are pulled forward by high growth rates on the leading edge of the invasion. It turns out that there are other classes of biological invasions. These are invasions that, like a sand dune, are pushed forward by dispersal from the top of the wave, rather than pulled forward by growth at the bottom.

These pushed waves are probably quite common, and they have quite fundamentally different properties to the pulled waves we have been considering so far. Now that we have, in Chapters 2–5, unpacked the basic theory around the ecology and evolution of invasive populations, we can examine how this theory is modified as we consider the overtopping dynamics of pushed waves. We will start by explaining the fundamental differences between pushed and pulled waves, and why this difference matters. We will then explore the various ways in which pushed waves might emerge, and ponder their ecological and evolutionary dynamics.

The Ecology and Evolution of Invasive Populations. Ben Phillips, Oxford University Press. © Ben Phillips (2025).
DOI: 10.1093/9780191924910.003.0006

6.1 Waves, pulled and pushed

The Fisher–KPP quartet pointed out that invasions propagate forward as a result of two forces acting on a point in space, x: (1) the balance of comings and goings at x, and (2) the growth of the population at x (Section 2.1). In Fisher's model, growth was made to be negatively density dependent, such that per-capita growth increases as density decreases. As such, the maximum per-capita growth rates would be found out on the leading edge of the invasion (see Figure 6.1). With simple diffusion, we now know that this situation results in a pulled wave.

But other descriptions of growth and dispersal are possible. It turns out that if per-capita growth is not highest on the leading tip of the invasion, interesting and complicated things can happen. When growth rates are higher further back, into the bulk of the invasion front (Figure 6.1), we have a dynamic that is more like an advancing sand dune. In this case, individuals rapidly emerge in this bulk part of the wave, and the dispersal of these individuals contributes more to wave advance than does the growth and dispersal of individuals on the leading tip. We have a wave that is pushed from behind rather than pulled from the front.

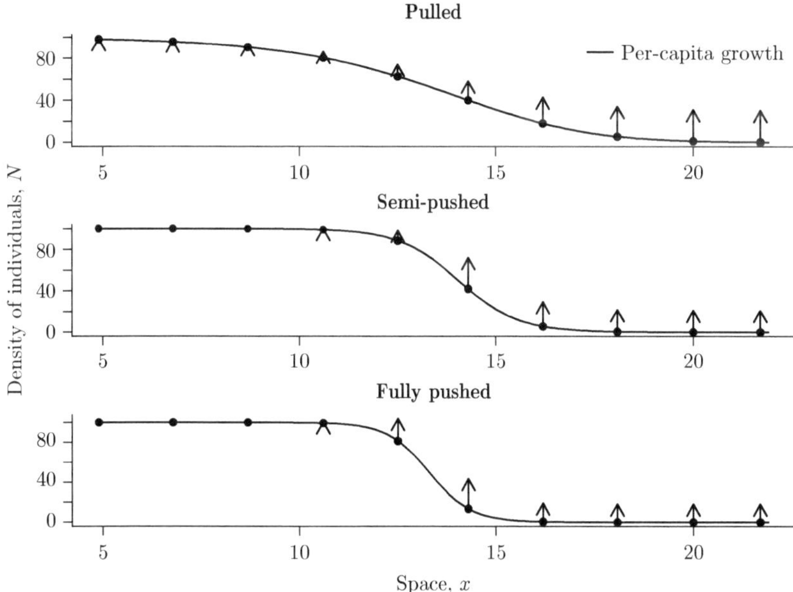

Figure 6.1 *Per-capita growth along pulled and pushed wavefronts. Effects have been scaled such that we can see the per-capita growth rate at each location relative to other locations along the wavefront. When dispersal is constant and maximum per-capita growth happens on the tip of the invasion front as $N \to 0$, we get a pulled wave. When per-capita growth is highest further back into the bulk of the wave, we get a pushed wave. This is the classic way of achieving a pushed wave, but as we will see, there are in fact many ways that waves can be pushed.*

This pushed dynamic can also arise when dispersal is density dependent. If dispersal rates are higher at high density (and, empirically, they often are; Bowler and Benton (2005)), then the same pushed dynamic can result: dispersal from the bulk of the population contributes more to forward propagation than does growth and dispersal of the leading edge (Bîrzu et al., 2019).

There are clear cases of waves that are considered to be fully pushed, and those that are considered fully pulled. But in between, we have a strange hybrid class of waves, the semi-pushed wave. These semi-pushed waves behave like a pushed wave when we consider invasion speed, but behave like a pulled wave when we consider the evolutionary and stochastic dynamics.

These different wave classes can have very different dynamics, and we will explore these differences in the coming pages. There is no doubt that this is a taxonomy that matters for anyone wanting to understand their particular invasive population. It is an important taxonomy, but it turns out to be quite tricky to tell these three wave species apart. To some extent, simply knowing about the possibility of pushed waves is the main thing, because knowing they exist helps explain patterns that wouldn't make sense against expectations built in previous chapters: for example, waves where dispersal doesn't evolve, or where heterozygosity doesn't appear to decline as the invasion progresses. Such outcomes are possible under a pushed dynamic, and the best way to identify such a dynamic (at least on paper) is to compare the observed wave speed with the speed you would expect from the Fisher–KPP model.

6.1.1 Relative wave speeds

One of the clearest situations that gives us a pushed wave is when the per-capita growth rate is much lower on the leading tip than it is in the bulk of the wave. So if we use only the leading-edge growth rate to calculate the expected speed of this wave, we will likely underestimate the realized speed. Those cascading sand grains push the wave forward faster than we would expect were we to just look at growth rate and dispersal on the leading tip.

You will recall Fisher's surprising result (Section 2.1) that the speed of a travelling wave is given by $v_F = 2\sqrt{rD}$, where r is the per-capita population growth rate, and D is the diffusion coefficient. In Fisher's (and later Skellam's) model, the speed of an invasion did not depend, in any way, on the carrying capacity. It matters not a jot whether your carrying capacity is 5 or 5 million; all that matters are those growth and dispersal parameters. It's a surprising result and, as it turns out, not correct for pushed waves. For pushed waves, the speed also depends on K. For pushed waves, higher carrying capacities can lead to faster spread (Haond et al., 2018); it matters a lot whether your carrying capacity is 5 or 5 million.

By the 1970s, theoreticians had been working on reaction–diffusion models for quite some time, and they had established that Fisher's result is true only so long as the maximum per-capita growth rate is found at the leading edge of the invasion (*e.g.*, Aronson and Weinberger, 1975; Fife, 1979). This condition was cooked into both Fisher and Skellam's models, so it took us a while to notice that it was even there. Let's consider the growth part of Fisher's model, for example. Here, we have.

$$g(N) = \frac{dN}{dt} = r_0 N \left(1 - \frac{N}{K}\right).$$

The per-capita growth rate of any such model can be seen simply by dividing both sides by the number of individuals:

$$r(N) = \frac{1}{N}g(N) = r_0 \left(1 - \frac{N}{K}\right).$$

Now we can see that the per-capita growth rate at any time, $r(N)$, is a function of the number of individuals at that time. This is the negative density dependence that makes the logistic model what it is. It is straightforward to see that this is the equation for a straight line with a negative slope and an intercept at r_0 (Figure 6.2), given that $N \geq 0$, $r(N)$ has its highest value at $N = 0$. And when we think of an invasion front, $N \to 0$ describes the situation right at the leading tip of the invasion (see Figure 6.1). So, cooked into Fisher's model is this quiet assumption that per-capita growth is highest right on the leading edge of the invasion.

Another quiet assumption in Fisher's model is that diffusion is constant. That is, the dispersal rate of the organism in question is the same, regardless of the environment. One obvious way to relax that assumption is to make dispersal also density dependent, *i.e.*, $D(N)$. When we do this, we discover that positive density dependence in dispersal (*i.e.*, dispersal increasing with density) also causes

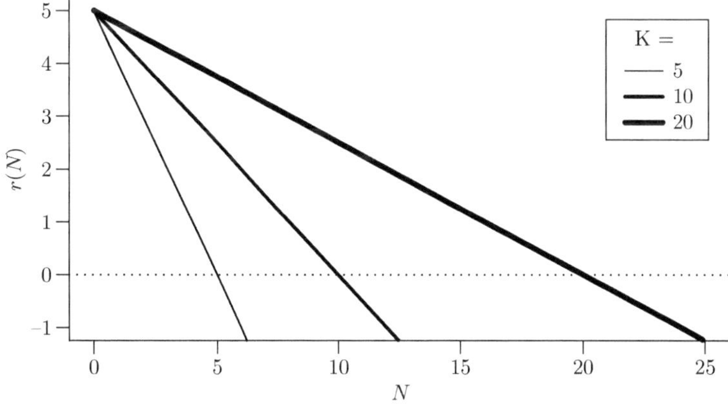

Figure 6.2 *The logistic population model has per-capita growth, $r(N)$, maximal as $N \to 0$. Examples here show $r_0 = 5$ for a range of carrying capacities ($K \in \{5, 10, 20\}$). In all cases, the growth rate at carrying capacity, $r(K) = 0$.*

pushed waves (Altwegg et al., 2013; Kawasaki et al., 2017); again, dispersal from the bulk of the population contributes more to the spread rate than does growth and dispersal at the leading tip.

So it turns out that the invasion speed Fisher derived is best thought of as the 'linearly determined' speed (Lewis et al., 2016). It uses the growth rate and diffusion values that pertain when the population is at very low density, on the leading tip of the invasion.[1] So Fisher's spread rate is better denoted as

$$v_F = 2\sqrt{r(0)D(0)}, \tag{6.1}$$

and it is the correct speed for deterministic invasions where growth and dispersal are either constant or highest at the leading tip. Here v_F is called the 'Fisher speed', and $r(0)$ and $D(0)$ are the per-capita growth rate and diffusion coefficient at a population size of zero. In the case of logistic growth, for example, we have $r(N) = r_0(1 - \frac{N}{K})$, which, evaluated at $N = 0$, gives $r(0) = r_0$.

Any invasion with simple diffusion and with per-capita growth highest on the tip of the invasion front has this linearly determined speed and is a 'pulled' wave. But, of course, we can imagine other ways that a population might grow and disperse, and when we start elaborating on the simple growth and dispersal functions that are in Fisher's model, we soon discover waves that are pushed; waves that travel faster than we would expect, given the growth rate on the leading tip.

Pushed waves generally do not have a simple function that describes their speed. The speed is no longer independent of conditions in the bulk of the population, but depends on the entire shape of the invasion front. This is a mildly vexing situation, but we have to live with it.

The Fisher speed is still important to us, however, because it turns out to be a very useful baseline against which to judge invasion speed. If we know the actual velocity that a wave is travelling, v, and we also know the Fisher velocity, v_F, then we can tell what kind of wave we are dealing with. Just over 60 years after Fisher's paper, in a series of research papers, Wim van Sarloos (reviewed in van Saarloos, 2003) showed that the ratio v/v_F allows us to perfectly separate pulled from pushed waves. Building from this idea, Bïrzu et al. (2018) then went on to discover semi-pushed waves and showed that these also fell into a nice neat pile based on v/v_F.

The key theoretical result is this: when $v/v_F = 1$, you have a pulled wave (your wave is travelling at the Fisher velocity); when $1 < v/v_F < \frac{3}{2\sqrt{2}}$, you have a semi-pushed wave (your wave is travelling slightly faster than the Fisher velocity); and when $v/v_F \geq \frac{3}{2\sqrt{2}}$ (faster still), you have a fully pushed wave (Table 6.1). To date, this relative wave speed is our best method for telling these different wave types apart. It's great that we have such a clear taxonomic character, but in practice, telling the difference between actual speed and expected speed may actually be quite difficult. Given that $\frac{3}{2\sqrt{2}} \approx 1.06$, the domain of semi-pushed waves is those waves travelling only about 6% faster than the Fisher velocity. For anyone who has ever worked on a real invasion, the precision required here is

---

[1] These come from the linearization of the growth functions $g(N)$ and $D(N)$ at $N = 0$, hence 'linearly determined'.

daunting, and suggests that we may ever only be able to use this criterion to identify semi-pushed invasions in laboratory systems, where such precision may be possible. Fully pushed waves, however, may move substantially faster than the Fisher velocity, so we may be able to discern these in a field setting, particularly when they are strongly pushed (see Section 6.2.1.1).

We may also be able to discern pushed waves not from their relative velocity, but by the ecological and evolutionary outcomes that emerge as the invasion progresses.

6.1.2   Mixing across the invasion front

We have just seen that the speed of the invasion is one clear axis across which our three classes of invasion segregate. Those invasions that propagate faster than the Fisher velocity are pushed. Fair enough, fully pushed waves have that cascading sand grain dynamic, and they travel faster than you would expect from $r(0)$, but really, who cares? Why does any of this even matter?

It matters a lot, because fully pushed waves sample from across the entire invasion front. Fully pushed waves draw the next generation's invasion front from across the entirety of this generation's invasion front. Pulled waves, by contrast, draw the next generation from the relatively tiny population at the leading tip of the invasion. As a consequence, there is much greater mixing across the invasion front in pushed waves than what occurs in pulled waves (Roques et al., 2012). In Section 3.9, we characterized the invasion front as a strangely isolated population, rolling across the landscape under its own set of evolutionary pressures, and largely isolated from the core of the population. This is a fair characterization of pulled dynamics, because with pulled dynamics there is very little gene flow from the bulk of the invasion front. Pushed dynamics, however, with their much greater levels of mixing, result in invasion fronts that remain more connected with the bulk of the population.

This connection with the bulk of the population means that we are in a world with high levels of gene flow. It is no longer reasonable to ignore the ecological and evolutionary dynamics occurring in the bulk of the invasion front as we did in earlier chapters (*e.g.*, Sections 3.3 and 5.2). All of the selective pressures we expect on invasion fronts—spatial sorting and natural selection—still play out, but the evolutionary response is now diluted by high levels of gene flow from parts of space that are experiencing different selective pressures.

6.1.3   Relative stochasticity

Another consequence of the greater connectivity across the invasion front is that pushed waves have a much larger effective population size. As we have seen in Chapters 4 and 5, stochasticity matters in interesting ways. From a purely ecological viewpoint, highly stochastic invasions—those with a small effective population size on the invasion front—are very difficult to predict. From an evolutionary

Table 6.1 *A taxonomy of invasion waves, comparing how each wave class differs in terms of its relative velocity, stochasticity, and connectivity across the invasion front.*

	Pulled	Semi-pushed	Fully pushed
Velocity	$v/v_F = 1$	$1 < v/v_F < \frac{3}{2\sqrt{2}}$	$v/v_F \geq \frac{3}{2\sqrt{2}}$
Stochasticity	High	High	Low
Connectivity	Low	Low	High

perspective, small effective population size drives serial foundering and all that comes with it: mutation surfing, the loss of heterozygosity, and the evolution of expansion load. Pushed waves have a much larger effective population size on the invasion front, so in pushed waves all of these stochastic effects are strongly mitigated.

Semi-pushed waves, as you might expect, are a strange hybrid of these two situations. They move faster than the Fisher velocity (so are pushed), but they are only *just* pushed. As a consequence, they still have quite small effective population sizes and so effectively suffer the full effects of invasion stochasticity. Semi-pushed waves are pushed waves with none of the benefits.

6.2　What causes pushed dynamics?

Most of the theoretical work on pushed dynamics has come about from exploring different growth functions. It is now well established that relatively low per-capita growth on the leading tip of the invasion will cause pushed dynamics. But pushed dynamics is an emergent outcome of this low per-capita growth on the leading tip; there are other ways to achieve pushed dynamics that aren't directly about the growth function, such as density-dependent dispersal. Fundamentally, pushed dynamics occur when dispersal contributes more to population growth on the leading tip than does reproduction. In this section, we will examine scenarios that naturally give us pushed dynamics.

6.2.1　Positive density dependence in growth

When per-capita growth is highest somewhere behind the leading edge of invasion, we have a pushed wave (this is not strictly true, but for our current purposes it is close enough). Another way of thinking about this is that pushed waves emerge whenever we have an initial positive density dependence in the growth rate (Figure 6.3).

When we see things this way, we can set our minds to thinking about the various reasons that a population might have a higher per-capita growth rate at intermediate density. And there are many such reasons that naturally emerge in ecological and evolutionary dynamics.

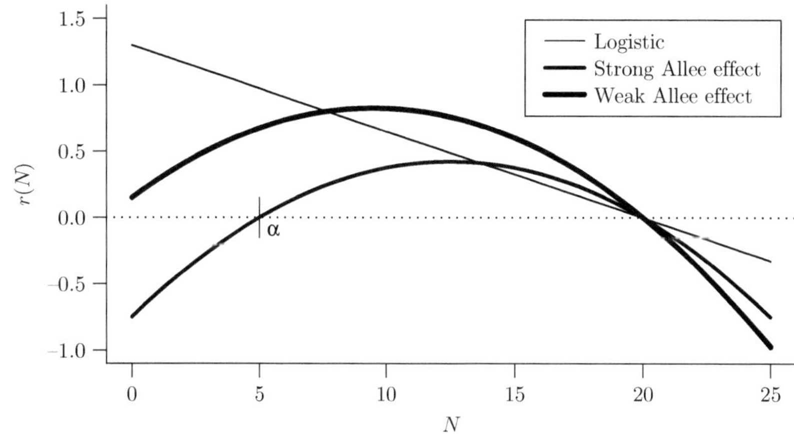

Figure 6.3 *Per-capita growth as a function of density for several alternative models. We compare the standard logistic model to models in which weak and strong Allee effects are implemented. Both Allee effect models show an initial positive density dependence before taking on negative density dependence as N increases. The strong Allee effect differs from the weak in that the strong effect has negative per-capita growth below a critical density, the Allee threshold, denoted* α. *All models have a carrying capacity* $K = 20$.

6.2.1.1 Allee effects

Allee (1938) in his 'Social life of animals' first pointed out the possibility of positive density dependence, when he considered that it might be very hard to find a mate if conspecifics are thin on the ground. He grew up in a very small country town, so perhaps had personal experience of precisely this kind of issue.[2] The idea that low density might result in lower survival and reproduction has since become known in ecology as the 'Allee effect'. It is now known to be caused by a wide array of ecological circumstances (Courchamp et al., 1999; Berec et al., 2007). In addition to difficulty finding mates—an effect that applies to obligate sexual species everywhere, including pollination in plants or external fertilization in broadcast spawners such as coral—plants and animals often facilitate conspecifics in complex ways, including outright cooperation (Stephens and Sutherland, 1999). For example, wolves hunt in packs, vigilance is more effective when there are many sets of eyes looking for predators, and yeast cells facilitate nearby cells when they break down complex sugars and their neighbours hoover up the by-products. Humans are, of course, also an excellent example of a cooperative species; put a lone human out into the world and, despite the fantasy of the rugged individualist, he typically won't last long.

Scholars of such effects are careful to point out that many of the examples of Allee effects are examples of how some component of individual fitness is reduced at low density. These effects—hunting and vigilance being more effective in large groups, that sort of thing—are termed 'component Allee effects', because they are examples of how some components of individual fitness

[2] Others have argued, much more persuasively, that his belief in the value of community for individual fitness actually stemmed from his upbringing as a Quaker.

decline as we head towards low density. Such component effects are necessary, but not always sufficient, to cause population-level 'demographic Allee effects' (Courchamp et al., 1999). It is entirely possible, for example, that one component of fitness (*e.g.*, pollination rate) declines at low density while other components (*e.g.*, germination and growth rates) improve at low density. Thus, we need to be careful in inferring demographic Allee effects from observations of component Allee effects. Component Allee effects are quite common, but demographic Allee effects are less so (Berec et al., 2007; Kramer et al., 2009). In the context of invasion dynamics, of course, it is the demographic Allee effect that matters.

Demographic Allee effects have been extensively studied by theoreticians (Boukal and Berec, 2002), and a useful distinction has been drawn between 'weak' versus 'strong' Allee effects. A weak Allee effect exhibits positive density dependence at low density, but does so in a way that growth always remains positive. In the context of Figure 6.3, a weak Allee effect always has $r(N) > 0$ for $N < K$. By contrast, a strong Allee effect has negative growth at low density. In the context of Figure 6.3, a strong Allee effect has $r(N) < 0$ for $N < \alpha < K$, where alpha is the critical density below which growth becomes negative.

Strong Allee effects cause perhaps the clearest example we could imagine of a fully pushed wave. On the leading tip of an invasion subject to a strong Allee effect, growth is actually negative (*cf.* the examples in Figure 6.1 where growth is always positive). Here, the leading tip of the invasion would actually head off to extinction were it not for the constant arrival of individuals from the bulk of the population. Invasion waves moving forward despite strong Allee effects are like a jogger tripping herself on a loose shoelace; she progresses forward, but not by lifting her feet.

6.2.1.2 *Bistability and strong Allee effects*

Strong Allee effects are a manifestation of a more general idea: 'bistability' in the growth function. 'Bistability' refers to the fact that the population has two stable equilibria (see Figure 6.4): there are two stable states for the system, and which one the system runs to depends on where it starts. In the case of a strong Allee effect, the population is stable either at carrying capacity or when it is extinct. If the population starts above a critical population size (usually denoted α), the population will grow to carrying capacity. If it starts below α, it will run to extinction.[3]

The idea of bistability is a nice general one. The first description of a pushed dynamic in biology came from Aronson and Weinberger's (1975) extension of Fisher's model (although arguably it came earlier, with theoretical work on the propagation of electrical impulses through the nerves of squid). Fisher's model was originally a description of an advantageous allele at a single locus. Extending the model to a diploid locus, and imagining a situation in which the heterozygote was less fit than either homozygote (a situation known as 'underdominance') yielded a bistable growth function: one in which the advantageous allele would actually be selected against when it was at low frequency. The reason for this is

[3] This dependence on a critical population size is why this idea is also sometimes termed 'criticality'.

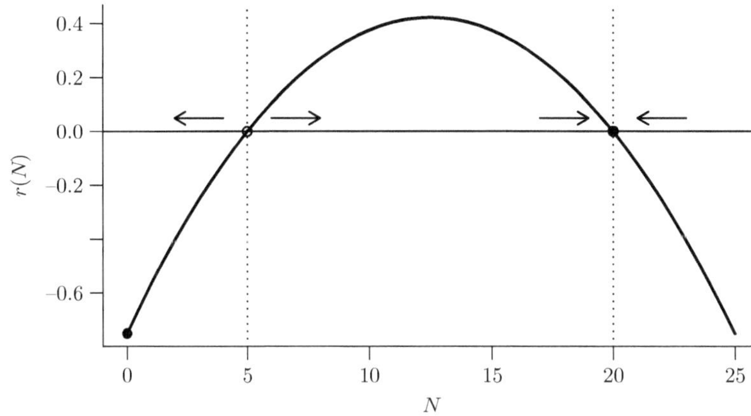

Figure 6.4 *Bistability in the growth function. Per-capita growth is zero when population size is either N = α = 5 or N = K = 20. When N = 20, any small perturbation in population size will cause the population to come back to N = 20 because growth is negative when N > 20 and positive when N < 20. When N = 5, however, the situation is reversed, and any small perturbation will cause the population to run off to either N = 0 or N = 20. Thus, this population has two stable equilibria: one at N = 0, and one at N = 20.*

that when the allele is at low frequency, most copies of the allele in the population will be in heterozygote form. Thus, if the advantageous allele started at low frequency, it would tend to go extinct. If it started at a high enough frequency, however, it would be favoured by selection (because now most copies of the allele will be in homozygotes).

Bistability in population dynamics also often appears as a natural outcome of two species models. For example, *Wolbachia*, an endosymbiont in many insect species, requires a certain threshold prevalence, before it can increase in frequency in its host population, but once this threshold is breached, it runs to high prevalence and will remain at high prevalence thereafter (Barton and Turelli, 2011; Hoffmann et al., 2014). This bistability has important consequences for the spread of *Wolbachia* through host populations.

Bistability in the growth function will typically cause a pushed wave to form, regardless of whether we are talking about the propagation of a population of dandelions, alleles, or endosymbionts. Bistability—by causing negative growth on the leading tip—causes an invasion to trip over itself. As a consequence, wavefronts subject to strong Allee effects will tend to be very steep.

6.2.2 Density-dependent dispersal

Another source of pushed waves is density-dependent dispersal (Newman, 1980). So far, all the theory and models we have looked at in this book have considered dispersal to be density independent. But there is considerable evidence, across a broad array of taxa (from cancer cells to birds), that dispersal is often positively density dependent (*e.g.*, Bowler and Benton, 2005). There are also good evolutionary reasons why we would expect this to be the case; at least one of the functions of dispersal is to escape competition (Van Valen, 1971; Hamilton and May, 1977). Thus, it may often be the case that a given genotype is more likely to disperse if it finds itself in a high-density population than if it finds itself in a low-density population.

With positive density-dependent dispersal, we see all the features we have come to expect from pushed waves (Altwegg et al., 2013; Haond et al., 2018; Bîrzu et al., 2019). Whether density-dependent dispersal results in a pushed wave will, of course, depend on the form of the density-dependent response. If, for example, dispersal rate only increases after the population reaches carrying capacity (and dispersal distances are smaller than the length of the invasion front), then density dependence will have no real effect on invasion speed. If dispersal increases steadily with increasing density, however, we will often see pushed dynamics emerge (Haond et al., 2018; Bîrzu et al., 2019).

6.2.3 Other scenarios resulting in pushed waves

While almost all the theory on pushed waves has come about as a consequence of examining density-dependent growth and dispersal functions, it seems very likely that pushed waves might also result from other processes, beyond those we have looked at already.

6.2.3.1 *Expansion load*

First is expansion load with hard selection. In Section 5.6, we saw that serial foundering on invasion fronts could cause deleterious mutations to accumulate. Where these mutations affect absolute fitness (*i.e.*, reproductive rate), they cause the invasion to slow down. Ultimately, what emerges is an equilibrium speed reflecting a balance between the accumulation of deleterious mutations, and the arrival of advantageous alleles (Peischl et al., 2015). Most of these advantageous alleles invade the invasion front as it slows down.

Expansion load seems to exhibit several of the ingredients we associate with a pushed wave. First, the growth rate is lower on the tip of the invasion (driven by the accumulation of mutations in this case, rather than density dependence *per se*). Second, the slowing down of the wave seems to allow greater mixing (which is why advantageous alleles can catch up, and invade the front). This greater mixing is perhaps most evident in the case of expansion load without recombination; here, Peischl et al. (2015) found that the invasion essentially stops as the front goes into mutational meltdown, and it only gets going again as functional variants finally invade the invasion front.

For these reasons, I suspect expansion load generates pushed waves. If true, this would be a class of pushed waves that results not from a deterministic demographic process, but from a stochastic evolutionary one; a kind of genetic

Allee effect (Luque et al., 2016). If true, expansion load would cause initial-ly pulled invasions (experiencing all the stochasticity that accrues to that class) to evolve into pushed invasions (experiencing substantially reduced stochastici-ty). It seems likely, even, that the wave might cycle between pushed and pulled states.

6.2.3.2 *Heterogeneous habitat*

Another contender for a scenario causing a pushed wave is the situation in which habitat is highly heterogeneous (Pachepsky and Levine, 2011). Imagine a situation in which habitat is fragmented into patches and there is some fixed probability, d, that an individual might disperse between patches. For a single individual, the probability of dispersing is d, but for a population of individuals, the probability that at least one individual disperses is $1 - (1 - d)^N$, where N is the population size. Now we can see that the probability that an adjacent empty patch is colonized is an increasing function of N; dispersal is positively density depen-dent, for stochastic reasons (Morel-Journel et al., 2016; Dahirel et al., 2021). The bigger the population, the greater the chance it will send propagules into the next vacant patch. As we have seen above, positive density dependence in dispersal can lead to pushed waves; a result that was both empirically and mathematically confirmed in a nice model system in which bacteriophages (a virus that infects bacteria) were invaded across a 'lawn' of bacteria (Hunter et al., 2021); the dis-crete nature of the host, and the fact that the virus had to stop to reproduce in the host caused pushed invasions to emerge.

This idea might extend to variation in habitat quality; a continuous habitat with strongly varying carrying capacity, for example. In this case, when the invasion hits a set of low-quality patches, the invasion will slow down (growth rate is slower in the low-quality patch, and the juxtaposition between high- and low-quality patches will have much of the population growth in the low-quality patch driven in fact by dispersal from the high-quality patch).

One scenario that has been neatly described is the situation in which a pop-ulation is invading into a space that is opening up before it. The analogy here is to climate change, in which the region of space that is suitable for the species is moving (as, for example, the climate changes). In this case, and under particular velocities of environmental shift, it has been shown that pushed waves can result (Garnier and Lewis, 2016).

6.2.3.3 *Competition*

We have already made the point that a strong Allee effect is a bistable system. The final outcome depends upon where you start. Another common situation in which bistability is expected is when two species compete. This expectation falls out of one of ecology's most famous models; the Lotka–Volterra competition model. We develop the issue of competition in much greater detail in Chapter 8, but for now, here is a brief description of how competition might cause a pushed wave. Under Lotka– Volterra competition, two species are considered, and their growth rate is made density dependent. The trick is that the density a species experiences is calculated using the density of the other species as well as its own. The equations look like this:

$$\frac{dN_1}{dt} = r_1 N_1 \left(1 - \frac{N_1 + a_{12} N_2}{K_1} \right) \tag{6.2}$$

$$\frac{dN_2}{dt} = r_2 N_2 \left(1 - \frac{N_2 + a_{21} N_1}{K_2} \right). \tag{6.3}$$

The values of K set the carrying capacity for each species, and lower values of K denote a species that experiences strong intraspecific competition. The values of a are exchange rates—a_{12} can be read as 'the number of one's per two'—and these determine the strength of interspecific competition. A larger value of a_{12}, for example, denotes a larger competitive effect of species 2 on species 1.

Depending on the settings we choose for the Ks and as, one of three outcomes will eventuate (see Chapter 8 for details). These are: competitive exclusion (one of the species goes extinct); stable coexistence (both species persist); or bistability (one or the other species will go extinct, depending on initial densities).

Bistability occurs whenever interspecific competition is stronger than intraspecific competition. In the terms of the model, this is whenever $K_1 < a_{12} K_2$ and $K_2 < a_{21} K_1$. Under these circumstances, whichever species gets the upper hand (in terms of density) will go on to drive the other species extinct.

From here, it is easy to see how this bistability might apply to an advancing invasion (Figure 6.5; Shigesada and Kawasaki (1997)). If the invasion is pushing into space already occupied by a strong competitor, then the invasion will fail unless the invader has a higher carrying capacity than the resident (and starts at high density). But if we imagine an invader with high carrying capacity pushing into a space occupied by a strong competitor (that has lower carrying capacity), we can see that the leading, low-density, tip of the invasion is in a world where the invader is being outcompeted by the resident. The high-density bulk of the wave, however, is in a world where the resident is being outcompeted by the invader. Here is a clear case of a pushed wave, analogous to a strong Allee effect (Roques et al., 2015), but in this case being driven by external (biotic) factors.

6.3 The dynamics of pushed waves

Surprisingly, it took nearly 20 years between the discovery of pushed waves and their first arrival in the literature on invasive populations, with Mark Lewis and Peter Kareiva's (1993) foundational paper. That it took so long is particularly surprising when we consider that the idea of Allee effects had been around for a long time. But air travel was expensive back then, and the internet didn't really exist; ideas travelled more slowly in the 1980s.

Given the boggling variety of circumstances in which they might arise, and the fact that there is no simple expression for the speed of a pushed wave,

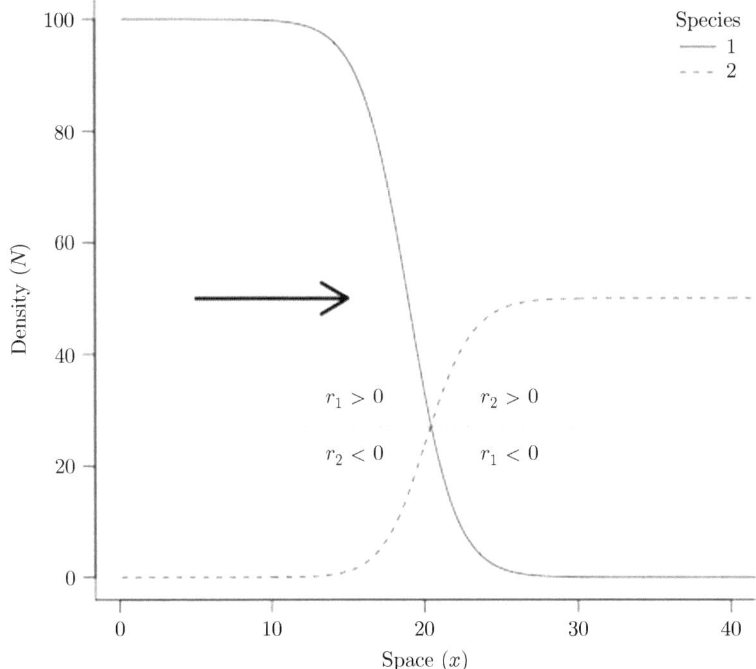

Figure 6.5 *A pushed wave resulting from competition. Here, species 1 is invading from left to right into the habitat of a less competitive species. In this example, interspecific competition is stronger than intraspecific competition, so there is a bistability in population growth. Whichever species has the lowest density at a location will have a negative growth rate. Because species 1 attains higher carrying capacity, it propagates forward as a pushed wave, while species 2 collapses. The dotted horizontal line passes through the point at which the two populations are at equal density. To the right of this point, the growth rate of species 1 is negative; a classic pushed wave, moving forward despite a bistability in the growth function.*

extracting general rules for the dynamics of pushed waves might seem challenging. Nonetheless, there are some key ideas that flow quite intuitively from our understanding of pushed dynamics.

6.3.1   Slower speeds, steeper waves

Our neat taxonomy of wave types (Table 6.1) makes the very clear statement that pushed waves travel faster than we would expect from a pulled wave parameterized with growth and dispersal rates at the leading tip. Thus, it is mildly confusing that many theoretical treatments in which pushed waves are generated by the growth function claim that pushed waves travel *slower* than the equivalent pulled wave. Both claims are correct, but they differ in how we define an equivalent pulled wave. In Bîrzu's (2018) classification, the equivalent pulled wave is one where the growth dynamics is exponential and the growth rate is given by growth at low density, $r(0)$. In many other treatments (*e.g.*, Lewis and Kareiva (1993)), the equivalent pulled wave is given by exponential growth dynamics at the maximum growth rate, $r(\hat{N})$.

If we characterize our population by the *maximum* per-capita growth rate it can attain, then it is intuitive that when this maximum is attained deeper into the bulk of the wave, the wave is likely to travel slower than if that maximum growth rate is attained out into the leading tip. So, standardized by maximum per-capita growth rate, pushed waves do travel slower than pulled waves.

This standardization by $r(\hat{N})$ is such a biologically sensible way to proceed that is often not stated explicitly. The per-capita growth rate is, after all, the balance of births and deaths, and if we imagine a particular organism—elephants, for example—it is fairly clear that there is some maximum value of r that such an organism can physically attain. Female elephants are simply unable to have five offspring per year. From a biologist's perspective, density dependence just messes with the population density at which this maximum per-capita rate might be achieved. Organisms have some intrinsic capacity for reproduction, and the harsh world works against them achieving that. This implicit understanding was so deeply entrenched that it took a team of physicists (less attached to the biological detail) to suggest an alternative measuring stick.

So while Bîrzu *et al.*'s taxonomy is very useful, it is still good to remember that if you take an organism and impose an Allee effect on its growth dynamic, you will actually slow that organism's invasion speed. This braking effect imposed by an Allee effect also applies to invasions that would otherwise accelerate as a consequence of fat-tailed dispersal kernels (Section 2.5.2). In some cases, the effect is sufficiently strong that there is, in fact, no acceleration, and a fat-tailed kernel results in a constant rate of spread (Kot et al., 1996; Wang et al., 2002).

The other outcome of a pushed dynamic is that the shape of the travelling wave tends to become steeper. This is most intuitively understood when we consider bistable dynamics, where the invasion is tripping over its own shoelaces. And from there, we can see a continuum of wave shapes all the way back to a pulled state. Indeed, this tendency for pushed waves to be steeper has also been put forth as a potential taxonomic character; another means by which we might identify pushed from pulled dynamics.

6.3.2 Failure to thrive

When we have a bistable growth dynamic, an intuitive outcome stemming from this is that invasions might fail to establish. For simplicity, let us consider a strong Allee effect. Below the critical density threshold, α, the growth rate will be negative and the population will head to extinction (*e.g.*, Figure 6.4). So if we do not introduce sufficient founders to start the invasions (*i.e.*, $N(0) < \alpha$), then the population simply goes extinct rather than establishing an invasion.

A more interesting extension of this idea is that if at any time the maximum density across space drops below α (*i.e.*, $\max(N(x)) < \alpha$), the invasion wave will collapse and the population will go extinct (Goodsman and Lewis, 2016).

This means that we can start with an initial density, $N(0) > \alpha$, but as diffusion spreads individuals out through space, the density at the introduction location drops (see, e.g., Figure A.3). If growth is not sufficient to keep density above that critical threshold, the invasion will collapse before it attains its equilibrium wave shape.

Thus, with a pushed dynamic, for a given spatial domain, there is a critical number of founders required before the invasion will establish (Lewis and Kareiva, 1993; Goodsman and Lewis, 2016). The greater the diffusion rate, or the stronger the Allee effect, the larger this critical number of founders will be (*e.g.*, Morel-Journel et al., 2016). This might sound like an abstract idea, but it can have very real consequences. There are many applications (*e.g.*, biological control, conservation reintroductions) where our aim is to establish a successful invasion. And here is a very common way in which we might fail to do so.

Intriguingly, recent theoretical work (Alfaro et al., 2020) is now considering the possibility that the Allee threshold itself might evolve. If it does (and this seems reasonable), then it is reasonable to treat the Allee threshold as a variable trait, to consider what might happen to that trait early in the establishment and spread of an invasive population, and what this might mean for that population's fate. Alfaro and colleagues' work implies that evolutionary rescue is possible, and that populations that we might expect to fail may actually thrive. In short, when evolutionary processes are taken into account, the landscape defining which populations will and will not thrive becomes quite complicated.

This idea of critical number of founders also extends to a critical area in two-dimensional invasions, but we will examine that result later in the book when we look at the geometry of invasions (Chapter 7).

6.3.3   Pinned invasions

This idea of failure to thrive due to bistability can be extended to situations with multiple discrete habitat patches. Let's imagine the one-dimensional lattice we have become familiar with (Section 3.3). We might start with a population above the critical threshold in a patch, and so this population will grow to carrying capacity. But unless dispersal is sufficiently high, there may never be enough dispersers arriving in the next patch to push that patch over the critical threshold (Keitt et al., 2001). Thus, in discrete spaces, it is possible for invasions to become 'pinned'; unable to move forward, because the critical threshold cannot be achieved in the next patch.

When dispersal is rare, there is a wide range of Allee thresholds that will result in range pinning (Keitt et al., 2001). As dispersal rates increase, however, only the strongest Allee effects will result in pinning. It is interesting to note that there must be a sweet spot in dispersal rates for an invasion to establish and propagate. If dispersal rates are very high, a population risks collapse by diffusing faster than the growth rate can keep up. If dispersal rates are low, the population might establish, but then fail to propagate, because of range pinning.

Interestingly, Morel-Journel et al. (2022) extended the idea of range pinning to another situation, beyond Allee effects. They found evidence for range pinning in invasions that did not exhibit Allee effects, but that did exhibit positive density-dependent dispersal. In their simulations and experimental work (both in discrete spaces), range pinning was observed in invasions in which dispersal rate was very low at low density. The range pinning they observed was often transitory—eventually the population did grow sufficiently to send dispersers into the next patch—but was also often remarkably stable due to a combination of low dispersal and demographic stochasticity.

It is important to note that, regardless of mechanism, pinning is a phenomenon of heterogeneous space. In a continuous and homogeneous space, pinning cannot happen: invasions only propagate forward, or collapse. Pinning becomes possible, however, when space becomes heterogeneous. Discrete patches are an obvious manifestation of a heterogeneous space, and it is clear how pinning might happen in that case. But when space is continuous and heterogeneous (*e.g.*, habitat quality, or connectivity varies through space), then pinning is also possible, though harder to conceptualize.

6.4 Evolution in pushed waves

Given that pushed waves have greater mixing, larger effective population sizes, and steeper wavefronts than pulled waves, we can expect an effect on the evolutionary processes we established in Chapters 3 and 5. In our earlier theory on pulled waves, we established (Section 3.3) that spatial sorting is a spatial analogue of natural selection. We revealed that, in a pulled wave, a conspiracy of natural selection and spatial sorting drives invasion front populations towards higher values of rD, causing the wave to accelerate. In this chapter, we have seen that both r and D might be density dependent (*i.e.*, $r(N)$ and $D(N)$), and that the form of this density dependence matters: positive density dependence can cause pushed waves.

The theory on evolution in pushed waves is still in its infancy, but the same forces are at play as in a pulled wave. When we consider pushed waves, we might consider that a more general outcome of spatial sorting and natural selection on invasion fronts is that populations are driven towards higher values of $r(0)D(0)$. That is, whatever favours high growth and dispersal at low population densities is favoured. In a pulled wave, rD is equivalent to $r(0)D(0)$, but in a pushed wave, it is not.

We have also established that, in a pulled wave, the leading tip is essentially isolated from the rest of the population: it experiences very little gene flow from the bulk of the invasion wave, and so can play by its own rules. Pushed waves are less the lone maverick, however: they are well connected, and experience strong gene flow across the entire invasion front (Roques et al., 2012). As a consequence, the evolutionary trajectory of a pushed invasion front is likely to be muted compared with that of a pulled wave. Another way of thinking about this is that pulled waves are adapting to the conditions right on the leading tip of the invasion: strong r-selection, and strong spatial sorting. Pushed waves, on the other hand, are adapting to conditions across the entire wave front (and some distance behind it). This is a mix of conditions, from high- to low-density populations, and from populations experiencing a range of density gradients (from steep to non-existent). Because of this, we might expect the evolutionary pressures on a pushed front to be towards some kind of compromise between these varying conditions (*e.g.*, Williams et al., 2016b).

Another consequence of this high connectivity is that the effective population size of a pushed wave is relatively large. Thus, the stochastic processes at play on an invasion front may also be strongly mitigated in a pushed wave (Bîrzu et al., 2019) (Table 6.1). Pushed waves should be both slower, and more predictable. Thus, we now stumble upon another reason why expansion variability might decline as an invasion progresses (Section 5.8): it might decline if the invasion evolves from a pulled state to a pushed state (through the accumulation of expansion load, for example).

In the remainder of this chapter, we will look at the detail that has emerged around evolution in pushed waves. We will consider how our three main processes—spatial sorting, natural selection, and serial foundering—may play out on pushed invasion fronts.

6.4.1   Spatial sorting

A few simulation models over the years have explored the effect of a strong Allee effect on the evolution of dispersal. They have found that strong Allee effects can retard the evolution of dispersal on invasion fronts. For example, Justin Travis and Calvin Dytham (2002) in their early simulation model demonstrating the evolution of dispersal on an invasion front also examined a simple Allee effect (in which individuals who found themselves alone in a patch were unable to reproduce). Their conclusion: 'Allee effects have a huge impact on the results we obtain.' Invasion speed was dramatically reduced, and dispersal rates showed only a mild evolutionary response.

This is more or less where things sat until Kirill Korolev turned his formidable mind to the problem. Korolev focused specifically on the Allee effect, and examined a range of effects from non-existent, to weak, to strong (i.e., bistable) effects (Korolev, 2015). His work showed something fascinating: when the Allee effect becomes strong enough, dispersal rates should actually evolve downwards. Korolev's model pitted two genotypes against each other, in competition; essentially the Lotka–Volterra equations (equation (6.2)), but with an extra parameter allowing an Allee effect to be threaded in. With standard negative density dependence, all went as expected; the most dispersive genotype came to dominate the invasion front. This effect continued to hold as a gentle Allee effect was introduced, and even as the strong Allee effect came into play.[4] But once the Allee threshold moved beyond a certain point (about $\alpha \geq K/4$), the world was turned upside down. Now, the highly dispersive type started to disappear and the invasion decelerated: evolved to *lower* invasion speeds. Under some circumstances, the invasion would eventually evolve itself to a complete standstill! The basic reason for this result is that with the strong Allee effect, the faster dispersers have been sorted into that part of the invasion front that is actually experiencing negative growth. These fast dispersers carry their genes out into the wild blue yonder, and they never come back. As a consequence, the less dispersive genotypes come to dominate.

We can also think about this result in terms of maximizing $r(0)D(0)$. In Korolev's model, under a strong Allee effect, $r(0)$ is in fact fixed to be negative. As a consequence, the only way to maximize $r(0)D(0)$ is to evolve $D(0)$ to be zero ($D(0)$ cannot be negative).

Another nice nuance of Korolev's work here is that this process acts a little like a ratchet: an evolutionary shift in the population facilitates the next evolutionary shift to happen. This is most simply seen with those values of the strong Allee effect in which high dispersal is still favoured. In this case, if a mutant arises that is a lot more dispersive than its population, it will end up way out on the tip of the invasion front, experience very low density, and suffer the consequences of negative growth. If, by contrast, it is only slightly more dispersive than the rest of the population, it will often end up dominating the positively growing part of the invasion front. Thus, the fitness of a new mutant is relative to how different that mutant is to the population mean; mutants with large effects will tend to be selected against even if, ultimately, the population will evolve to that mutant's dispersal rate.

The idea of Allee effects leading to the evolution of lower dispersal on invasion fronts was also discovered, in the same year, by a team of theoretical ecologists, Shaw and Kokko (2015). Their model was a lot less abstract than Korolev's, but mapped the idea to some real ecological effects, including resource limitation, and the difficulties of finding a mate (if you are an individual of a sexual species). Shaw and Kokko's work also shows that, if Allee effects are sufficiently strong, dispersal might evolve downwards on the invasion front. Their work was, in turn, abstracted to simpler dispersal functions and applied in an integrodifference framework, again showing that dispersal might evolve downward on invasion fronts subject to strong Allee effects (Lutscher et al., 2023). So this is an idea with a decent amount of theoretical support; empirical support is pending.

Allee effects are, of course, just one way that we can get pushed waves. They can also result from positive density-dependent dispersal (Section 6.2.2). In a nice experiment, Van Petegem et al. (2018) investigated invasions in a laboratory mite population that exhibits strong density-dependent dispersal. As we

---

[4] At this point, we are probably in a situation similar to Travis and Dytham's model in which lone individuals failed to reproduce: a strong Allee effect, but in their model only *just* a strong Allee effect.

might expect from the pushed dynamic that would result from density-dependent dispersal, evolution of dispersal on the invasion front was a relatively minor outcome from their experimental invasions: they found evidence for the evolution of higher r, but not for higher D following 10 generations of spread.

Spatial sorting may also affect the way that density-dependent dispersal itself evolves. Spatial sorting should collect individuals whose dispersal rate is most sensitive to density and place these on the invasion front. In the absence of a strong Allee effect, this may cause the density dependence itself to evolve. Justin Travis et al. (2009) built a simulation model to explore this question and confirmed this expectation. In their model, they consistently saw the density dependence of dispersal evolve in such a way that dispersal rates become higher on the leading tip of the invasion (*i.e.*, $D(0)$ increased). They observed that this led to acceleration of the invasion front. This observation is once again consistent with the general idea that invasion fronts operate to maximize $r(0)D(0)$. Although Travis and colleagues did not place their work in the context of pushed and pulled waves, it seems likely that selection on density dependence in their simulations may well allow invasions to evolve from being pushed waves, to being pulled (an idea we will come back to shortly). The evolution of density-dependent dispersal is, of course, a difficult prediction to test empirically, but it has been tested a few times: by Fronhofer et al. (2017), by Weiss-Lehman et al. (2017), and by Dahirel et al. (2023), all of whom, using experimental invasions, found some evidence to support the basic idea that we should see the emergence of negative density-dependent dispersal on invasion fronts.

6.4.2   Natural selection

In Section 3.2, we discussed how natural selection on an invasion front should favour individuals with higher reproductive rates. This remains true in a pushed wave, of course, but in a pushed wave, that higher reproductive rate is best achieved by reducing the strength of the Allee effect. In terms of most models, it is not r that is under selection, so much as the Allee threshold, α. By continually exposing a population to low-density conditions, natural selection should favour rugged individualists; individuals who don't need to lean on the help of others, and so the evolutionary response should result in populations that are less susceptible to Allee effects. To explore this possibility, Phil Erm and myself developed a simulation model in which the Allee threshold could evolve (Erm and Phillips, 2020). Given the high levels of mixing across a pushed invasion front, it wasn't at all clear to us that the population could evolve towards lower Allee thresholds. But it turns out that it can. In our simulations, we consistently saw populations on the invasion front evolving towards a lower Allee threshold. Given sufficient time, our populations would evolve to the point that there would no longer be an Allee effect: our invasions evolved from being fully pushed to being pulled. Our model was a simple simulation model, and the results are restricted to the particular conditions we simulated. It remains to be seen how general our result is, but we at least know now that it is possible for an invasion to evolve from being pushed to being pulled; again, we see the increase of $r(0)D(0)$ through evolution on the invasion front.

In this context, it is interesting to note that many plants (and occasionally animals also) appear to have evolved either a capacity for self-fertilization or asexual reproduction in marginal populations (Baker, 1955; Peck et al., 1998; Barrett, 2002). Sexual reproduction comes with many overheads (Felsenstein, 1988), but one obvious cost at low population density is that it might be hard to find a mate. Thus, sexual reproduction usually imposes an Allee effect; the evolution of selfing or of asexuality removes that Allee effect. Where it is possible for an organism to evolve selfing or asexuality, then we might very well expect conditions on an invasion front to favour such a transition (Baker, 1955; Pannell, 2015). Removing the need to find a mate is likely to increase $r(0)$.

6.4.3   Serial foundering in pushed waves

One of the subtler, but more important, effects of the pushed dynamic is that the pushed invasion front is well-mixed (Roques et al., 2012). There is a lot of gene flow across the invasion front, and the leading tip tends to grow from a relatively larger founder size, so the front has a larger effective population size (Hallatschek and Nelson, 2008). This situation has really only become clear recently (starting with van Saarloos, 2003), but it means that stochasticity is a much weaker effect on pushed invasion fronts. This, of course, has consequences for both the ecological and evolutionary dynamics. In both cases, a pushed wave should be more predicable; less prone to unexpected outcomes.

In a pushed wave, all the outcomes of serial foundering—mutation surfing, expansion load, declining heterozygosity—should be strongly mitigated. It is, for example, regularly observed that invasive populations have more genetic diversity than we might naively expect them to have (reviewed in Estoup et al. (2016)). Here, now, is a reason why we would expect that to happen: not all invasions are pulled.

In 2015, I put together a simulation model examining serial foundering and its effect on invasion speed (Phillips, 2015). I gave my simulated population variation in the rates of both dispersal and reproduction, and allowed it to spread for a small number of generations. As discussed earlier (Section 5.8), this resulted in a quite surprising variation in invasion speed. The genotypes that arrived on the invasion front in the first few generations (by chance or otherwise) had a very large impact on the resultant invasion speed.

In a subset of simulations, I introduced a strong Allee effect (in which $r(0) \approx 0$). This caused a dramatic decline in stochasticity: invasions were much less variable across simulated realizations. I knew this was because the Allee effect was generating larger effective population sizes on the invasion front, but that was about all I could say at the time.

Thanks to the work of Bîrzu et al. (2019), however, we now have a more general view of what is going on in these situations (Figure 6.6). Bîrzu and colleagues found that when waves are pulled (or semi-pushed), the effective population size on the invasion front scales only weakly with carrying capacity, K. By contrast, when waves are fully pushed, the effective population size scales linearly with K. Another way to put this is that beyond a certain point, it really doesn't matter what the carrying capacity of a pulled invasion is: you can't reduce stochasticity much by increasing the carrying capacity. For a pushed invasion, however, every additional individual added to the carrying capacity reduces the stochasticity of the invasion.

6.4.4   Evolution of pushed waves, in summary

Evolution playing out under pushed dynamics is an active area of theoretical work. I suspect that it is a place where 'the math gets interesting', and we can expect many new developments in our theoretical understanding in the coming years. A particularly fruitful approach will likely come from combining the generalized reaction–diffusion equation of Bouin et al., (2012), with functions generating Allee effects, and from there stepping out to stochastic reaction– diffusion equations. Empirical work will also constitute a formidable challenge.

It is already clear that evolution on pushed invasion fronts should be, like the waves themselves, slower, and more predictable than evolution on pulled invasion fronts. Interestingly, however, there is a strong hint that evolution on invasion fronts (pushed or pulled) acts to maximize $r(0)D(0)$. If this contention is true, then pushed waves will tend to evolve into pulled waves. The sure, steady, and conservative progress of a pushed wave might, given sufficient time, transmute itself into the wildfire of a pulled invasion.

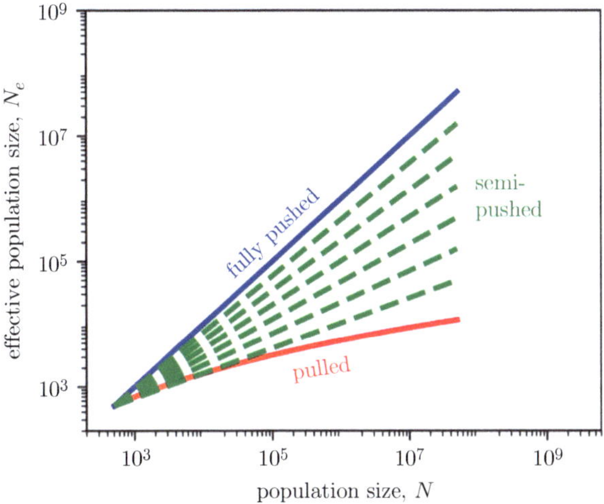

Figure 6.6 *Effective population size (N_e) as a function of carrying capacity, for pulled, semi-pushed, and fully pushed waves. Reproduced from Bîrzu et al. (2019).*

6.5 The wrap

In this chapter, we introduce the idea of pushed waves. Although challenging to identify in nature, these waves are probably quite common, and their dynamics are more complex than the pulled waves we had been focused on until now. Pushed waves result from density dependence in growth and dispersal. Negative density dependence in dispersal and/or positive density dependence in growth can yield invasions that are pushed forward from high-density parts of the invasion front rather than pulled forward from the low-density parts.

Pushed invasions may result from a wide range of situations, from competition through to genetic Allee effects. Pushed waves change our expectations, built from an understanding of pulled waves, in important ways. Pushed waves propagate forward by sampling broadly across the invasion front. As such, there is much greater genetic mixing across the invasion front, and much lower stochasticity (both demographic and evolutionary) on the invasion front. Pushed waves thus have muted evolutionary responses relative to pulled waves, and will exhibit much less variation in invasion speed. Pushed waves can also be pinned; held stationary by their own dynamics.

Although we are getting towards the edge of our current understanding, it is likely that evolutionary processes can cause invasions to transition between pushed and pulled states and there is a hint that evolution might work, in various ways, to maximize the product of growth and dispersal at low density, $r(0)D(0)$.

Some interesting wrinkles

7

It was a good friend and colleague, Stuart Baird, who got me hooked on wrinkles. Stuart loves nothing more than a good bit of speculative theorizing. Give him a problem—anything from the evolution of linkage disequilibrium on an invasion front, to how to start a recalcitrant chainsaw—and you are likely to witness a brilliant, clearly laid out, and entirely extemporized piece of theorizing about how one might understand the issue. If you are keeping up, and interject with some concern or another, your point will be considered for a few moments, carefully weighed, and then put aside. 'An interesting wrinkle! Let's come back to it.'

Stuart sees a perfectly clean landscape of logic before him, and it will not stand for elaboration until its broad contours have been described. But once that is done, we come back to explore the wrinkles in detail. It's an excellent habit of thought, and an exercise in memory. Sometimes the wrinkles grow to replace the old landscape, sometimes they solidify into new contours, and sometimes they get sanded flat. But they're always interesting.

This chapter is in the spirit of Stuart's wrinkles. It explores some facets of the story that we have deliberately put aside while we surveyed the broad contours. Many of these wrinkles are actually fundamental parts of the story; they are unlikely to be sanded flat. But I have included them here in this post hoc grab bag of wrinkles because, well, I heard Stuart's voice while I was writing.

7.1 The geometry of space

Nearly all of the theory we have so far examined (and many of the experimental examples) have been placed in just a smidgen of space; one dimension of it, to be precise. Other than Skellam's model (Section 2.2), we've largely ignored the fact that most of the invasions we care about happen in two or three dimensions. Does this extra space matter? Below, we step things out to two dimensions.

7.1.1 Invasion speed in two-dimensional space

When we step into a two-dimensional space (or three, for that matter), we have another piece of geometry to consider. While an invasion in one dimension depends only on the density gradient of the advancing wave, an invasion in

The Ecology and Evolution of Invasive Populations. Ben Phillips, Oxford University Press. © Ben Phillips (2025).
DOI: 10.1093/9780191924910.003.0007

two dimensions also depends upon the shape and orientation of the invasion front. When the invasion front is a straight edge, we can expect the invasion speed to be the same as in the one-dimensional case (Skellam, 1951). But when the invasion front is some other shape—say, a circle, or some other curved shape—we will see departures from the one-dimensional spreading speed.

In the simplest case of an invasion that starts at a point in 2D space and spreads in a perfect, growing circle, it is worth noticing that the curvature of the invasion front changes over time. When the invasion is just starting, the population inhabits a small circle; when the invasion has been progressing for a long time, the population inhabits a large circle (for example, the spread of bacteria in a Petri dish; see Figure 2.1). Small circles have much more strongly curved boundaries than do large circles. It turns out that this strong convex curvature acts to slow the invasion down. Thus, circular invasions will tend to show accelerating radial spread simply because their curvature is decreasing as the invasion progresses (Lewis et al., 2016). Such an invasion will eventually asymptote at the one-dimensional spread speed.

Why does the curvature of the invasion front change the invasion speed? Intuitively, it is straightforward to see that the curvature of the invasion front modifies the density of dispersers that arrive ahead of the invasion front (see Figure 7.1). Effectively, a convex front has a much larger empty space in front of it than does a concave front, so dispersers from a convex front are diluted in comparison to dispersers from a straight front. The reverse is true for a concave front.

Intuitively, then, we can see that parts of an invasion front that are concave will tend to move faster than a straight front, and parts of an invasion front that are convex will tend to move more slowly than a straight front. The picture that emerges, then, is one in which a straight edge is a difficult shape to perturb; if an invasion has attained a straight edge, it is likely to maintain it, despite the occasional wobble. This intuitive hunch is now well supported theoretically (Berestycki et al., 2009).

This shape dependency can also have a profound impact on pushed waves. In Chapter 6, we saw that pushed invasions resulting from a strong Allee effect could be pinned, or could collapse if the threshold density was not attained (Section 6.3.2). Now we can see that when an invasion has only just commenced, it has a strongly curved invasion front, and so its dispersers are strongly diluted. This strong curvature makes it much more difficult for an early stage invasion to climb above the threshold density. As a consequence, for invasions subject to a strong Allee effect, there is not only a critical number of founders, but also a critical area of occupancy that these founders must attain before the invasion can propagate. Below this critical area, the invasion will fail to establish (Lewis and Kareiva, 1993).

So, yes, moving from one to two (or three) dimensions can make a substantial difference to the dynamics of an invasion.

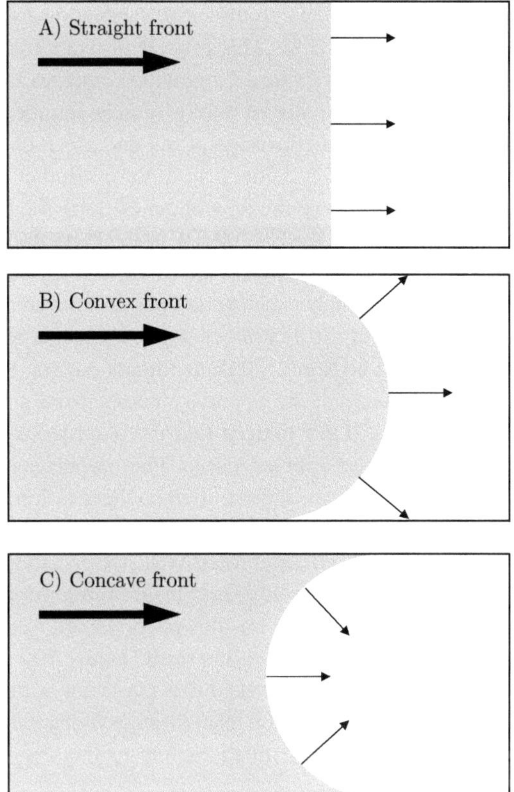

Figure 7.1 *The curvature of an invasion front alters the density of dispersers arriving ahead of the invasion front. Here we compare a straight invasion front with a convex and concave front (invasion moving from left to right in each panel). The figure imagines individuals dispersing from regularly spaced locations along the invasion front. The density of arrow points is a visual representation of the density of dispersers ahead of the invasion front.*

7.1.2   Evolution along an invasion front

We have now developed the idea, intuitively, that the curvature of the invasion front matters. Concave parts of an invasion front—divets in the front—will tend to expand a little faster than straight edges. The reason for this is that, relative to straight regions, divets are receiving proportionately more dispersers from neighbouring parts of the invasion front; they are being invaded from the side. This invasion from the side should bring with it genes from other parts of the invasion front. Thus, divets should not only expand faster, but they should also be prone to genetic swamping: alleles from neighbouring regions come to fill in the divet.

In Chapter 5, we learned how serial foundering can lead to fixation of genotypes on invasion fronts (Section 5.2). This process of fixation can lead to the sectoring of an invasion such that genetic differentiation emerges along the length of the invasion front; differentiation driven purely by serial foundering. This idea was demonstrated beautifully by the invasion of red and green bacteria across a Petri dish (Hallatschek et al. (2007); Figure 5.1). It turns out that these bacterial invasions are great places to observe the flow of genes along the invasion front. The reason for this is that the sector boundaries in the Petri dish are set down on the invasion front and remain static after the invasion front has passed; they are a record of how the boundary between genotypes has moved as the invasion progressed; they are a record of the flux of genes across the boundary.

In one of Hallatschek and Nelson's (2007) landmark papers, they meditated on this boundary between sectors; the boundary between red and green parts of their bacterial populations. They noticed that the boundaries were not perfectly straight lines, but that they wobbled a little. These wobbles must represent slight deviations to the left or right each generation on the invasion front. Perhaps this wobble was just a simple random walk—small 'errors' in the location of the boundary on the invasion front each generation? Well, it turns out that the boundaries wobbled a little more than we would expect from a random walk; something else was going on. That something else was this phenomenon of genes invading from the side in convex regions of the invasion front (Figure 7.2).

Hallatschek and Nelson's analysis revealed that the sector boundaries moved in response to the curvature of the invasion front through which the sector boundary passed. If the invasion front was tilted slightly to the left of the average

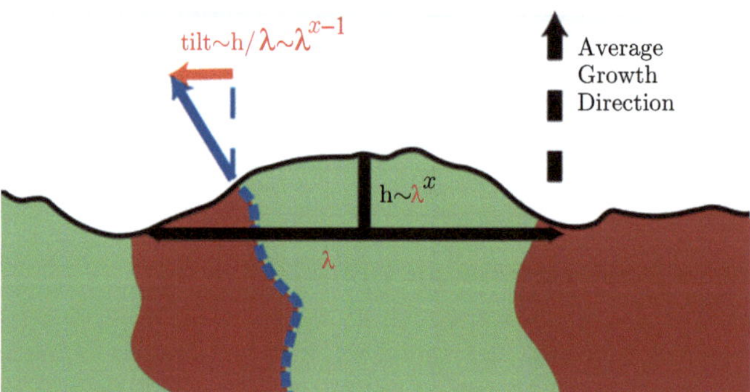

Figure 7.2 *Wobbles in the position of a sector boundary between red and green bacterial genotypes are biased by the curvature of the invasion front. This schematic (from Hallatschek et al. (2007)) shows the sector boundary (blue dashed line) growing at an angle that deviates from the average direction of the invasion. The size of this deviation depends upon the angle of the invasion front at the sector boundary relative to the average direction of the invasion. Thus, the random curvature of the invasion front causes the sector boundaries to wobble more than they would under simple Brownian motion.*

direction of invasion, then the sector boundary would move to the left; if it was tilted to the right, then the boundary would move right. In this way, tiny, random, fluctuations in the shape of the invasion front led to variation in the position of the sector boundary. These sector boundaries move along the invasion front as the invasion progresses. If two sector boundaries bump into each other, the genotype that was in between them is no longer on the invasion front. It has been squeezed out of the invasion front and will likely never make it back.

In essence, Hallatshek and Nelson (2007) showed that this movement of sector boundaries was akin to another form of genetic drift operating on the invasion front. Drift resulting purely from the geometry of the invasion front. The collision of sector boundaries, in this case driven purely by random wobbles in their position, causes the loss of alleles from the invasion front. On an approximately straight-edged invasion front, given enough time, this process would eventually result in the loss of all diversity from the invasion front. Interestingly, however, on an invasion front whose length is growing rapidly over time (for example, the circular invasion front on a Petri dish, or in the fields of the Czech Republic), the growth in the length of the invasion front drives the sector boundaries apart faster than they can collide. It is like the classic experiment of placing dots on a balloon and blowing it up. Except in this case, the expansion of space acts to prevent boundary collisions and so preserve genetic diversity on the invasion front. Thus, invasions that are expanding in all directions at once will tend to preserve the genetic diversity on the invasion front; those that are expanding in a relatively narrow habitat, by contrast, will rapidly lose genetic diversity on the invasion front, through geometry-enhanced drift.

7.1.3 Variation in space, and geometry-enhanced drift

All of this theory has been built assuming that an invasion progresses through a homogeneous space. All of this complexity emerges without any variation in the quality of habitat. But what if habitat quality does vary? Wolfram Möbius et al. (2015) set out to explore this possibility, and their results follow very nicely from what we have already learned. Let's imagine, they said, that an invasion flows around some obstacle in its path (Figure 7.3). The invasion is a pulled invasion, and so it is rapidly differentiating along the length of the invasion front. An obstacle emerges—a dispersal barrier, or a region of unsuitable habitat—that acts to slow down (or halt) the impacted part of the invasion front.

The result of hitting this barrier, of course, is that the invasion front now has a concave region, a divet. Once the invasion spreads around the barrier, this concave region will tend to travel faster than the rest of the invasion (Section 7.1.1), and it will do so because it is receiving individuals (and their genes) from the

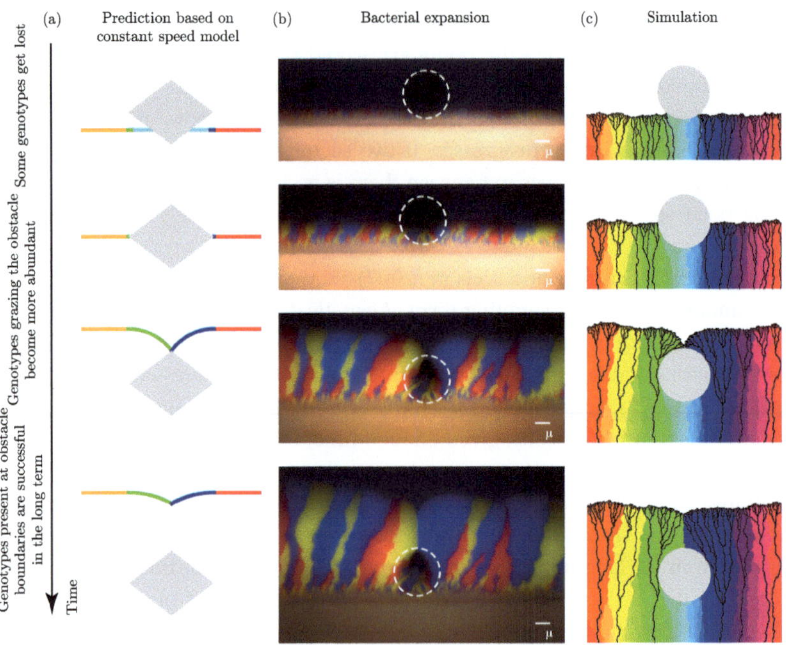

Figure 7.3 *Geometry-enhanced drift, in which lineages surfing the invasion front are lost after encountering a barrier to spread. This figure, reproduced from Möbius et al. (2015), shows results of an analytic model, an experimental invasion involving different bacterial genotypes, and a simulation. All clearly demonstrate the loss of affected genotypes as they are slowed down, and neighbouring genotypes take up their position on the invasion front.*

nearby parts of the invasion front that are further advanced in space. Genes will flow into the divet region in essentially the same way outlined by Hallatschek et al. (2007) (see Section 7.1.2). The net result is that spatial heterogeneity in the environment will act to annihilate some of the lineages that have been happily surfing. Which lineages are lost, and which preserved, depends entirely on which ones happen to run into obstacles, and which ones do not. Möbius and colleagues dubbed this process 'geometry-enhanced drift'. I have applied their moniker to the earlier result of Hallatschek and Nelson because it is an excellent name, and it is really the same process at play, albeit under a different mechanism.

It turns out that by tweaking the heterogeneity of the environment, it is possible to effectively increase the strength of drift, relative to selection, on the invasion front. In a strongly heterogeneous environment, the additional stochasticity provided by the environmental heterogeneity can eventually cause drift to overwhelm selection on the invasion front, even in dense bacterial colonies composed of millions of individuals (Gralka and Hallatschek, 2019). Spatial sorting and natural selection can, it seems, be overwhelmed if the environment is too crazy.

Interestingly, this process of geometry-enhanced drift probably emerges in situations subtler than that of a complete obstacle. The 'obstacle' need not completely halt dispersal and population growth to have an impact. It is enough that it reduces either of these rates (Möbius et al., 2019). The necessary condition—a concave section of the invasion front—results as soon as one part of the invasion front moves more slowly than the rest, regardless of why it has moved more slowly.

We might also imagine the opposite situation: in which one part of the invasion front gets a boost, by hitting a part of space that favours growth and/or dispersal. This would cause a bulge in the invasion front at this location. Here, we would expect neighbouring regions of the invasion front to be invaded by the genotypes lucky enough to have found themselves in the bulge (Möbius et al., 2019). Again, even though the effect in this case bulges the invasion front, it still leads to a loss of diversity on the invasion front.

7.1.4   Sorting and selection along an invasion front

Back in a homogeneous space again, we have now built a picture in which our invasion front consists of concave and convex regions; bulges and divets. We can think of these as high- and low-pressure systems: genes tend to flow outwards from the bulges, and inwards towards the divets. This flow is a pure consequence of geometry: divets tend to be invaded from the side, by parts of the invasion front that are further advanced in space.

We can now add to this mental picture the possibility that bulges might not just happen for stochastic reasons. Imagine a mutation cropping up on the invasion front that increases rD. Such a mutation will be strongly favoured on the invasion front (Section 3.3), and as it increases in frequency it will cause its part of the invasion front to move faster than the average. In other words, a mutation that increases rD will rapidly manifest a bulge in the invasion front. You can see where this is going, right?

We now have the situation of a mutation that is favoured on the invasion front, and that is also favoured by pure geometry. Not only will such a mutation dominate the invasion front in the direction of invasion, it will also rapidly spread along the invasion front. It will be driven along the invasion front not only by selection and sorting, but also by the low- and high-pressure systems of bulges and divets.

Hallatschek and Nelson followed up on their earlier work looking at the random walking of sector boundaries to examine exactly this situation: sector boundaries in which a favourable mutant interacts with a wild type (Hallatschek and Nelson, 2010). In addition to an elegant mathematical description, they once again furnished us with some beautiful images from experimental invasions (Figure 7.4). Here, we see something more reminiscent of an elegant gingko leaf than the awkward and wobbly sector boundaries resulting purely from drift.

Their mathematical treatment of this phenomenon described the sector boundary on the invasion front as being subject to both drift and selection.

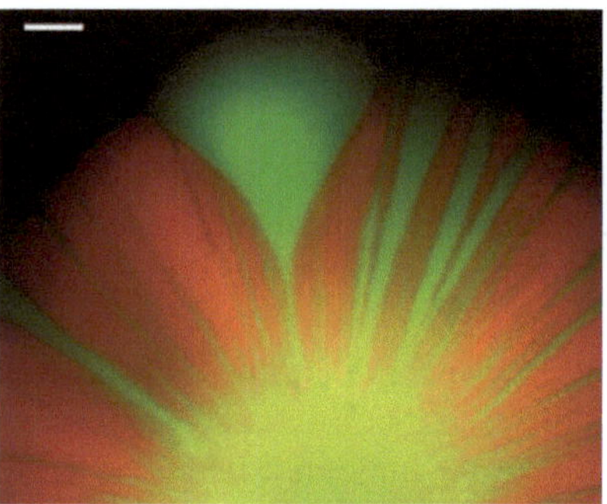

Figure 7.4 *Mutants that are favoured on an invasion front will generate a bulge in the invasion front, and rapidly invade along the invasion front. Here, an image from Hallatschek and Nelson (2010) shows the naturally arising sectoring between red and blue genotypes dominated by a naturally arising mutation with a higher growth rate. The mutant green genotype bulges the invasion front and steadily invades the front to either side of where it arose.*

This results in a biased diffusion of the sector boundary: it still wobbles a little, but has a clear tendency to invade along the invasion front. Interestingly, their mathematical treatment ignored the bulge-divet geometry and instead described only the selective advantage. As a consequence, their mathematical model describes the growth of neat triangular wedges, whereas the empirical observations show sector boundaries growing as graceful arcs. One interpretation of the empirical result is that the invasion along the front has accelerated over time, and one potential explanation for that would be the changing geometry of the bulge. As the bulge grows further ahead of the wild-type front, the tilt at the sector boundary should increase. Thus, it seems likely that the invasion of advantageous alleles along the invasion front happens faster than would be predicted by the selective advantage alone.

Regardless of the speed with which an advantageous mutant invades along the invasion front, it is clear that it will. Thus, an advantageous allele that comes to dominate any point along an invasion front will eventually come to dominate the entire invasion front. This is interesting because it increases the chances of an advantageous mutation occurring on the invasion front. We know that, to be successful, these advantageous mutations need to happen in the relatively small population at the leading tip of the invasion (Section 5.4.2, Hallatschek and Nelson (2008); Deforet et al. (2019)). In one dimension, this implies a reasonably small population, and so a reasonably small chance of such a mutation arising. But in two dimensions, along a broad invasion front, the absolute size of the leading tip population might actually be quite large. Thus, in two dimensions, the chance of an advantageous mutation—one that will eventually affect the entire invasion front—scales with the length of the invasion front. Longer invasion fronts have greater evolutionary potential.

Another implication of this tendency for genes to flow along the invasion front is that deleterious mutants—those that tend to slow an invasion—have a limited tenure on the front. We know that, because of serial foundering, such mutations can drift to high frequency on an invasion front. In a one-dimensional space, this allows such mutants to dominate the invasion front for very long periods of time, and to surf across space (Section 5.4). In a two-dimensional space, the brake that these mutants apply to their part of the invasion front rapidly has them being swamped by wild-type alleles from neighbouring parts of the invasion front. Thus, the mutational meltdown of an invasion caused by expansion load (Section 5.5) is much harder to achieve in a two-dimensional space. Mutational meltdown of the invasion front is still possible, but it requires a very high rate of mutation to offset the steady removal of deleterious mutations caused by gene flow along the invasion front (Hallatschek and Nelson, 2010).

7.1.5   Dented fronts

Above, we have built the argument that invasion along the 2D front will follow the basic geometry: moving from bulges and towards divets. In 2021, however, a beautiful exception to this was found. Lee et al. (2022) found a system where v-shaped 'dents' appear on invasion fronts and continue to grow (Figure 7.5).

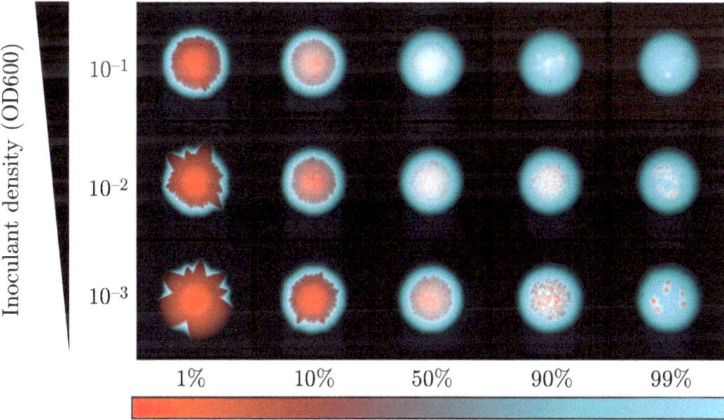

Figure 7.5 *Dented invasion fronts; formed by competition between two bacterial strains. The red strain has a higher invasion speed than the blue strain when grown in pure cultures. When grown together, however, the blue strain outcompetes the red strain on the invasion front. Thus, when a founder event causes the red strain to arrive on the invasion front, it causes rapid advance of that part of the front until it is invaded from the side by the competitively superior blue strain. The panels show the effect of changing initial population density and the ratio of blue:red strains. Image reproduced from Lee et al. (2022).*

The dented invasion front turns out to be the result of competition between two strains of bacteria. When a slower invader is also a superior competitor at low density, it is possible that, if the slow invader arrives on the front (spatial drift), it can dominate its section of the front despite having a lower invasion speed. Here, invasion from the side is undone by the superior competitive ability of the strain occupying the dent.

This observation sparked a nice piece of purely geometrical theory showing that bulges and dents are possible with two strains in competition, and that the shape that forms is entirely dependent on the ratio of the velocity of the two types when grown in pure culture, as well as the speed with which the mutant type invades along the front. It is a nice piece of theory, but it is worth noting a peculiarity of this system (and similar non-motile bacteria/yeast) systems: there is no real movement of cells. This feature makes these systems perfect for watching sectors emerge, and for mapping back through time how these sector boundaries have changed along the invasion front. But we need to remember that the sector boundaries are in some sense fossilized. In systems where the players are motile, those sector boundaries will usually move after the invasion has passed. What happens to those sector boundaries behind the invasion front will obviously depend on the fitness of the two types at high density (Perkins et al., 2016).

The other thing worth noting is that this observation does not undo the argument we have been developing: that high values of rD are selected for on the invasion front. In this case, the biotic environment has changed the value of r that each strain achieves. If we look at the growth rates when grown in isolation and we denote the growth of the fast red strain as r_R and the slow blue strain as r_B, then we can see that when grown in isolation, $r_R > r_B$, but in this situation, of course, there is nothing for selection to work with; there is no variation. When grown in competition, however, selection has something to work with, and $r_R < r_B$. Thus, in a mixed population on the invasion front, selection favours the blue strain, despite the fact that it is the slower strain when grown in isolation. This points to the fact that there are important nuances for invasions when competition is afoot, which we come back to in Chapter 8.

What is surprising is that we now have a case where we might see the evolution of a slower invasion speed than is theoretically possible. The only way the red strain can dominate on the invasion front is if it achieves fixation across the entire length of the front. Otherwise, it will eventually be replaced by the stronger competitor (the blue strain). So the invasion can be seen to, on average, slow down over time. In Lee et al.'s study, the red strain mostly arrived on the front when the invasion was initiated with a low-density population (Figure 7.5). Under this setting, stochastic effects were strong, and there was a chance that the red strain could arrive on the invasion front. When they initiated the invasion with a high-density culture, stochastic effects were dialled down and more often than not, the red strain was completely left behind: the populations rapidly unmixed and the red strain found itself excluded from the invasion front. The other way they found they could get the red strain onto the invasion front was simply by manipulating the ratio of strain 1 to strain 2 in the initial population (Figure 7.5).

7.1.6 Additional dimensions matter

Overall, it is very clear that the space through which an invasion propagates matters in interesting and non-trivial ways. Moving into two- and three-dimensional spaces doesn't overturn any of the theoretical mechanisms developed in a one-dimensional world, but it does provide additional axes along which the ecological and evolutionary dynamics can play. These additional axes can sometimes lead to very different results than in the one-dimensional case: invasions expanding in two dimensions, for example, can preserve all genetic diversity along the expanding front, whereas genetic diversity is rapidly lost in the one-dimensional case. Similarly, moving from a homogeneous to a heterogeneous space also adds interesting wrinkles. In both cases, of course, we are moving towards a more realistic description of how real invasions—from tumours to toads—might play out. When we consider real-world invasions, it is clear that space matters.

7.2 The geometry of traits

Until now, we have largely been imagining the evolution of traits in isolation: the evolution of r or the evolution of D. This is the trait equivalent of imagining an invasion in one spatial dimension. It's a very useful starting point, but it is a simplification of reality. Organisms have numerous traits, and so individuals and populations might best be characterized as sitting somewhere in a multidimensional trait space. When we consider just one trait, we are compressing this multidimensional space into just a single axis, just as when we consider just one spatial dimension we are (usually) compressing the physical, three-dimensional space into just a single axis. We have just seen how expanding the geometry of physical space can add some interesting wrinkles to our story. What about when we expand the geometry of trait space?

7.2.1 What is a trait, anyway?

Early evolutionary theory was essentially an elaboration of Mendel's model of inheritance: particulate and complete inheritance of discrete traits (e.g., wrinkled vs. smooth peas). There was quite some confusion as to how we might apply such a model to understand the more common case of continuously varying traits (e.g., the mass of a pea). It was Fisher (1918) who cracked this problem. He showed that particulate inheritance was consistent with continuously varying traits if we allowed traits to be under the influence of many genes, each with a small effect. The additive effect of all these varying alleles across all these contributing genetic loci could give rise to a normal distribution of phenotypes (see Appendix A): continuous variation arising from particulate inheritance.

Fisher extended his model to include interactions within and between contributing loci (dominance and epistasis, respectively), and to include the possibility for environmental effects. Body size, for example, might be affected not just by your genes, but by what you eat. The end result of all these effects was a pretty complicated (but now well-evidenced) picture of how traits come to be expressed, and how they are inherited. In recent years, there is a growing emphasis on epigenetics (essentially, inheritance of states of gene regulation) as yet another layer of complexity to be added atop.

So, traits are quite complicated things. They are more complicated when we consider that what we humans see as a trait is not the same thing that natural selection sees. Natural selection (plus or minus spatial sorting) does not select upon body size, nor upon the wrinkliness of a pea, the hairiness of an ape, nor the colour of a flower; it selects upon the entire phenotype, the entire organism. In the terms of this book, natural selection sees only those masterpieces of minimalism r and D. But r and D are very difficult for humans to measure. We might observe selection upon flower colour, for example, but such observations are never the full picture: selection is operating on many traits simultaneously; the many traits (including flower colour in this case) that determine an individual's reproductive rate.

So, selection operates simultaneously on many measurable traits, and each of those traits might be controlled (loosely) by many genes. Just to really make things complicated, we also have to contend with pleiotropy: the fact that each of these genes might well have effects on more than one of our measured traits. There are many-to-one and one-to-many relationships everywhere.

I think I mentioned this already: traits are quite complicated things. This complexity brings with it a marvellous blend of resilience and flexibility, but it also carries constraint. Traits are not always free to respond to selection, and these constraints can, of course, affect the course of evolution and the dynamics of invasion.

7.2.2 Genetic correlations

One of the consequences of pleiotropy is that, within a population, trait values become correlated. Individuals with relatively long thumbs tend to have relatively long toes; that kind of thing. Why? Because there is a set of genes common to both traits.

In the world of invasions, we are clearly interested in the possibility that we might have a correlation between rates of reproduction (r) and dispersal (D). In fact, we see this correlation quite often in nature: with examples ranging across insects (Hughes et al., 2003), bacteria (Fraebel et al., 2017), and cancer cells (Boddy et al., 2018). When we think about the world this way (i.e., genetic correlations between r and D), it is important to remember that we are compressing a high-dimensional trait space into one with only two dimensions; a dimension for r and a dimension for D. But as we learned in Section 7.1, two dimensions are quite a lot more than even a pair of ones. Interesting things can happen in two dimensions.

We have already seen in Section 4.10.3 that a correlation between r and D can affect invasion speeds (Lewis et al., 2016). Positive correlations speed up invasions, and negative correlations slow them down. It doesn't really matter whether this is a genetic correlation, or one driven by the environment; the effect is the same.

But genetic correlations do make a big difference to the evolutionary trajectory of an invasion. We have developed the idea that evolutionary processes on invasion fronts tend to select for higher values of rD. So it is straightforward to see that if there is a positive genetic correlation between r and D, selection on the invasion front will be aligned with the dominant axis of trait variation. When the correlation between r and D is negative, by contrast, selection is offset from the main axis of variation (Figure 7.6).

It is a fundamental rule of evolution that the response to selection is stronger when there is more genetic variation to play with (Fisher, 1930; Price, 1970). If we view genetic variance along the axis of selection, we can see schematically from Figure 7.6 that with a positive correlation, selection has a lot more genetic variation to play with than with a negative correlation. Evolution on invasion fronts will proceed faster when there is a positive genetic correlation between r and D (Perkins et al., 2013; Ochocki et al., 2019).

While a positive genetic correlation between r and D has a positive effect on both invasion speed and evolutionary response, it may often be the case that the correlation between r and D is negative. Dispersal, for example, often carries

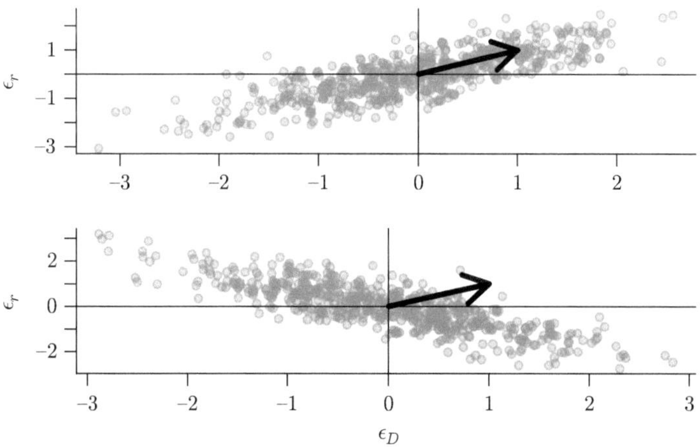

Figure 7.6 *Hypothetical distribution of trait values in populations subject to a genetic correlation between r and D. Values of ϵ express deviations from the trait mean in units of trait standard deviations. The top panel depicts a positive genetic correlation ($\rho = 0.8$), and the lower panel a negative genetic correlation ($\rho = -0.8$). Blue arrows show the direction of selection on an invasion front.*

costs associated with energetic investment or reduced survivorship. A very clear example here comes from winged insects, in which investment in the thorax favours dispersal (larger wings and bigger wing muscles), while investment in the abdomen reduces dispersal (more weight per gram of thorax) but allows greater reproduction (greater egg production capacity). Thus, many insects are seen to face a fundamental trade-off between reproduction and dispersal (Hughes et al., 2003).

In the face of a strong genetic trade-off between r and D (i.e., a negative genetic correlation), selection on the invasion front may have very little effect. With a strong trade-off, there may be little scope to increase rD, and this evolutionary constraint might place a strong constraint on the ultimate speed of the invasion. Well, maybe. Evolution often finds a way, and in this case there are two possibilities to be aware of.

The first possibility is scaling. Unlike dimensions of physical space, dimensions of trait space may be measured in very different units. In this case, r is a rate (something per time), and D is measured in space squared per time; these are different units. In Figure 7.6, I have changed the units to trait standard deviations, but this obscures the true scale. If, for example, D has a much larger absolute variance than does r, then quite large changes in D might be possible on an absolute scale, even if it cannot move very far in terms of standard deviations. Thus, we might still see a large shift in invasion speed despite a negative correlation between r and D.

The second way that evolution might proceed despite the negative correlation is simply by changing the correlation structure itself. If there is any genetic variation for the correlation structure itself, it might evolve. Given a positive correlation results in faster invasion speeds, we can certainly expect strong selection towards a positive correlation structure on an invasion front.

7.2.3 Genetic correlations and pushed waves

One interesting possibility, noticed in passing by Ochocki et al., (2019), is that a strong negative genetic correlation between r and D can result in pushed waves. The mechanism is a simple result of spatial sorting. The most dispersive individuals are sorted out into the leading tip of the invasion front, as usual. But with a strong negative correlation, these individuals will tend to be individuals with relatively low reproductive rates. Thus, we see a gradient in per-capita population growth rate emerge as a consequence of spatial sorting.

This gradient in per-capita growth is functionally very similar to the gradient in per-capita growth that emerges from an Allee effect. The mechanism is completely different, of course, but the result is the same: higher per-capita growth rates in the bulk of the invasion. As we now know (Chapter 6), this is one of the conditions that causes a pushed dynamic.

Thus, a strong enough negative genetic correlation seems likely to not only slow an invasion, but place quite strong constraints on the evolutionary response on the invasion front. Constraints not due solely to the trait geometry but also to the strong gene flow across the invasion front that is a consequence of the pushed dynamic (Section 6.4).

7.2.4   Arbitrary relationships between r and D

One quiet assumption we have been making is that there is some straight line relationship between spatial and temporal fitness. We have been considering that the relationship might be either a positive or negative straight line. But of course nature revels in curve balls. Maybe, in crushing the trait space into only two dimensions, we can actually expect some pretty wonky relationships to emerge between r and D. Maybe, straight lines are simply too much to hope for.

We can get an intuitive feel for what might happen under an arbitrary relationship, if we define a spatiotemporal fitness function, $w = rD$, that claims (based on arguments in Sections 3.3 and 6.4) that fitness on the invasion front is defined by an individual's dispersal and reproductive rate at low population density. We can draw contour lines onto our r-D space that connect regions of equal spatiotemporal fitness (Figure 7.7). If we then plot our arbitrary curve, $D(r)$, onto this map of fitness, we can easily see where local and global optima occur.

This is a long way from a formal analysis, but it nevertheless allows us to see that there may be one or more optima, a result that has been shown to emerge in integrodifference models in which both r and D can evolve (Poloni and Lutscher, 2023). Where a population evolves to in this space will depend upon the shape of $D(r)$, and upon where in r-D space the population begins. If the second differential $D''(r) < 0$, for example, then we will have a convex trade-off function laid over our fitness surface, and there will be a single evolutionary optimum. Conversely, if the second differential $D''(r) > 0$, we will have either one or two evolutionary optima (one when $0 < D''(r) < 2w/r^3$).

In the case of a function that has both local and global optima, the population might get stuck on a local optimum and persist there for a long time until some force (for example, drift, serial foundering, or geometry-enhanced drift) pushes it over to another evolutionary attractor. Thus, complicated relationships between r and D may well result in invasion dynamics that show stepped invasion speeds.

7.3   Anomalous invasion speeds

We have just learned that the geometry of traits—the genetic architecture of r and D—might cause interesting wrinkles. Trait geometry can affect invasion speed both directly, and via the evolutionary response on the invasion front. We have been imagining a continuum of phenotypes, but it might also be instructive to imagine the simplest situation in which trait variation can be admitted: two phenotypes. Elizabeth Elliott did just this as part of her PhD research (Elliott and Cornell, 2012), and what she found was fascinating.

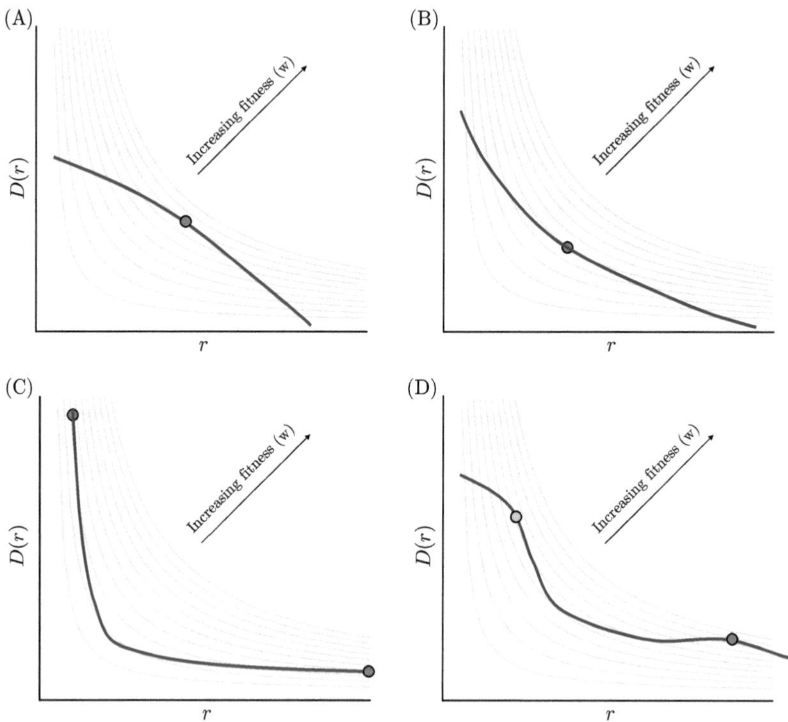

Figure 7.7 *The spatiotemporal fitness landscape, and evolutionary constraint. Here, light grey contours show isoclines of increasing fitness. Darker lines show hypothetical relationships between r and D. When D(r) is convex or weakly concave (panels A and B), there is a single point at which highest fitness can be attained along the trade-off curve (shown by the dot). When D(r) is strongly concave, or non-monotonic (panels C and D), there may be two or more fitness peaks. Panel D also shows the possibility of local and global fitness optim (lightly and darkly shaded dots, respectively).*

Elliott imagined two genotypes: a disperser (high D, low r), and an establisher (low D, high r). Two genotypes that would, together, describe a negative trade-off between dispersal and reproduction. She stuck these two genotypes into a pair of coupled reaction–diffusion equations (similar to those in Section 6.2.3.3) and instated a small mutation rate that switched individuals between the two geno-types. Did the resultant mixed invasion wave move at the speed of the establisher, or of the disperser?

It turns out that the resultant mixed wave often moves faster than the wave of either of its constituent genotypes. The invasion speed of this mixed population is anomalous: how can it be possible to have a mixed invasion moving faster than either of its component parts? It is a difficult result to explain. We might think of the mixture as having a weighted average value for r and D, and that this average would fall somewhere on the straight line connecting the two phenotypes in our two-dimensional trait space (Figure 7.8; Keenan and Cornell (2021)). If we think of things this way, and we lay our straight line over a surface describing the Fisher velocity, then we can see visually that in some circumstances (but not others), the weighted average will yield a faster spread velocity than either of the constituent genotypes.

But this graphical explanation doesn't explain all cases of anomalous invasion speeds. The pair A and D in Figure 7.8, for example, will yield an invasion speed higher than that of D alone, even though the weighted average of these two genotypes suggests a speed less than that attained by D alone. It seems that anomalous speeds are best thought of as a kind of inadvertent population-level cooperation between the genotypes (Elliott and Cornell, 2012); it occurs even though the genotypes are actually in strict competition with each other. Weird, eh?

The existence of anomalous invasion speeds suggests that, rather than a single genotype coming to dominate the invasion front, it might be possible to have 'coalitions' of genotypes that come to dominate. By playing together, these coalitions can move faster than any individual genotype. It's an interesting idea, but it turns out that anomalous invasion speeds require quite a few conditions. First, the two genotypes need to be quite different from each other; second, there needs to be mutation—some kind of switching—between

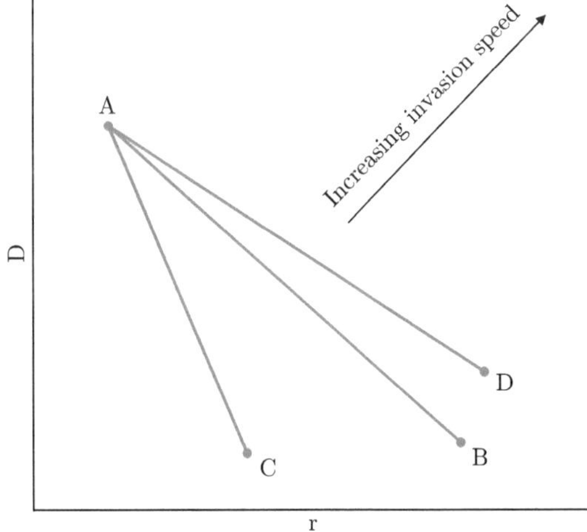

Figure 7.8 *A weighted average between two genotypes (denoted by straight lines on this plot) will often yield invasion speeds (speed isoclines shown in grey) that are faster than the invasion speed of either genotype alone. This is the case between genotypes A and B. Some pairs of genotypes (A and C, or A and D, for example) will have weighted averages that always yield speeds intermediate between those of the two genotypes.*

the genotypes; third, they require a trade-off between r and D that is either straight or concave; finally, anomalous invasion speeds largely disappear in the face of demographic stochasticity (Elliott and Cornell, 2013; Keenan and Cornell, 2021). If all these conditions are met, then an invasion might progress substantially faster than we would predict from either genotype alone; a result that has also appeared in integrodifference models (Poloni and Lutscher, 2023).

It is good to keep our eyes open to the possibility of anomalous invasion speeds. But perhaps the most general message to emerge from work on anomalous speeds is that trade-offs matter; the geometry of the trait space once again has a large bearing on both the ecological and evolutionary dynamics that are possible. If we want to predict what an invasion will do, it is not sufficient to go and measure the mean phenotypes. That is a good start, but the next most important piece of information will be the relationship between r and D; the shape of that relationship as well as the genetic variance and covariances at play.

7.4 Spatial heterogeneity

Most of our survey thus far imagines a nice world in which it doesn't matter where you live. One location is as good as another. With a few exceptions, so far we have largely treated space as homogeneous, and uniformly favourable. It's a nice, safe imaginary world, but quite a boring one. The real world is not so boring. There are beaches and mountains; rivers and roads; shady and sunny sides of a tree; there is bark beside cambium; muscle around blood vessels.

Spatial heterogeneity is a fact of life. It is also a subtle and scale-dependent thing. We could spend considerably more than the length of this book pondering it. But let's not. Instead, we will consider three very general sources of spatial heterogeneity: discrete habitat, gradients, and biotic interactions. Much work has been done in these realms, and that work is typically not thought of as being work about invasions. So in venturing into these realms, I mostly hope to open doors and admire the view, as well as help us make a few links across fields.

7.4.1 Discrete patches

We experience space as a continuum, but there are many cases where it might be useful to think of it as being discrete. In a discretized space, we imagine locations that may be inhabited by a well-mixed population, with the possibility of the exchange of individuals between locations.

In this sense, we can find many examples in the nature of discrete populations. Many parasites live in a discretized space: their hosts are locations in space within which they can form populations; transmission between hosts is an exchange of individuals between locations. Similarly, we might think of an archipelago of islands, with populations of butterflies on each island. We might think of lakes and an aquatic crustacean; rock outcrops and a saxicolous orchid; and so on.

Discrete patches are so common that an entire field of population biology has sprung up around them. We call populations living in such spaces, metapopulations: a large population made up of many sub-populations. Most work on metapopulations—and there has been a lot (for reviews, see Hanski and Gaggiotti (2004))—is concerned with 'established' populations, rather than invasive ones. But the basic dynamics of a metapopulation involves the extinction and recolonization of patches. If we took a metapopulation model and started with all but one of the patches being extinct, then we would have a model of an invasion.

With few exceptions (and these have mostly already been discussed in Sections 6.2.3.2 and 6.3), we can expect invasion dynamics across a discrete space to be similar to that across a continuous space. Growth at each location is still due to local growth and the balance of comings and goings; natural selection and spatial sorting are still at play; drift and serial foundering still matter; and pushed dynamics can emerge.

The metapopulation worldview tends to think of extinction-recolonization dynamics: sub-populations blink out of existence and are then recolonized by individuals from other patches. It is possible for a metapopulation to be at equilibrium, despite all this extreme turnover happening at the sub-population level: the total number of sub-populations remains constant despite continuous extinction and recolonization of sup-populations. But from the perspective of invasion dynamics, extinction-recolonization looks like lots of 'sub-invasions' occurring : transitory moments in time and space in which the population is invading. Thus, even in a metapopulation at equilibrium, invasion dynamics are happening. Spatial sorting and foundering are happening continuously in metapopulations, but in a fragmented way: first happening at location x, then at location y, z, and so on. Invasion fronts blinking in and out of existence.

If a recolonization event is subject to spatial sorting (as it must be), then we would expect recently colonized sub-populations to have slightly higher dispersal rates than long-colonized sub-populations. There is, in fact, some nice empirical evidence (from butterflies, voles, grasshoppers, and bugs) that this does happen (Hanski et al., 2002; Forsman et al., 2011; Berggren et al., 2012; Comerford et al., 2023), and the idea may also explain the emergence of metastatic cells in the interior of tumours (see Section 9.1). In aggregate, then, these moments of spatial sorting should cause the dispersal rate of the entire metapopulation to be higher than it otherwise would be. This is something that metapopulation theorists anticipated, before spatial sorting even had a name. Elegant modelling work by Isobele Olivieri and collaborators (1995) showed that the dispersal rate in metapopulations should be higher than what we would predict, given the costs of dispersal. This was such an interesting observation that they called this shift in dispersal rate the 'metapopulation effect' (Olivieri and Gouyon, 1997).

There are, of course, similar parallels with genetic drift. We learned in Chapter 5 that founder events can be seen as genetic drift happening through space rather than time. In metapopulations, recolonization will cause a founder event; thus, metapopulations at equilibrium will also experience greater evolutionary stochasticity (drift + foundering) than we would anticipate simply from a raw accounting of sub-population sizes (Gilpin, 1991; Harrison and Hastings, 1996).

In short, there are very clear parallels between invasion theory and metapopulation theory. Metapopulations at dynamic equilibrium will contain many brief instances of invasions. Metapopulations do not, in general, exhibit invasions in which there is a persistent invasion front, but there is no doubt that the processes we have observed in persistent invasion fronts are also playing out in metapopulations, they are just doing so in a fragmented and locally transient way.

7.4.2 Gradients

Another simple way we can introduce spatial heterogeneity is to imagine a gradient in some aspect of the environment. Such gradients are commonplace in nature: mean temperature decreases with distance from the equator; variation in temperature increases rapidly as we move from inside your skin to outside of it; a bird poop on a windscreen represents a glorious gradient in nutrient availability to your average bacterium; and so on. Sometimes, the gradient we see in some condition or resource strongly affects organismal fitness. Tropical species are ill-equipped for life at the poles; internal parasites tend not to live for long outside of bodies; bacteria require an appropriate nutrient substrate if they are to thrive. Thus, it is an intuitive step to imagine that environmental gradients often generate fitness gradients. What happens, then, to an invasion that progresses along a fitness gradient?

Well, to some extent, this is a trivial question. If environmental conditions change in some steady way across space, x, and the environment determines rates of birth and death, we have $r(x)$. Our invasion will advance through space until it hits the situation at which $r(x) = 0$, at which point the speed of the invasion equals 0. If the environment is deteriorating in a steady way towards this point, we can expect the invasion to gently decelerate to this standstill. That is, some kind of stable range edge will emerge. We can imagine various ecological wrinkles around this idea, but they are wrinkles.

The question rapidly becomes non-trivial, however, when we give the population some capacity to adapt to this changing environment. As often happens, admit a little bit of heritable variation and suddenly things get very interesting indeed. With a little bit of heritable variation, the last paragraph does not end with a fullstop, but it ends with a question: why doesn't the population adapt to conditions on the range edge, and so continue to expand its range?

One possible answer to this question was first pointed out by that most rakish of early contributors to the modern evolutionary synthesis, J. B. S. Haldane.

Haldane was an afficionado of cigars and double-breasted pin-stripe suits; he was a larger than life character who enjoyed asking large questions about life. In the mid-1950s, there was a fierce (and occasionally acrimonious) debate raging in ecology between those who felt that all populations must be density regulated and those who felt that density-independent processes were more important. Haldane, in a surprisingly diplomatic move, wrote a short paper pointing out that, for many species, both situations might be true (Haldane, 1956). In the core of the range, wrote Haldane, environmental conditions are good: density is likely high, and so the population is likely density regulated. On the edge of the range, however, environmental conditions were less suitable and so here, he suspected, population density would be low and density-independent processes were probably the ticket. So far, so good; all very diplomatic. But then he notices something interesting. Haldane realizes that these density differences between core and edge imply a net flow of genes from core to edge. High-density populations, argued Haldane, produce more dispersers than do low-density populations. For a given rate of migration, then, the low-density population receives a greater fraction of immigrants than does the high-density population. In Haldane's view, this meant that low-density populations on the edge of the range were doomed to be maladapted. Any adaptation in the small population at the edge of the range is immediately swamped by the immigration of less well-adapted individuals.

This is an interesting idea, and may help explain why we very often see clear geographic limits to populations. It is an interesting idea, but it took 40 years for it to be formalized into a mathematical model. This was finally achieved by Mark Kirkpatrick and Nick Barton (1997). Why so long? Well, I have every suspicion that it is because the mathematics is tough. The math is tough, but we have already wrapped our heads around the reaction–diffusion model, so let's have a go at seeing how Kirkpatrick and Barton described Haldane's idea. They first started with an equation describing how the mean of the trait determining fitness, $\bar{z}(t, x)$, would change at any location, x, in a one-dimensional space:

$$\frac{\partial \bar{z}}{\partial t} = \frac{\sigma^2}{2} \frac{\partial^2 \bar{z}}{\partial x^2} + \sigma^2 \frac{\partial \ln(N)}{\partial x} \frac{\partial \bar{z}}{\partial x} + G\frac{\partial \bar{r}}{\bar{z}}. \tag{7.1}$$

This is quite a mouthful, but you will recognize that first term as a diffusion term (Appendix A). It is the change in the trait mean caused by the coming and going of individuals at x. The last term is also straightforward; it is the effect of natural selection on the trait mean: the selection gradient (the gradient between fitness and the trait) multiplied by the genetic variation in the trait, G. The middle term is one we haven't seen yet, but it is important. This middle term is the advection term. It captures the idea that there might be asymmetry in gene flow (of the kind that Haldane envisaged), caused by the gradient in population density, $\frac{\partial \ln(N)}{\partial x}$. Together, these terms completely describe why the mean trait value might change over time: the (asymmetrical) coming and going of individuals, and natural selection.

They next need to define a gradient in the environment. Well, that's easy: let's just say that the environment is changing at some constant rate with respect to space, and we can jump straight to describing the environment as defining some trait optimum, θ, that changes through space: $\theta(x) = bx$, where b is the steepness of our environmental gradient.

Now all that is left is to describe how deviation of the trait mean from this optimum affects fitness. This is slightly more complicated, but not too bad. For this, Kirkpatrick and Barton chose a pretty standard form of stabilizing selection:

$$r(x) = r_{\max} - \frac{(\theta(x) - \bar{z})^2}{2V_s},$$

where V_s controls the strength of selection towards the optimum.

With these descriptions of the environment and how it relates to fitness, the last term in equation (7.1) becomes $G\frac{bx-\bar{z}}{V_s}$. Now we have a partial differential equation that captures the mechanisms that Haldane had imagined (asymmetrical gene flow, selection, and, implicitly, some gradient in the environment that affects fitness). You will note, however, that while we now have a model for how a trait might change, we still don't have an explicit model for population size at x, $N(x)$. To do that, we simply need a model that links fitness, $r(x)$, to a demographic model.

Kirkpatrick and Barton introduced a simple density-dependent demographic model (a modified version of logistic growth) to develop a joint model describing changes in population size and changes in trait means. Analysing this model suggested that (a) if b was small enough, the population could be perfectly adapted at all locations; (b) if b was large enough, the population would simply collapse to extinction; and (c) at intermediate values of b, a stable range edge would emerge. This last outcome is the interesting one. It suggests that an invasion along a gradient will naturally come to a halt somewhere along that gradient, and remain stable there, pinioned by the opposing forces of natural selection and maladaptive gene flow.

It's a nice answer, and one Haldane undoubtedly would have enjoyed. It's a nice answer, but it's not the end of the story. On the other side of the Atlantic, Richard Gomulkiewicz and Bob Holt were also pondering this problem, and were reaching a subtly different conclusion. Gomulkiewicz and Holt, in a series of papers, played with the idea of a simple two-patch system, where one of the patches is unsuitable for the species. They had a source patch, in which growth rates were positive, and a sink patch, in which growth rates were negative. Here, they argued, is an environmental gradient boiled down to the bare necessities: two places connected by dispersal but which differ in their environment. They were interested in whether the population might, by constantly sending immigrants into the unsuitable patch, eventually adapt to that second patch and so come to fully occupy both. Their work, over several papers (Holt and Gomulkiewicz, 1997; Gomulkiewicz et al., 1999; Holt et al., 2003), seemed to suggest that such adaptation would eventually occur if there was sufficient genetic variation in the source patch. Essentially, if a variant is produced that has an absolute fitness greater than 1 in the sink patch and that variant arrives in the sink patch, the sink patch will be colonized; the population will have invaded across the gradient. In this view of things, we have a range edge emerge that is only stable temporarily: yes, there is maladaptive gene flow, but this gene flow also provides a constant source of migrants, any of which might be fit enough in the new habitat to kick off a new population.

So what's going on? How can these two models be reaching such different conclusions? It was difficult to compare the two approaches, because they really were quite different. On one hand (Kirkpatrick and Barton, 1997), we have a deterministic reaction–diffusion model (continuous time and space) coupled with a model from quantitative genetics, and on the other hand (Holt et al., 2003) we have a stochastic individual-based model (discrete time and space) with a single-locus population genetic model. Theoreticians got busy building various models that fell between these two extremes. It's a complicated flotilla of models, and the result is also quite complicated. The answer as to whether stable range edges will emerge often ends up being, 'it depends'. But probably the main reason for the different results is how genetic variance was handled. In the original Kirkpatrick and Barton model, genetic variance was treated as a constant, but it is clear that genetic variation will evolve along the fitness gradient. It turns out that when we allow genetic variation to freely evolve in the deterministic model, the ultimate result is typically adaptation across all space—no stable range edge (Barton, 2001; Polechová et al., 2009).

This is all a little vexing, because we regularly observe sharp and apparently stable range edges in nature. It is more vexing given that the latest empirical synthesis of natural systems finds little evidence for Haldane's idea (Kottler et al., 2021), while recent lab experiments do (Moerman et al., 2022). We often see sharp range edges, but now we can't explain them. Well, it's not that bad. The truth is that genetic variation can rarely be considered to evolve freely as it does in these models. There are natural limits to genetic variation: it takes time to accumulate, and organisms face fundamental constraints. When population size is limited, for example, stochastic effects and limited genetic variation can seriously impede adaptation along environmental gradients (Bridle et al., 2010). We may well be in the situation where the mathematical equilibrium is perfect adaptation across the entire gradient, but the time taken to achieve such adaptation is sufficiently long that, for practical purposes, we can consider that a stable range edge has emerged. Rather than thinking about ultimate outcomes, then, it may often be more useful in this case to think about how long it will take for the population to spread across the gradient. If progress is sufficiently glacial, it is going to appear to be a stable range edge across most timescales that we care about. Seen in this light, the initial results of Kirkpatrick and Barton are still a very useful guide.

So, when a fitness gradient is steep enough, it is going to be hard for a population to invade across that gradient, because maladaptive gene flow will work against the evolutionary response. But what do we mean by a steep gradient? 'Steepness' in this sense only makes sense when measured against the dispersal rate of the organism in question (Kirkpatrick and Barton, 1997). A gradient where fitness declines by 10% per kilometre will appear quite steep to a bird or to a flying insect that regularly moves more than 1 km but will essentially be flat as far as a bacterium on a Petri dish is concerned. Gradient steepness is only defined relative to dispersal rate, but we know that dispersal evolves on range edges (Chapter 3). As an invasion progresses, and dispersal rates evolve upwards on the invasion front, fixed environmental gradients become steeper.

So now we have an interesting situation. Let's say we have a gentle gradient and invasion proceeds. The process of invasion selects for higher dispersal, which has the effect of making the environmental gradient steeper, which has the effect of slowing (or even halting) the invasion. The likely picture that emerges is an interesting interplay between dispersal evolution, invasion speed, and gradient steepness. In 2012, I put a simple simulation model together to play with this idea (Phillips, 2012). The model space consisted of two differing homogeneous environments connected to each other by a gradient. I imagined a population in which there was heritable variation in two traits: one trait for dispersal and one allowing adaptation to the environment. Across simulations, I allowed the population to spread for some period of time before it encounters the gradient. Sure enough, the longer the population had to spread before it hit the gradient, the longer it took to adapt across the gradient. Here is a case in which a long run-up actually impedes your ability to jump.

7.4.3   Cliffs, and asymmetrical dispersal

Let us now indulge in a thought experiment, with a squeeze of lemming. Lemmings are a group of small rodents found in the arctic. They are cute, but even taking this into account they hold a strange fascination with popular culture. Every now and then lemming populations explode in numbers and (here we transition to urban legend) engage in mass suicide events by throwing themselves en masse off cliffs and into the ocean. Let us now imagine a desolate arctic island, a fortress of cliffs upon which lives a population of lemmings that, every now and then, indulge in this strange cliff-hurling behaviour. Let's imagine that the tendency to throw oneself off a cliff is actually a form of dispersal (closer to the truth) and that there is a genetic basis to it. What happens to the frequency of this trait over time?

It is pretty clear that, because the cliff-hurling phenotype removes itself from the population, the only ones left are those less prone to this behaviour. Our island population should evolve, quite rapidly, towards a race of relatively sedentary, sensible lemmings with a healthy respect for steep drops. We could think of this as a straightforward application of natural selection.

Now, let's imagine that it isn't actually suicide and that, in fact, most of the lemmings taking the jump actually swim to the mainland and live long and prosperous lives. Let's imagine that no lemmings are harmed in the making of this thought experiment. The island population still evolves towards an aversion to jumping. But the mechanism is no longer natural selection, it is spatial sorting. Arguably, it was spatial sorting before we imagined the swim to the mainland.

When we think about what is going on here, it is clear that traits are being sorted through space. The mainland is receiving a lot of cliff-jumpers, and the island is losing its cliff-jumpers. The result, on the island, is evolutionary change. In this case, spatial sorting is causing a decrease in dispersal, rather than an increase. The reason for this is that dispersal is asymmetric; it is only happening in one direction (away from the island).

We may have just indulged in a fun thought experiment with lemmings, but in the process we have noticed something interesting; spatial sorting can also cause dispersal rates to decline. And the island example is not all that far-fetched. Indeed, there are countless examples of insular populations with greatly reduced dispersal ability relative to their mainland relatives. The Galápagos Islands, for example, have flightless cormorants; mountaintop-restricted dung beetles in Australia's Wet Tropics have lost their wings; Rails—a group of birds that look like small chickens—have colonized islands throughout the world's oceans, and then become flightless. Often, the argument for such evolutionary shifts is disuse—the dispersal traits are no longer needed, and so are selected against on the basis that maintaining them is costly. With spatial sorting, however, we have a much more direct argument: all the most dispersive phenotypes have left town, and only the most sedentary are left. With spatial sorting, we have a mechanism that can drive dispersal rates down very quickly (e.g., Baxter-Gilbert et al., 2020); it helps explain, for example, why some island populations maintain all the capacity for dispersal, but simply never exercise that capacity: spatial sorting on behaviour has occurred, while the loss of capacity through disuse has not (Jessop et al., 2018).

We can generalize the idea of the lemming cliffs to any sharp change in the environment that a population encounters. If the environmental change is an absorbing boundary—in the sense that individuals dispersing out there don't come back—then we can expect dispersal rates in parts of the population close to that boundary to be driven down by spatial sorting. Indeed, we can generalize further to any environment in which there is strong spatial heterogeneity. In such environments, dispersal rates will tend to evolve downwards simply because most individuals that disperse are coming from high-quality environments and are likely to end up in lower-quality environments (Hastings, 1983). As a consequence, dispersal is asymmetric in the same way as it is in our lemming story.

Asymmetric dispersal can also occur in other situations in nature, particularly when dispersal is a passive exercise. Water tends to flow only one way down a stream; prevailing winds and ocean currents generate similar biases; blood flows only one way down the veins. What happens if, instead of imagining a discrete island losing dispersers, we imagine a continuous space in which dispersal is biased in one direction, like a stream? Well, the result from simulations is as you would expect: a cline in dispersal rate emerges along the space, with higher dispersal rates accumulating as you travel in the direction of the dispersal bias (Allgayer et al., 2021).

7.5 The wrap

In this chapter, we came back to a few wrinkles that we skipped over in earlier chapters. We looked at the additional interesting things that emerge when we allow more than a smidgen of space (i.e., moving into two- and three-dimensional spaces). Here we have the curvature of the invasion front to contend with. This curvature can affect both the ecology and evolution of the invasion front. Strongly convex fronts move more slowly than concave fronts, and, in pushed invasions, convex fronts can cause the invasion to collapse altogether. In regard to gene flow, however, convex parts of an invasion front are like high-pressure systems, sending genes outwards into concave parts (the low-pressure systems). It is possible to reverse this flow in situations where the genes in the convex section are fitter in competition. Regardless, the interplay of these two pressure systems along an invasion front can increase the speed with which genetic diversity is lost from an invasion front.

We then looked at two-dimensional traits. We allowed r and D to share some kind of genetic correlation. We developed the idea of an invasion fitness surface and graphically developed the idea that we may see one or more evolutionary attractors on such a surface, depending upon the relationship between r and D. We saw that strong negative correlations may even be (yet) another way to manifest a pushed invasion.

Finally, we looked at spatial heterogeneity and how it might influence the dynamics of invasion. Here we made links to broader fields of study: metapopulations, and the study of stable range edges. We see that, internally, metapopulations can be thought of as invasions that are patchy in space and time.

There are, of course, many more wrinkles we could explore.[1] Perhaps given the grab-bag nature of this particular chapter, there is no obvious single further resource to point you towards, but for those interested in the evolutionary dynamics of metapopulations, Hanski and Gaggiotti (2004) is a broad-ranging and valuable resource.

---

[1] I am developing a suspicion that wrinkles are fractal in nature. We came back to look at some of the bigger ones, anyway.

Biotic interactions

8

8.1 A quick primer on biotic interactions

Biotic interactions are of immense importance in biological invasions. The vast majority of invasions fail before they really get underway, and in many cases this will be because of biotic interactions. This was something that Charles Darwin (1859) simply reasoned to be true; I, on the other hand, learned the principle indirectly from HG Wells's *The War of the Worlds*. Wells's *The War of the Worlds* has aliens arriving en masse with weapons that are so superior to ours that human civilization is rapidly overthrown. It is game over within days. Well, it would be except for the biologically satisfying denouement. Humans are conquered, but the aliens get sick and are all killed by everyday terrestrial bacteria to which their bodies have no resistance. The war of the worlds is not won by the technology of which we are so proud, but by the small and common things that we routinely ignore; a few novel biotic interactions.

To a first approximation, there is nothing interesting on earth except for biotic interactions. The number of potential biotic interactions vastly outnumbers the number of species in an ecological community; the potential number of biotic interactions scales with the square of the number of potentially interacting species in a system. We need to look no further than our bodies to see how ubiquitous these interactions are. Each of us supports thousands of species of microbes, some of which are dangerous to us but many of which are critical for good health; they help us digest food, synthesize vitamins, and ward off parasites. Indeed, the very cells of our body represent ancient partnerships between species (our mitochondria were once free-living bacteria), and important parts of our genome have their origin in viruses. It is increasingly clear that what we think of as our body is actually a complex community of biotic interactions.

The sheer, boggling number of biotic interactions requires us to come up with a classification scheme; some way to make sense of them. Ecologists have long used a functional classification in which we consider a pair of species and their fitness effects on each other. Effects can be either positive, negative, or neutral. When we see the world this way, it is clear that there can be only five possible categories of interaction.

Seen this way, it is clear that *The War of the Worlds*—in which humans and aliens are trying to kill each other—is an instance of competition between two species; the interaction between you and the mitochondria that live in your cells and generate energy is an instance of mutualism. There are also some surprises. The common cold—a virus that uses your cells to produce more copies of itself— is engaging in a form of predation; as is the dairy cow as it chews up grass.

The Ecology and Evolution of Invasive Populations. Ben Phillips, Oxford University Press. © Ben Phillips (2025).
DOI: 10.1093/9780191924910.003.0008

Table 8.1 *Ecologists classify biotic interactions by examining the effect of the interaction on each partner. Here, columns represent the effect on species 1, and rows the effect on species 2. Table entries below the diagonal just repeat the classification and so have been left blank.*

Effect on Spp 1/2	Negative	Neutral	Positive
Negative	Competition	Amensalism	Predation
Neutral	–	No interaction	Commensalism
Positive	–	–	Mutualism

The other thing that becomes apparent from this table is that it contains two quite different evolutionary scenarios. Table entries with a neutral row or column (i.e., amensalism and commensalism) describe an essentially one-way interaction; only one of the partners experiences a fitness effect from the interaction. Thus, any adaptation to these interactions will be shouldered by only one of the pairs: grass might evolve to be robust to the footfall of horses, but horses do not evolve soft feet to avoid damaging the grass. Adaptation in these one-way pairs is very similar to adaptation to the abiotic environment.

In the remaining parts of the table, however, we have a very different evolutionary dynamic at play. Here, we have a fitness effect accruing to both partners. As a consequence, both partners will tend to adapt to the interaction. This is now adaptation to a moving target, and we can expect to see coevolution in which evolutionary shift by one partner causes evolutionary shift in the other partner. In these three categories—competition, predation, and mutualism—we can expect to see rich eco-evolutionary dynamics. If we place the actors into space, these rich dynamics will likely lead to interesting spatial dynamics also. In this chapter, we will do a quick survey of these possibilities, and see how the pieces we have learned so far play out when we have two species engaged in a coevolutionary dance.

8.1.1 Modelling approach

We tend to think about biotic interactions as in some way affecting the growth rate, r, of our population. The growth rate is now no longer just a function of the density of the focal species, $r(N)$, but of all the species with which it interacts, $r(N_1, N_2, \ldots, N_n)$, where n denotes the total number of interacting species in the system. Moreover, each of these Ns is itself a dynamic variable, with its own growth rate depending on the species with which it interacts, $r_i(N_1, N_2, \ldots, N_n)$. In short, biotic interactions can get complicated quickly, and this is before we even admit space, or the possibility of evolution.

Typically, though, we start with just two species—N_1 and N_2—and we consider how their population dynamics might play out. We typically set this up as a system of two differential equations, one for species 1, and one for species 2. To examine the dynamics in space, we simply include a diffusion term. Thus, models can be easily specified that describe the dynamics for each of our classes of interaction from Table 8.1. They will tend to look like this:

$$\frac{\partial N_1}{dt} = D_1 \frac{\partial^2 N_1}{dx^2} + N_1 r_1(N_1, N_2) \qquad (8.1)$$

$$\frac{\partial N_2}{dt} = D_2 \frac{\partial^2 N_1}{dx^2} + N_2 r_2(N_1, N_2), \qquad (8.2)$$

where N_1 and N_2 are the densities of species 1 and species 2 at some location, x, $r_1()$ and $r_2()$ are functions returning the per capita growth rate at location x and time t.

8.1.1.1   *The phase plane*

One of the nice things about dealing with only two species is that we can use a phase plane to observe how the system changes over time. The phase plane is simply a Cartesian plane defined by our two state variables, N_1 and N_2, in this case. If we know the number of each species at a particular time, we can plot that as a point on the plane. We can then watch where this point moves during the next instant of time, and so on. In this way, the phase plane allows us to see where the system will run to from a given starting position. Indeed, it is useful to think of the phase plane as a topographic map, with the hills and valleys on the map determining where a ball will roll to if placed at a given point on the map. Figure 8.1 shows a couple of example phase planes, with little grey arrows pointing to where the ball will roll to if placed at the start of the arrow.

8.2 Competition

8.2.1   Lotka–Volterra competition

We have already touched on competition in Section 6.2.3.3, where we argued that it could cause a pushed wave to occur. We introduced the Lotka–Volterra competition equations (Volterra, 1926; Lotka, 1932) and talked briefly about the fact that these have three possible outcomes. We will unpack these outcomes a little further here. For reference, here are the equations again, but this time with a spatial diffusion term thrown in:

$$\frac{dN_1}{dt} = D_1 \frac{\partial^2 N_1}{dx^2} + r_1 N_1 \left(1 - \frac{N_1 + a_{12} N_2}{K_1} \right) \qquad (8.3)$$

$$\frac{dN_2}{dt} = D_2 \frac{\partial^2 N_2}{dx^2} + r_2 N_2 \left(1 - \frac{N_2 + a_{21} N_1}{K_2} \right). \qquad (8.4)$$

You can see immediately that they are a manifestation of the general form given in equation (8.1). Here we track change in the density of two species

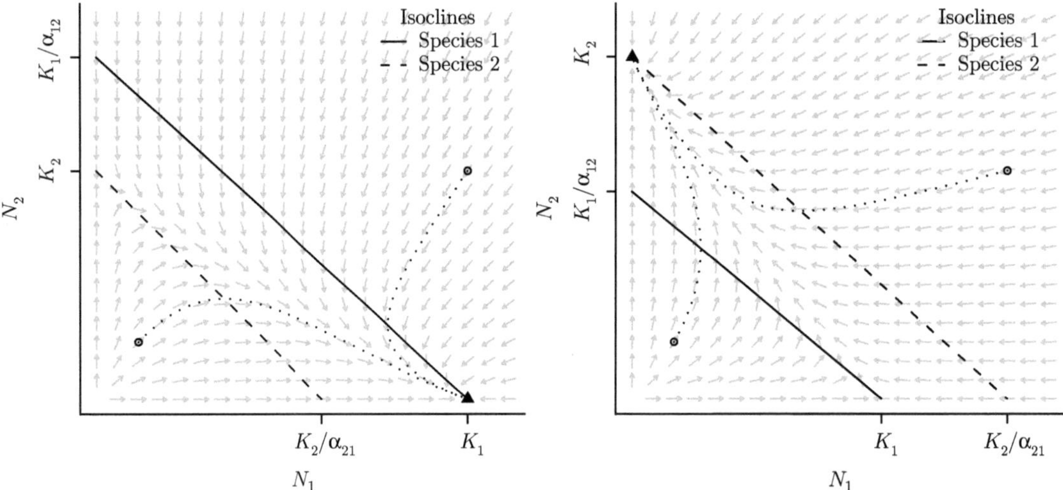

Figure 8.1 *Phase plane for the Lotka–Volterra model, where one species overwhelms the other. Grey arrows are vectors showing the velocity of the system at various combinations of* (N_1, N_2). *Solid and dashed straight lines are isoclines showing the set of states in which one or the other species is not changing. Dotted lines show trajectories of the system from two different starting positions (given by the open circles). These panels show the situation in which one or the other of the two competing species will go extinct. Regardless of initial starting densities, the left-hand panel will result in the extinction of species 2; the right-hand panel will result in the extinction of species 1. The filled triangle shows the stable equilibrium towards which the system moves.*

(N_1 and N_2) at some location x. Each species has its own intrinsic growth rate (r_1 and r_2), carrying capacity (K_1 and K_2), and diffusion rate (D_1 and D_2). There are also exchange rates (a_{12} and a_{21}) that determine the contribution that each individual of species 2 makes to the effective local density experienced by species 1, and vice versa.

8.2.1.1 *Local ecological dynamics*

Before we embark on thinking about how competition might play out in space, it is worth understanding what happens in an aspatial model, where we set $D_i = 0$. Some simple analysis of this model reveals that each species has a zero isocline on the phase plane. That is, for each species, there is a straight line on the phase plane that describes a set of $\{N_1, N_2\}$ values where there is no change in the density of that species. If these two lines ever intersect, then, we have a point on the phase plane where both species stop changing. We can think about these isoclines as either valleys or ridges on a landscape, and we can think of these intersections as either mountain tops or basins.

The local dynamics of competition have three well-known outcomes. In the first outcome—let's call this outcome 'competitive exclusion'—we have either one or the other species going extinct, regardless of starting conditions. This occurs when one of the species is simply a superior competitor, regardless of density. This is captured when the model's isoclines are arranged according to Figure 8.1.

In both these scenarios, it doesn't matter where you start on this phase plane. It doesn't matter if you start with a density ratio of 1000:1 for species 2 to species 1, or vice versa; the same outcome will eventuate. More formally, species 2 will always go extinct when $K_1/\alpha_{12} > K_2$ and $K_1 > K_2/\alpha_{21}$; and species 1 will always go extinct when $K_1/\alpha_{12} < K_2$ and $K_1 < K_2/\alpha_{21}$. These inequalities describe the situations in which one of the species is simply an overwhelmingly superior competitor.

The second possible outcome—'bistability'—occurs when interspecific competition is strong. Now the outcome depends upon the starting position; the relative density of the two competitors. If we start with sufficiently more of species 2, that species will go on to drive species 1 extinct, and vice versa. This situation is shown in Figure 8.2, with isoclines that intersect under the conditions that $K_1/\alpha_{12} < K_2$ and $K_2/\alpha_{21} < K_1$.

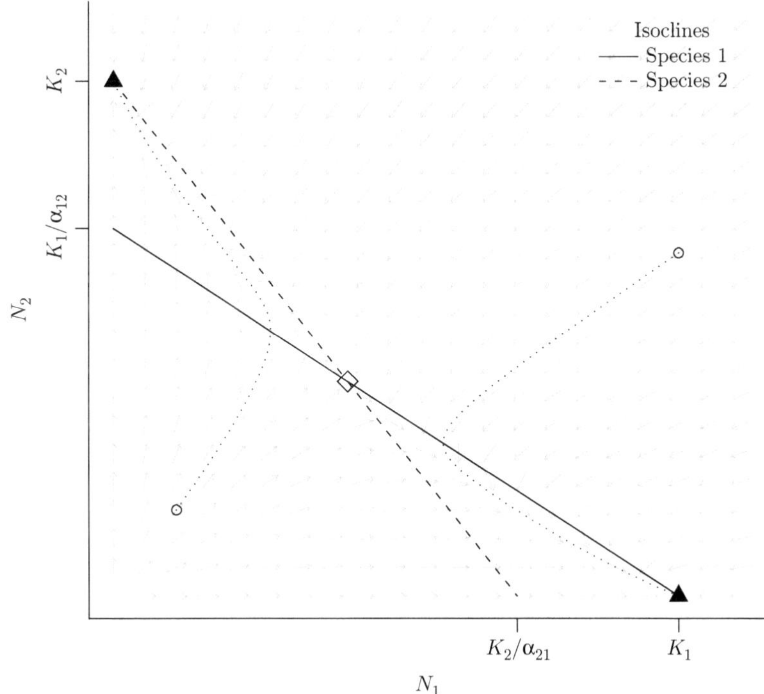

Figure 8.2 *Phase plane for the Lotka–Volterra model in bistability mode, where one or the other of the two competing species will go extinct depending on the initial densities. Grey arrows are vectors showing the velocity of the system at various combinations of (N_1, N_2). Solid and dashed straight lines are isoclines showing the set of states in which one or the other species is not changing. Dotted lines show trajectories of the system from two different starting positions (given by the open circles). The open square shows the unstable equilibrium at the intersection of the two isoclines. Closed triangles show the stable equilibria corresponding to one or the other species extinct.*

This is the situation that we discussed in Section 6.2.3.3 in which there is a bistability: the growth rate of an invader pushing into territory occupied by a competitor may swing from negative to positive as we move from the low density tip of the invasion front deeper into the bulk of the population.

Finally, we have the third outcome—'stable coexistence'—in which both our competing species can persist indefinitely. This situation occurs when within-species competition is a stronger force than between-species competition $K_1/\alpha_{12} > K_2$ and $K_2/\alpha_{21} > K_1$. This situation is depicted on the phase plane in Figure 8.3.

This section on competition will largely focus on the theoretical world of the Lotka–Volterra competition equations. There is good reason for this, and there is plenty enough to explore. It is worth noting, however, that the Lotka–Volterra system is a *very* simplified world. There are quite a few implicit assumptions in

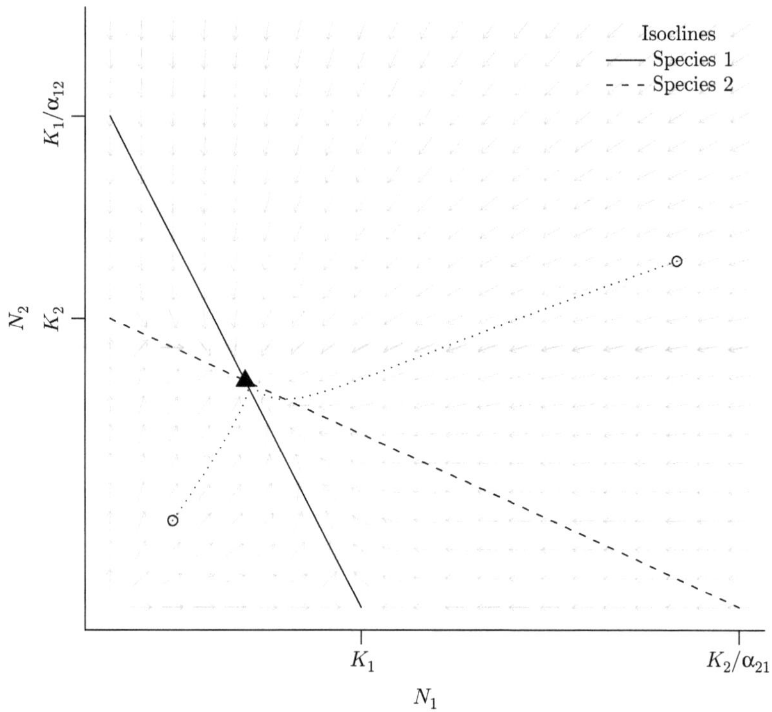

Figure 8.3 *Phase plane for the Lotka–Volterra model in stable coexistence mode, where both species will coexist regardless of the initial densities. Grey arrows are vectors showing the velocity of the system at various combinations of (N_1, N_2). Solid and dashed straight lines are isoclines showing the set of states in which one or the other species is not changing. Dotted lines show trajectories of the system from two different starting positions (given by the open circles).*

the model. To see an important one, we simply need to ask: what is it that the two species are competing over? Well, we don't really know; the model doesn't say. Are the two species competing for a single resource, or multiple resources? If so, what are these resources, and why aren't they depleted over time? Perhaps, the two species are competing for local space? There are abundant wrinkles to be unearthed here. All we really know is that there is some equilibrium number of organisms that can be supported, and as we move towards this number, growth rates drop. This might apply to competition for local space. Or (maybe) it might apply to an environment in which new resources appear at some fixed rate, like manna from heaven. Ecologists have, of course, spent a lot of time thinking about competition (Schoener, 1983) and there are assumptions under which a dynamic resource can neatly be factored out, returning us to the simpler model (e.g., Macarthur and Levins, 1967; Schoener, 1974). Thus, while there are other, more complex, models, the Lotka–Volterra system remains a pretty robust place from which to build our understanding.

8.2.1.2 *Why so many species?*

One of the striking things about the results of the Lotka – Volterra competition is that stable coexistence appears to be an unusual result. Coexistence requires the situation in which each species is its own worst enemy, and where interspecific competition is relatively weak. On the face of it, this theory seems to suggest that coexistence might be rare and that competitive exclusion should be commonplace. But we are routinely faced with rich diversity in nature, despite clear evidence for interspecific competition. So what's going on?

There are two strong contenders here that might resolve our paradox; the first is an evolutionary explanation, the second ecological.

8.2.2 Evolutionary dynamics under competition

Before we even start, it is probably useful to step back a little and remember that competition is defined as an interaction in which both participants are negatively affected. Both players suffer a fitness consequence when competition is afoot. If there is any heritable variation in traits affecting the competitive interaction, then we would expect one or both partners to evolve in response to competition. There are two possible ways in which this coevolution might play out; species can either evolve to become fiercer competitors—a kind of convergence to each other in niche space—or they might evolve to avoid competition—a divergence. Either of these options might improve a species' fitness in the short term, but given that competition is, by definition, costly to both parties, it is easy to see that divergence is the most globally optimal outcome: you can spend your whole life grinding away at the competition, or you can simply find ways to avoid each other.

Evolution, of course, doesn't always find the most optimal outcome, so it is heartening that in many models in which a pair of competitors compete along a niche axis, we typically see divergence; competitors evolve so as to minimize their niche overlap and so minimize competition. This outcome was first referred to as 'character displacement' by Brown and Wilson (1956). Their paper set off about 50 years of good, albeit at times acrimonious, debate and research around the idea. It was worth debating because it was a big idea, and it's a big idea because it may help explain, among other things, much of the diversity we see in nature. It now seems clear that character displacement—adaptation to avoid competition—is a common outcome, with substantial theoretical and empirical support (Dayan and Simberloff, 2005).

Interestingly, however, theoretical work also uncovered situations in which we might see, not divergence, but convergence. There exist various situations in which we might see species converge upon each other's niches, and become fiercer competitors as a consequence. One such circumstance arises when we consider several species all competing along the same niche axis. Here, it is possible that adaptation away from one competitor drives a species into fiercer competition with another competitor; the niche axis is a busy place, and there isn't really room for everyone. In this circumstance, we can expect pairs of species so affected to converge in niche space, and to become fiercer competitors (Macarthur and Levins, 1967). The net result of such jostling is a limit to the number of species that can be maintained upon a niche axis.

Other circumstances also encourage convergence, and it is a possible outcome in many of the various models exploring evolution under competition (reviewed in Dayan and Simberloff (2005)). Despite its persistence in the theoretical literature, convergence has only been rarely observed in nature. There are two perfectly good reasons why this should be so. First, convergence brings with it the very real risk that one or the other species will go extinct. This follows directly from the arguments presented in Section 8.2.1. Of course, if one of the competitors has gone extinct, we cannot observe it, so our observations will be biased towards the divergent case (where everyone lives happily ever after, and can thus be observed).

A second reason for so few studies finding convergence under competition is that convergence is unlikely purely because the niche is such a complex thing. Most theoretical models imagine competition along a single niche axis; this is an axis that defines variation in some essential and non-substitutable resource. Let's make an example and say that our niche axis describes variation in the body size of insects that a couple of bird species use as their sole food source. Divergence would have the two species specialize on different ends of this axis: species 1 would become a specialist on small insects, and species 2 on the larger insects. Happily ever after. But what if there really wasn't sufficient variation in insect body size for divergence to occur? Our theoretical models would likely have our birds converge in this case. But nature is almost always much more complicated than this. There are many more niche axes available, along which species might diverge (see Figure 8.4). Maybe there are other prey species that can be

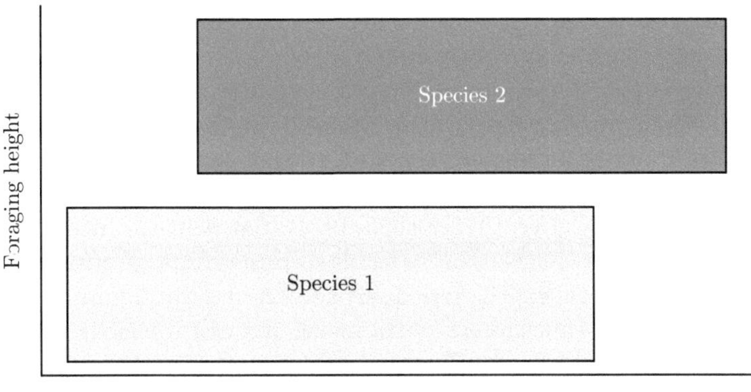

Figure 8.4 *A two-dimensional niche space with the niche of two hypothetical species depicted on it. There is clearly a strong overlap in niche when viewed from the prey size axis, but when viewed from the foraging height axis, it is clear that there is no niche overlap at all. Most species niches will have many more than two dimensions.*

utilized; maybe one can specialize to forage on the ground, and the other in the trees; maybe they can diverge in foraging times, with one foraging in the morning and one in the evening. The fact is that in very many real ecological settings, the niche is so complex that there are many ways to find a room of your own. The sheer number of alternative rooms means that the theoretical circumstances leading to convergence will often seem contrived when looked at from the field. It is nice to know that convergence is possible, but it seems likely to be a rare outcome.

8.2.3   Competition in metapopulations

The other contender for explaining the persistence of diversity in the face of competition is a purely ecological one. It requires two things, space and disturbance, and was first pointed out in a general way by G. Evelyn Hutchinson.[1] Hutchinson was a deep thinker and had a lovely turn of phrase. At the very end of a paper entitled 'Copepodology for the ornithologist' (1951), he pointed out the possibility of 'fugitive species'. A fugitive species is one that is, in all regards, competitively inferior. Wherever it happens to live, it will eventually be overwhelmed by competition from some other species. We would expect it to rapidly go extinct, but for one reason: it happens to be a much better colonizer than the other species. Now all we need is a dose of space: a world in which there are occasional disturbances that cause local extinction of the dominant species. Now, argued Hutchinson, we have a recipe for persistence, albeit of a harried kind. As Hutchinson put it, fugitive species 'are forever on the move, always becoming

[1] The same character who gave us the n-dimensional niche discussed above.

extinct in one locality as they succumb to competition, and always surviving as they reestablish themselves in some other locality'.

Hutchinson had a lovely turn of phrase, but also excellent mathematical intuition. When metapopulation theory arrived on the scene in the 1970s, Hutchinson's idea of the fugitive species found its way into various models, where it was found to be well supported by more rigorous theoretical analysis (e.g., Levins and Culver, 1971; Slatkin, 1974). Entertainingly, none of these theoretical papers cite Hutchinson. It is hardly surprising really, given the obscure title of Hutchinson's paper. This was clearly another idea whose time had come.

Let us now take Hutchinson's verbal model and cast it into the terms we have been using in this book. His contention was that in what we now call a metapopulation, good colonizers will be the first to exploit empty patches and so stay ahead of the competition. In other words, species with a relatively high value of $r(0)D$ are likely to be the first to colonize these new spaces. Fair enough, but why would we expect there to be species whose life history has evolved to sacrifice competitive ability for colonization ability? We learned in Chapter 3 that invasion fronts select for precisely this kind of life history. We also learned in Chapter 7 that metapopulations—populations characterized by extinction-recolonization dynamics—manifest the same set of evolutionary forces that play out on invasion fronts: spatial sorting, and r-selection. Thus, Hutchinson's fugitive species spring into existence as a consequence of the spatial disequilibrium characteristic of metapopulations and invasion fronts. Once their fugitive life history exists, of course, they can remain in coexistence with competitively superior species by being better able to exploit empty patches.

It is worth pointing out that there is, again, a very real space–time analogue here: the ecological definition of competition is very much a competition playing out in time; Hutchinson, by pointing out the fugitive species, gave us a competition playing out in space, also. Fugitive species are not, in fact, competitively inferior: they tend to lose the competition for time, sure, but they tend to win the competition for space.

Another observation to take from this discussion is that the selection processes playing out within a species can also often be seen playing out between species. Nanako Shigesada and Kohkichi Kawasaki (1997), in their now classic book on biological invasions, spent some time in their chapter 7 on competition for space. They used the two-species reaction–diffusion model in equation (8.3) to investigate which of the two species would win a competition for space if species 2 was introduced somewhere inside the invasion front of species 1. They found the very clear answer that the species that would come to dominate the joint invasion front (and so win the competition for space) would be the species with the highest value of $r_i D_i$. More than 20 years later, and in innocent ignorance of Shigesada and Kawasaki's result, Deforet et al. (2019) used the exact same equation and derived the exact same result to argue that evolution on invasion fronts should select for the genotype with the highest value of $r_i D_i$.

8.2.4   Invasion despite competition

Notions of competition (and predation and mutualism) gave rise to the idea of 'biotic resistance'—that an ecological community might be hard to invade because it is already full of competitors and predators, or might lack mutualists on whom the invader depends (Levine et al., 2004). In the case of competition, it is clear from the foregoing that invading a space already occupied by a competitor will require a few conditions to be met. Let us first attach the status of 'resident' to species 1, and 'invader' to species 2. It is clear that if the invader is an inferior competitor (in the sense of Figure 8.1), then the invasion will not progress. But what happens in the other situations detailed in Section 8.2.1.1?

First, we have the situation in which there is competitive exclusion, but the invader is the superior competitor (Figure 8.1). In this case, the growth rate of the invader is positive for any density $< K_2$, so we can expect it to invade, and in doing so displace the resident species. We might also expect invasion to be a clear outcome in the 'stable coexistence' scenario (Figure 8.3) because, again, at low density, the growth rate is positive. These intuitions turn out to be supported by analysis (Volpert et al., 1994). In fact, these two situations, despite the different consequences for the resident species, manifest the same invasion speed for the invader. In situations where $\alpha_{12}\alpha_{21} > 1$ (Hosono, 1998), it turns out that the invader will invade at a speed given by (Okubo et al., 1989; Volpert et al., 1994)

$$v_2 = 2\sqrt{r_2 D_2 (1 - \alpha_{21} K_1 / K_2)}.$$

This is always going to be slower than the speed that would be achieved in the absence of competition. In the absence of competition, we would expect the Fisher velocity, $v_F = 2\sqrt{r_2 D_2}$, so we can see that competition reduces this velocity by an amount determined by the ratio of the two species' carrying capacities. $\alpha_{21} K_1 / K_2$ is a measure of the force of interspecific competition applied by the resident on the leading tip of the invader's wave. If that force is large enough, the invader's speed of invasion might be very much slower than it would achieve otherwise. Competition can really slow things down.

While we now have a sense of how invasions might propagate under three of our four parameter settings, we have yet to examine the 'bistability' setting (Figure 8.2), in which one or the other species will dominate, depending upon the starting densities. This scenario, it turns out, is complicated. As mentioned in Section 6.2.3.3, this scenario likely yields a pushed wave, and pushed waves are theoretically more challenging than pulled waves. On one hand, analogous to the Allee effect (Section 6.2.1.1), it seems clear from the local dynamics that the invader will need to be introduced at a sufficiently high density, or else it will fail to establish and propagate. We can also imagine scenarios in which the invader is introduced at sufficient density, but then, because of diffusion, drops below some critical density across all space and again fails to propagate. Again, analogous to Allee effects, we might expect this minimum critical density to also translate into a minimum critical area in a two-dimensional space. Beyond such broad expectations, however, it is difficult to say much. More than a hundred years after the competition equations were proposed, the behaviour of this system in the bistable setting in space remains an active area of research (e.g., Guo and Lin, 2013; Tang and Chen, 2022).

Stepping back from the equations for a moment, however, it is worth pointing out that the maximum density of the invader looks to be an important determinant of outcomes here. In many situations, it will be the case that if the maximum density of the invader is higher than that of the resident species, the invader will be successful. A higher maximum density for the invader will either put it into the category of a superior competitor or increase the speed of its invasion wave (whether that wave be a pulled wave or a pushed wave). In nature, maximum density is almost always set by more than just intraspecific competition. In particular, parasites can play a powerful regulatory role, and by responding rapidly to host density can cause negative density-dependent population growth. Thus, the 'carrying capacity' in a species native range might be set not just by intraspecific competition, but also by the density-dependent responses of parasites and pathogens. When species are introduced to new areas, they very often leave a large fraction of their parasites and pathogens behind (Torchin and Mitchell, 2004; Mitchell and Power, 2003). As such, we might often expect invasive species to reach a higher maximum density in their new range than what they will in their native range. We see here that this release from natural enemies might—by increasing the maximum density attainable—enhance a population's competitive ability and so make it more effective at spreading in the face of competition (Keane and Crawley, 2002).

8.2.5   Character displacement in space

We spent some time in Chapter 7 examining Kirkpatrick and Barton's (1997) model of adaptation to environmental gradients. It is a complicated model, but imagines a species invading along a linear environmental gradient, with the capacity to adapt to that gradient. They showed that stable range limits could emerge along such gradients, held in place by maladaptive gene flow. This outcome of stable range limits required a gradient that was steep relative to the genetic variation available for adaptation. A few years later, Case and Taper, (2000) went one step further and imagined not just a linear environmental gradient and adaptive capacity, but a whole other competing species in the mix as well. In their model, an individual's fitness is determined by a single trait, z, that determines not only an individual's interspecific and intraspecific competitive ability, but also how well adapted the individual is to the abiotic environment. In this model, z is doing a lot.

The Taper and Case model is only a little bit more complicated than the Kirkpatrick and Barton model, but some nice results emerge. It manifests clear character displacement in regions where the two species interact. Moreover, when a competitor is in evidence, stable range edges emerge on much gentler environmental gradients than is the case in the single species model of Kirkpatrick and Barton. The reason for this is that competition causes the invasion front to be steeper.[2] As a consequence, maladaptive gene flow becomes a stronger force; both character displacement and adaptation to the abiotic environment are reduced as a consequence and the population becomes sufficiently maladapted that the invasion front no longer moves forward.

It is interesting to consider that, to date, none of the models that incorporate evolution under competition have incorporated the possibility that dispersal might evolve. It is almost certain that there will be some fascinating eco-evolutionary dynamics here; these dynamics might help us understand invasions, and they may also give us new insights into the formation of stable range edges.

[2] It probably causes it to become pushed; see Section 6.2.3.3.

8.3 Predation

Predation, despite its tendency to gore, is a slightly less antagonistic relationship than competition. With predation, at least, one of the partners draws a benefit. When we think of predation, we typically think of lions, snakes, sharks, and so on. These are all predators, and the kind of interaction they engage in can be thought of as 'true' predation: a predator that kills many prey items during its lifetime. But the broad definition of predation—one partner wins, the other loses—includes quite a few other interactions that might initially surprise us. Parasitism—in which the predator (the parasite) consumes only a portion of the prey and impacts only a few prey in its lifetime—is also a form of predation. Grazing—in which the predator (think a cow) consumes only a portion but impacts many prey (think grass) but rarely kills—is also a form of predation (vampire bats engage in grazing, apparently). Diverse as these interactions are, they all involve situations in which one partner draws a fitness benefit from the interaction while the other partner pays a fitness cost.

8.3.1 Lotka–Volterra predation

As with competition, we will start with the basic Lotka–Volterra predator–prey equations, and we will put them into space. This system tracks the number of predators, P, and the number of prey, N, and takes the same general form as equation (8.1):

$$\frac{dN}{dt} = D_n \frac{\partial^2 N}{dx^2} + rN\left(1 - \frac{N}{K}\right) - aNP \tag{8.5}$$

$$\frac{dP}{dt} = D_p \frac{\partial^2 P_2}{dx^2} + faNP - \delta P. \tag{8.6}$$

Once again, we have the diffusion terms for each species. The prey are governed by logistic growth, but they also suffer explicit mortality caused by predation. The number of attacks causing death of prey scales according to aNP, where a is an attack rate (essentially the rate at which encounters between predators and prey cause death of the prey). The NP comes straight from statistical physics: it is a mass action model, which is assuming that predator and prey move about at random, so the number of encounters ('collisions') between predator and prey increases as the densities of both increase.

Predators, P, also diffuse through space according to their particular diffusion coefficient, D_P. The growth dynamics of predators are determined entirely by the rate at which they can convert successful attacks into offspring. This is denoted $faNP$, where f is the conversion efficiency of the predator. Finally, we have the death of predators happening at some fixed rate δ. By tuning the parameters f and a, we can dial up situations that sound like true predation (high attack rate, low conversion efficiency), parasitism (low attack rate, high conversion efficiency), or grazing (low attack, low conversion).

8.3.1.1   *Local dynamics of predation*

Once again, it is useful to ignore space for a few moments and spend a little time understanding the local dynamics of this system. It is a simpler system than the one we saw above for competition. In this system, there are really only two possible outcomes, and once again, we can see these if we find the isoclines on the phase plane and think about the various ways these isoclines might sit in relation to each other. The first outcome—predator extinction—occurs when the carrying capacity of the prey is below that necessary for the predator to sustain itself: when $K < \frac{\delta}{af}$. Here, it doesn't matter what our initial density of predator and prey is; the predator goes extinct (Figure 8.5).

The only other sensible arrangement of isoclines on this phase plane gives us a stable equilibrium in which predator and prey coexist. On the phase plane, the approach to this equilibrium follows a spiralling trajectory, suggesting damped oscillations of predator and prey numbers as this equilibrium is approached (Figure 8.6).

8.3.1.2   *Predator–prey invasions*

Let us bring back space. Let us imagine a situation in which the prey species is at carrying capacity across all of space, and we introduce a new predator species. What happens? Well, given the local dynamics, it will come as no surprise to you that if $K < \frac{\delta}{af}$, the predator fails to propagate forward in space and goes extinct. More interesting is the situation in which $K > \frac{\delta}{af}$. Here, we do see a travelling wave propagating forward, and it turns out that this will attain an equilibrium speed of $v = 2\sqrt{D_p(afK - \delta)}$ (Dunbar, 1983). Note that the condition $K > \frac{\delta}{af}$ ensures that this speed is positive; this is a wave that propagates forward. Of course, the

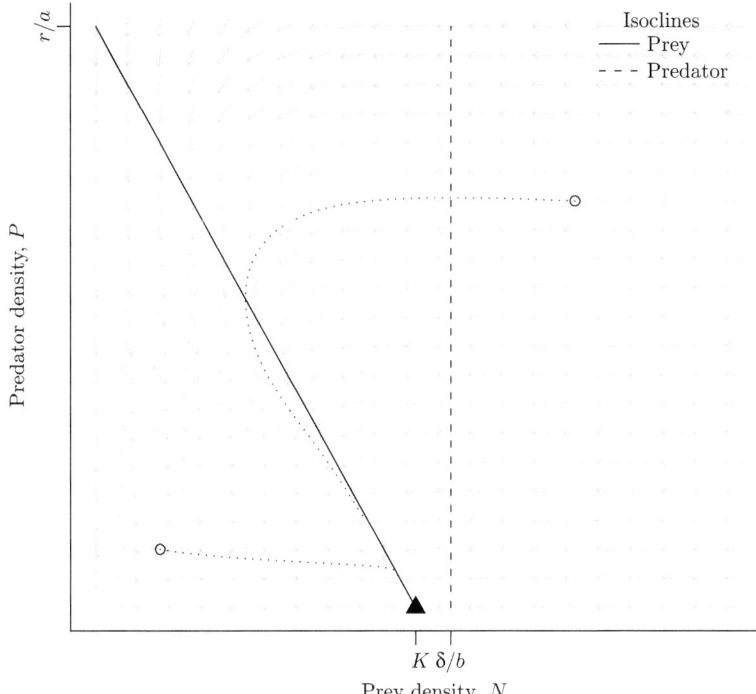

Figure 8.5 *The phase plane showing the situation in which prey density is insuffi-*
cient to support the predator population. Here, $K < \frac{\delta}{af}$, and the predator will always
go extinct. Grey arrows are vectors showing the velocity of the system at various com-
binations of (N, P). Solid and dashed straight lines are isoclines showing the set of
states in which one or the other species is not changing. Dotted lines show trajectories
of the system from two different starting positions (given by the open circles). The stable
equilibrium is shown with the filled circle.

invasion of the predator causes a wave of destruction to move through the prey
population also, dropping it from its previous equilibrium, at K to its new (lower)
equilibrium, somewhere below K.

We could try to imagine a situation in which the predator is at carrying capacity
across all space and we introduce a new prey species, but that situation really
doesn't make much sense with this model. In this model, there is only one prey
species and if it is absent, the predator goes extinct, so it doesn't make sense for
the predator to be at carrying capacity across space.

Another situation we can imagine is one in which both predator and prey are
introduced at the same time. This may seem a bizarre scenario to imagine, but
it actually happens all the time. Most species that spring to mind when you are
asked to name an animal or a plant carry parasites and pathogens, and when we

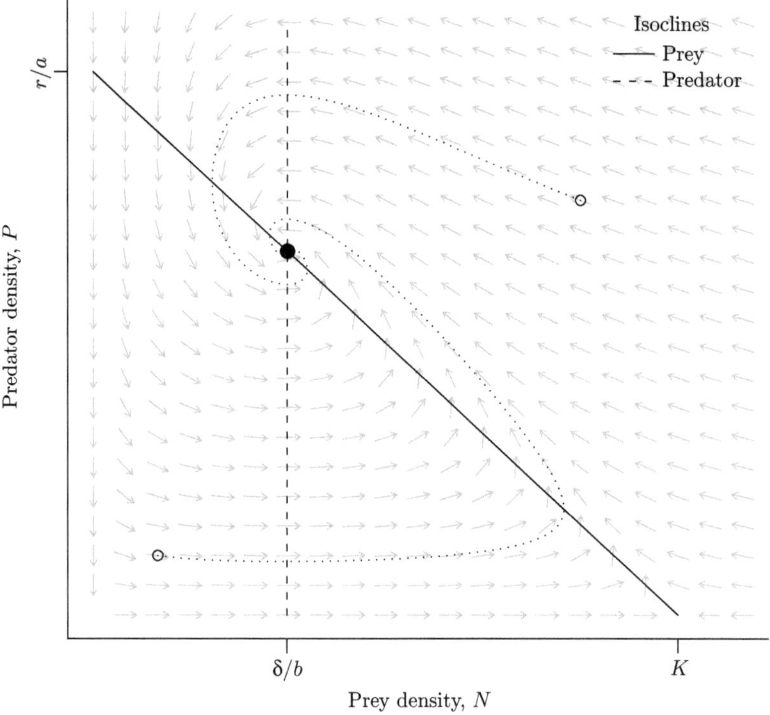

Figure 8.6 *The phase plane showing the situation in which predator and prey coexist. Here, $K > \frac{\delta}{af}$, and the predator and prey will always reach an equilibrium of coexistence, albeit oscillating on their way there. Grey arrows are vectors showing the velocity of the system at various combinations of (N, P). Solid and dashed straight lines are isoclines showing the set of states in which one or the other species is not changing. Dotted lines show trajectories of the system from two different starting positions (given by the open circles). The stable equilibrium is shown with the filled circle.*

introduce a species to a new location, we very often introduce a few of its parasites and pathogens with it (Torchin and Mitchell, 2004; Mitchell and Power, 2003). So, many situations in which a species is introduced also involve the introduction of a few parasites. We might also imagine a scenario in which a species is introduced, commences an invasion, and at some later time, a predator or parasite arrives; the predator is able to establish because now it has a prey species to exploit. In either case, what happens next?

Before we start, it is worth remembering that we have both deterministic and stochastic processes at play. For the deterministic processes, we can probably learn a thing or two from the Lotka–Volterra system (equation (8.5). Here, Owen and Lewis (2001) give us a thorough answer. If the prey invasion is a pulled wave (Chapter 6), then the speed of the prey wave is utterly unaffected by the predator (Figure 8.7). It doesn't matter if the predator eventually catches up with

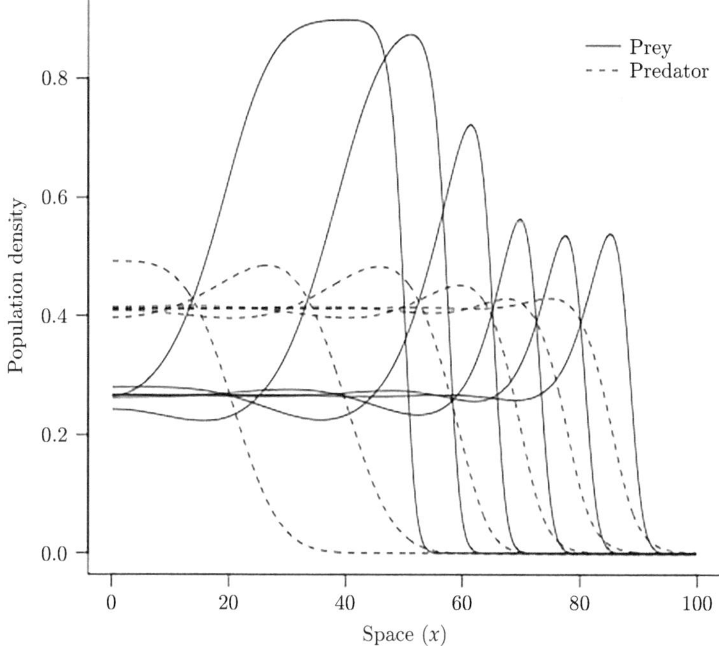

Figure 8.7 *Numerical solution for the predator–prey model in which a prey species is initially distributed across $x = [0, 40]$ when the predator is introduced at $x = 0$. Both invasions are travelling from left to right, and each series is a snapshot in time. The predator has a much higher dispersal rate than the prey and so catches up with the prey's invasion front. Note the very clear slowing and steepening of the predator wave as it converges upon that of the prey. Note also that the velocity of the prey wave is unaffected even though prey density is greatly reduced by predation.*

the invasion front of the prey or not; it doesn't change the speed with which the prey invades. By contrast, the invasion of the predator can be strongly affected by its prey. If the predator never catches up with the prey invasion front, then it invades as we would expect from the previous section, where we considered a predator invading a homogeneous prey population. If, however, we imagine a predator whose invasion speed is faster than the prey, eventually the predator will catch up with the prey, but it can proceed no faster than the prey, so the predator invasion slows down to match that of the prey (Figure 8.7). Why? Well, on one level, it is an obvious fact of life: if the predator relies on the prey species, it can't reproduce in places where there is no prey. But if we step back into invasion theory for a second, we can see that the deceleration of the predator's wave is because the predator's wave has become pushed (or more strongly pushed if it was already pushed). Thus, a predator whose invasion rate is constrained by prey availability can exhibit a fundamentally different invasion dynamic than one that is invading into an endless population of prey.

Owen and Lewis (2001) also examine the situation in which the prey species exhibits an Allee effect in its population dynamics. Now the prey wave is no longer pulled, it is pushed. In this situation, it becomes possible that the predator might not only catch up with the prey wave, but also slow it down. Indeed, it even becomes possible that the predator catches up and then causes the prey wave to stop, turn around, and march in the opposite direction. The end result, of course, would be the extinction of both species.

8.3.1.3   *Loss of parasites and pathogens from invasion fronts*

I mentioned earlier that these kinds of predator–prey races must often play out when a species is introduced to a new area and inadvertently brings with it a few of its natural parasites. Two invasions for the price of one. In many of these cases, it will only be parasites with direct life cycles that will likely establish and spread (parasites with intermediate hosts will likely find themselves with a severed life cycle in the new area). This was precisely the scenario that played out when cane toads were introduced to Australia. It turns out that they brought with them a lungworm, a nematode parasite that is directly transmitted between toads and that exists at relatively high prevalence in the toad population.

Given the considerations above (Section 8.3.1.2), we would expect this lungworm to have little effect on the spread rate of toads, and—given lungworms make use of toads as their dispersal vehicles—we would expect them to have a similar rate of dispersal as toads. Given lungworms are a small organism with a fast generation time relative to toads, and that, like many parasites, they produce a huge number of eggs, we might also expect their growth rate to be as high or higher than that of toads. Back-of-the-envelope considerations, then, suggest this parasite should keep up with its host invasion front. But when we looked at the prevalence of lungworms as the toad invasion front arrived and passed us by, it was very clear that the lungworms were lagging a few years behind the toads (Phillips et al., 2010b). What was going on?

First, it is worth noting that the toad invasion had been progressing for about 75 years at this stage. If the lungworms had a deterministic invasion speed slower than that of the toads, it is very unlikely that they would be so close behind them at this stage. Something must be happening on the invasion front to cause this lag. That something is very likely a few somethings (Phillips et al., 2010b). First, we can make an analogy here between parasites and alleles. In the same way that spatial drift on the invasion front causes loss of alleles purely by chance (Chapter 5), so too can we lose parasites. If the individuals that are on the leading tip of the invasion just happen to have arrived there without picking up parasites along the way, then the invasion front will not have parasites. If the wave is a pulled wave, then it will be hard for the parasite to reestablish on the front.

The second effect is a manifestation of spatial sorting. If having a lungworm reduces a host's dispersal ability, even if it reduces it a little bit, we will have spatial sorting by parasite load; hosts on the invasion front will tend to be free of the parasite. Thus, even if the parasite's maximum invasion speed (into a population of hosts at carrying capacity) is faster than that of the host, the parasite will tend to be lost each generation from the host's invasion front.

The third effect comes down to the dynamics of transmission. Many parasites have density-dependent transmissions such that low host densities make transmission between hosts difficult. If that is the case, then we have a third reason why parasites might be absent from an invasion front.

The overall picture for an invasive host population, then, is quite interesting. Not only does it lose many of its specialist pathogens and parasites when it is established in a new area, but once it begins invading that area, the invasion front loses the remaining parasites and pathogens. The end result then is a population on the invasion front that is relatively free of the selective pressures imposed by parasites. It is not at all clear what this does to an organism, but we can imagine that there might be some fascinating ecological and evolutionary dynamics at play here. In toads, at least, it is very clear that their immune system shows substantial differences between invasion front and core populations (Brown et al., 2015).

8.3.2   Predation, evolution, and invasion

We have already talked about two models that examine trait evolution in a continuous space. The Kirkpatrick and Barton model (1997, Section 7.4.2) examined invasion along an environmental gradient to which the species could adapt, and the Case and Taper model (2000, Section 8.2.5) added a competing species into the mix. Following along this tradition, in 2012, Alex Perkins (who we have already met) applied these models to the situation in which there is a predator and prey species interacting across a one-dimensional space (Perkins, 2012). Unlike the previous models, Alex did not impose an environmental gradient. But he did use an evolving trait, z, that was under stabilizing selection towards some optimum value. This same trait also determined the fitness of individual prey and predators when the two species were present in space. He imagined a world in which the prey species was already at carrying capacity across all space before a predator was introduced at one part of space. He examined the four cases where 'neither, either, or both the invader and native are evolutionarily labile'.

When neither species was evolutionarily labile (that is, there was trait variation, but that variation was not heritable), the predator invasion proceeded roughly as we might expect. Like the simpler model in which there is no trait variance (Section 8.3.1.2), the invasion proceeds according to the difference between the rate at which prey are converted to predators and the predator's death rate. Added to this, however, were some additional terms that accounted for the reduction in fitness due to genetic variation around the trait optimum plus any mismatch in trait between prey and predator (if we ignore the environment for a second, the predator's fitness is optimized when it has a value for z that closely matches that of the prey).

When only the predator is capable of evolving, the model shows a predator population steadily evolving towards an optimal phenotype that balances its fit to the abiotic environment with its match to the prey phenotype. On the predator's invasion front, prey are abundant and predators are scarce, whereas behind the invasion front, prey numbers have been suppressed. Because of this, the optimum phenotype actually changes between the invasion front and areas behind the invasion front. On the invasion front, the optimum phenotype is further away from the environmental optimum, because it can afford to be; prey are everywhere! The result of this is that the predator population of the invasion front evolves to closely match the prey; this increases predator fitness on the invasion front and causes the predator invasion to (transiently) accelerate.

The third case that Alex examined had the invasive predator unable to evolve, but the prey capable of evolving. What might happen here? This is actually a pretty complicated scenario. On one hand, we might just say, well, the prey will just evolve to be uncatchable by the predator, at which point the predator invasion will fail. That's more or less what would happen in an aspatial model. But of course whether that happens on the invasion front of a spatial model will depend upon how fast the predator's invasion wave moves relative to gene flow in the prey population. We might imagine a scenario in which the predator wave rolls across space even though it is going extinct behind the invasion front as the prey adapt. Overall, this spatial effect in Alex's model gives the impression that evolution by the prey population is unlikely to have a great deal of impact on the speed of the predator invasion. It might have an impact in theory, but the parameter values yielding that impact are pretty unlikely to happen in nature.

And what about the final scenario, where both predator and prey can evolve? Well, as we might expect from the foregoing, the predator evolves on the invasion front to better exploit the abundant prey. This causes the predator invasion to accelerate, making it even less likely that adaptations by the prey can keep pace. Behind the invasion front, the prey adapt, but this adaptation has essentially no impact on the speed with which the predator invasion plays out.

8.3.3 Parasites, pathogens, and the evolution of virulence

As we have mentioned earlier, parasites and pathogens are really just a particular kind of predator, so much of the foregoing discussion about predators applies to the spread of a novel pathogen also. And, of course, novel pathogens spring up all the time: rinderpest in the ungulates of Africa; chytrid fungus in frogs of the Americas; white-nosed syndrome in North American bats; HIV, Zika, Ebola, SARS-CoV-2, in humans. While pathogens are just another predator, we are often interested not just in how quickly they might spread through space, but also in the impact they might have on their hosts when they arrive.

The impact a pathogen has will depend to a large extent on its transmissibility and its virulence. These two factors—transmissibility and virulence—are very closely related to a pathogen's growth rate, r, and they also affect (indirectly) its dispersal rate, D. We know the process of invasion selects on rD, so it seems very likely that evolution during invasion will have some kind of important effect on the traits determining how dangerous a pathogen is. This realization was another one of those ideas whose time had come; several workers around the world were, once again, all working on it at the same time.

If we reconstruct history, the idea was first mentioned as a side observation by Alex Perkins in his 2012 paper (Perkins (2012); just discussed in Section 8.3.2). He pointed out that when his 'predator' was spreading into a novel 'prey' population, the predators on the invasion front were experiencing a much higher abundance of prey than their conspecifics behind the invasion front. This set up conditions for the predator to sacrifice other aspects of fitness in favour of maximal exploitation of the prey resource. That is, on the invasion front, the predator evolves to be quite voracious. Alex pointed out that we could swap 'predator' for 'pathogen' and the result would be a pathogen evolving higher virulence in order to exploit high host densities on the invasion front.

At about the same time as Alex's paper came out, I was working with a good friend, Rob Puschendorf, analysing a dataset he had compiled on the invasion of chytrid fungus across central America (Phillips and Puschendorf, 2013). I had also been pondering the invasion of lungworms into cane toad populations in Australia (Phillips et al., 2010b) and learning about the rapid shifts in lungworm life history that had resulted (Kelehear et al., 2012). Somewhere in the process of thinking about these systems, the basic arguments that Alex had made occurred to me also (embarrassingly, I had missed Alex's point in my first quick read of his paper). I built a statistical model to examine whether the time lag between chytrid arrival and host population decline had changed as the invasion progressed. Sure enough, there was very clear evidence that early in its invasion, chytrid took years to impact frog populations; after 30 years of spread, however, it was wiping out frog populations within months of arrival. This wasn't strong evidence for the evolution of virulence, but it was intriguing.

Meanwhile, unbeknownst to all of us, a group in the US were thinking along similar lines. While I was casting around for other systems with which to test this idea, they had already hit on a nice one. Dana Hawley and colleagues examined archival samples of a bacterium that infected house finches in the US and found rapid increases in virulence associated with the emergence of that pathogen into its novel (high-density) host population (Hawley et al., 2013). Erik Osnas and Andrew Dobson from this group then teamed up with Paul Hurtado and put together a reaction–diffusion model of an evolving host–pathogen system (Osnas et al., 2015).

The model of Osnas and colleagues takes a classic model from theoretical epidemiology—the susceptible-infected model—and places it into space with a classic reaction–diffusion framework. With a few neat tricks, they then boiled the system down to a one-dimensional version of Skellam's (1951) (see Chapter 2) original reaction–diffusion model. This made it simple to work out invasion speeds for different pathogen strains. The key aspect of this model boils down to the constraints that the pathogen strains need to work with. In this model, virulence, α, affects both local transmission between hosts, and dispersal of infected hosts through space. Increasing virulence increases local transmission, but decreases dispersal. In the terms of this book, they set up a clear trade-off between r and D, with virulence determining where a pathogen strain sits on this trade-off curve. Maximizing rD in the face of this trade-off will depend upon the shape of the trade-off functions (Section 7.2.4). With the functions they used, they found that in order to maximize invasion speed, virulence will decrease on the pathogen's invasion front, and in their model this was primarily driven by the fact that virulence reduces dispersal of the host.

Meanwhile, on the other side of the Atlantic, another team was working on the same theory. Quentin Griette, Gäel Raoul, and Sylvain Gandon also developed a reaction– diffusion instance of the susceptible-infected model (Griette et al., 2015). This team did not enforce a trade-off between virulence and dispersal, so in their case they found that virulence would increase on the pathogen's invasion front. There has, of course, been additional work in this area since, but the general message appears to be that, yes, invasion is likely to have an effect on the virulence and transmissibility of a novel pathogen. Precisely what the effect is, however, is going to depend upon the trade-offs between virulence, transmissibility, and dispersal. The details will matter here. Certainly, in pathogens that do not depend upon their primary host for dispersal (such as the case for chytrid), we may well expect virulence to increase on the invasion front. For pathogens that rely on host movement for dispersal, however, making their host sick may well reduce their dispersal rate and so cause less virulent forms to arise on the invasion front.

8.4 Hybrid zones as a collision of competitors

Let's leave predators behind for a while and switch our minds back to competitors. Let's imagine a scenario in which we have two competing species, but where our two competing species have the capacity to mate with each other and produce hybrid offspring. Let's have these two species invade from different directions and have the invasion fronts crash into each other. The result is a hybrid zone.

Hybrid zones are a fascinating and complex phenomenon and have contributed much to our understanding of speciation. They are where incipient species meet, and are natural laboratories where we can examine the evolutionary processes that enforce species boundaries. There is a very rich intellectual history in the examination of hybrid zones, and this small section really cannot do it all justice. We will look very briefly at the most successful model of hybrid zones—the tension zone model—and draw a few analogies to invasion theory.

8.4.1   Complete competitors, with benefits

We might think to understand hybrid zones by enlisting the Lotka–Volterra competition equations. We have two species after all, and if they are closely related enough to mate and form hybrids, we might expect them to have broadly equivalent niches. This might be a tempting way to start in on the problem of hybrid zones, but it rapidly falls over when we realize that the two species mate and form F1 hybrids. Not only that, but the hybrids might mate among themselves to form F2 hybrids, or backcross to either of the parental types to make F2 backcross hybrids, and so on. It quickly becomes clear that there are not just two classes of individuals in this system. Rather, there are parental species and a 'hybrid swarm' comprising individuals expressing various depths of hybridization and backcrossing. The Lotka–Volterra competition won't really cut it.

Our best foothold on the hybrid zone problem comes instead from an approach closer to Fisher's spread of an advantageous allele. Let us just look at a single genetic locus, and imagine an allele, p, from species 1 and another allele, q, from species 2. In a tour de force, Nick Barton (1979) developed a model on this premise. It was similar to Fisher's model, but Barton examined the situation in which heterozygotes (those carrying an allele from both parental types) suffer a loss of fitness; he also (similar to models we have seen in Section 7.4.2) added in the possibility of density and movement gradients causing asymmetrical gene flow.

8.4.2   The tension zone model

The resulting model has since become known as the 'tension zone model'. A slimmed down version of the model (assuming movement rates, selection, and density are constant) looks like this:

$$\frac{\partial p}{\partial t} = \frac{\sigma^2}{2}\frac{\partial^2 p}{\partial x^2} + spq(p-q). \tag{8.7}$$

σ^2 is the variance in the position of individuals per unit time; s is the strength of selection operating against heterozygotes. It is worth noting that the growth term $spq(p-q)$ implies a per-capita growth rate $(sq(p-q))$ that is actually at a minimum as $p \to 0$, and so we can expect a pushed wave in this system.[3]

[3] Indeed, it was this case of selection against heterozygotes that inspired some of the first serious work on pushed waves (Aronson and Weinberger, 1975).

Barton showed that the invasion wave has an equilibrium shape, given by

$$p = \frac{1}{2}\left(1 + \tanh\left(\sqrt{\frac{s}{2m}}(x - x_0)\right)\right), \tag{8.8}$$

where x_0 is the centre of the wave (the place where $p = q$). He also showed, quite clearly, that the movement of this invasion front was quite complex. If the p allele had a fitness advantage over the q allele, then this invasion front would propagate forward, but if there was no fitness advantage, the invasion front would just drift around in no particular direction.

If we focus on this 'no fitness advantage' scenario and bring back spatial variation in density, selection, or movement (captured in the full model, not shown here), things start to get interesting. It turns out that the invasion front will tend to move down density gradients; it will tend to move down gradients in dispersal; it will tend to move down gradients in selection. The overall picture that emerges is like a nicely shaped bead on a string: tilt the string and the bead won't change shape, but it will slide in the direction of the tilt. And of course if we let the string relax and sag in the middle, the bead will slide to the bottom of the string and stay there. Barton's analysis clearly showed that these tension zones tend to move into density troughs and then stay there.

8.4.3 Spatial sorting, and reinforcement

Although Barton included the possibility that dispersal might change across space, the tension zone model does not explicitly incorporate the possibility that dispersal might evolve. But of course all the conditions are there for that to happen. A hybrid zone is a kind of absorbing boundary: an individual dispersing into it has a high chance of mating with the wrong species and so suffering reduced fitness. We might think of a hybrid zone—and particularly one sitting in a density trough—as a kind of black hole that draws genes into it, never to return. Spatial sorting means that the most dispersive genotypes have a higher chance of falling into this black hole than less dispersive genotypes. As a consequence, we can expect dispersal rates around the margins of the hybrid zone to evolve to lower levels. This evolutionary dynamic can be expected to have complex outcomes for the shape of the tension zone, but may well cause it to reduce in width over time.

Although the question of dispersal evolution around hybrid zones remains an open one (Phillips and Baird, 2015), other traits, particularly those around mate choice, are known to evolve on the margins of hybrid zones. If mating with the wrong species is bad for fitness, but you can only really encounter the other species on the edge of a hybrid zone, we would expect better mate discrimination to evolve on the edges of hybrid zones. Indeed, this pattern of reproductive character displacement has been regularly observed in nature (Higgie et al., 2000). Here, at least, hybrid zones do give results similar to the character displacement arising in models of competition (Section 8.2.5).

Finally, it is worth noting that hybridization is a bit of a dice throw. Surprising things can happen in hybrids. It is often observed in plant breeding, for example, that hybrid varieties have higher fitness ('heterosis'). Bourret et al. (2022) showed that the idea of heterosis might also extend to dispersal. In their trout system, hybrids had much (much) higher dispersal probability than either parental line. In a situation like that, it is entirely possible that the 'black hole' model of a hybrid zone is reversed, and we might see net gene flow out of the hybrid zone.

8.5 Evolutionary responses from the invaded community

Most of the preceding discussion has focused on the dynamics of invasion fronts confronted with biotic interactions. We have largely ignored the possibility that the 'resident' population might adapt to the presence of the 'invader'. This is not to say that such adaptive responses are not routinely observed in nature. Indeed, invasive populations are very often potent forces of natural selection, and evolutionary responses by resident species have been observed in a wide array of taxa in the face of novel competitors, predators, and pathogens (Strauss et al., 2006; Le Roux, 2022).

Given the powerful role that biotic interactions have on fitness, it is not surprising that we should see such rapid evolutionary responses. Indeed, the impact that invaders can have on resident fitness is often sufficient that the impact is of existential concern; we should be concerned as to whether the resident can adapt before it is driven extinct (Gomulkiewicz and Holt, 1995; Mack et al., 2000).

While evolutionary responses of the resident population clearly happen, I have largely ignored them here because these adaptive responses are unlikely to have a major impact on the process of invasion. As pointed out by Alex Perkins (Section 8.3.2), among others, it requires a fairly special set of circumstances for gene flow from the long-invaded parts of the resident population (where adaptation has occurred) to catch up and overtake the invader's invasion front. These circumstances are not impossible, but they do seem likely to be rare in nature. This state of affairs might be frustrating for people trying to manage an invasion or its impacts, but it also opens up space for novel management interventions, as we shall see in the next chapter.

8.6 The wrap

In this chapter, we looked at how biotic interactions might affect the ecological and evolutionary dynamics of invasion. We set out the types of interactions that are possible and then focused on the interesting cases (competition, and predation) in which both interacting partners accrue a fitness effect from the interaction.

We developed arguments for why 'fugitive species' might evolve in metapopulations. In the process, we unearthed another spatial analogue of a temporal process, by re-casting fugitive species as taxa that, while poor competitors for time, are strongly competitive for space. We also noticed that precisely the same model that argues for the evolution of rD on invasion fronts was used by ecologists more than 20 years earlier to argue that competing species with high rD will come to dominate a joint invasion front. A nice example of the re-discovery of an idea and how processes can be recapitulated across levels of organization.

We found that invasions spreading into space occupied by a competitor will be slowed by that competition, and in some cases can become pushed invasions as a consequence. Invasions of this kind should also cause the evolution of character displacement between the two competing species. Invasions of this kind are also more prone to forming stable range edges where they progress across an environmental gradient.

The invasion of a novel predator into a naive prey population was also examined. This also includes the situation of a novel pathogen spreading through a host population. Here it is a clear result that the predator invasion progresses at a rate that depends, in part, upon the density of the prey. Given the effect of predators on prey, prey densities are typically highest on the predator's invasion front. Predators on the invasion front are faced with an abundance of prey and this abundance favours predators that are able to rapidly exploit an abundant prey resource (r-selection again). In the case of a spreading pathogen, then, we might expect increased virulence to evolve on the invasion front.

In cases where predators and prey are both spreading through space, the speed of the predator's invasion is ultimately regulated by that of the prey. If we examine the particular situation of specialized parasites and pathogens, a number of processes (both deterministic and stochastic) conspire to remove these parasites and pathogens from their host's invasion front. This can lead to a population of hosts on the invasion front that are evolving in the absence of specialized parasites and pathogens.

We then consider the interaction between two species that are able to hybridize. We can conceptualize this as a kind of competition, but we show that in this case it is more straightforward (slipping between levels of organization again) to consider the dynamics at the genetic level, rather than at the population level. This leads to a modified version of Fisher's original reaction–diffusion model in which the growth term is bistable. The result is that hybrid zones can be usefully described as a pushed wave resulting from the reduced fitness of hybrid genotypes.

One of the clear messages from this chapter is that we often see invasions nested within invasions. We see this nestedness not just when one species invades over the top of another, but whenever we see evolutionary waves sweep through populations. This nestedness often allows invasion models to slide quite neatly between levels of organization. A hybrid zone between two competing species can be conceptualized as an invasion of a population of genes into a world already occupied by competing genes; the evolution of r and D on an invasion front can be conceptualized as a competition between two 'species' competing for space ahead of the invasion front. There are waves within waves, here, and rich dynamics at play.

The discerning reader may have noticed that we never really go beyond pairwise interactions between species, but of course real ecological communities are substantially more complicated than that. Hui and Richardson's (2017) *Invasion Dynamics* is a good place to go if you would like to spend time unpacking some of the complexity of invasions and community dynamics (though primarily through an ecological rather than evolutionary lens).

<table>
<tr><td>

9

</td><td>

Management of invasive populations

</td></tr>
</table>

Invasive populations are everywhere and are often of high consequence, impacting our economic, environmental, and social well-being. A gene for herbicide resistance spreads; a tumour grows in a loved one's body; an agricultural pest sweeps across the country; a new pathogen spreads around the world. All of these are invasive populations; populations of genes, cells, or organisms spreading without control and causing massive impact. Regardless of scale and context, we have learned that all are governed by the same fundamental biological processes of reproduction and dispersal. We have also learned that there are rich eco-evolutionary dynamics at play here, and that these dynamics change the outcomes of invasion in important ways. Evolution can speed up invasions, slow them down, and even cause them to falter and collapse altogether. The new eco-evolutionary view of invasions potentially suggests novel management ideas for old problems.

In this chapter, we will examine several case studies. The intention here is threefold. First, to show the diversity of situations to which the ecological and evolutionary dynamics of invasions pertain. Second, to show how eco-evolutionary thinking is providing new and deeper insights into these problem invasions. And finally, to just have a little fun and speculate on ways that we might one day manipulate these invasions for management effect.

9.1 Tumours

In 2010, I was invited to give a talk at a workshop on 'evolutionary applications in conservation, medicine, and agriculture' organized by Scott Carroll from UC Davis. Truth be told, they had invited my postdoctoral supervisor, Rick Shine, but he couldn't go and suggested I might like to. I did indeed: the workshop was being held on Heron Island in the midst of Australia's great barrier reef, with a bunch of interesting people in attendance, so off I went. I had been thinking almost exclusively about the cane toad invasion for a few years, and having to talk to a bunch of folks interested in agriculture and medicine caused me to stop for a moment and think about the broader implications of what we were unearthing in toads. How might this stuff be interesting to a medical researcher, I wondered. One idea that struck me is that a tumour is very similar to an invasive species.

The Ecology and Evolution of Invasive Populations. Ben Phillips, Oxford University Press. © Ben Phillips (2025).
DOI: 10.1093/9780191924910.003.0009

If I simply visualized cancer cells as little toads, I suddenly had some interesting ideas about what might happen as a tumour grows.

But of course, cells are not toads and I am not a cancer researcher, so I rather timidly made this analogy (and a few others) at the end of my talk and left it at that. I was quietly relieved that no one told me it was a completely stupid idea. I have no idea whether the analogy travelled from that workshop; more likely, this was just another idea whose time had come. Regardless, I was delighted in 2013 to read a theoretical paper examining the eco-evolutionary dynamics of cancer cells on the expanding edge of a tumour (Orlando et al., 2013). This was rapidly followed by an insightful review article by Kiril Korolev and colleagues on the value of eco-evolutionary thinking for understanding cancer growth and evolution (Korolev et al., 2014). It wasn't just me seeing cancer cells as little toads, apparently. Then in 2018, in a very lucid review, Cindy Gidoin and Stephan Peischl (2019) pointed out that spatial sorting and spatial drift could prove to be very valuable in understanding tumour growth.

Variation between cells within tumours has of course long been known about. This 'intratumoral heterogeneity' can complicate both diagnosis (biopsies might not capture the full diversity of cell types in a tumour (López-Fernández and López (2018))), and treatment (a given therapy might not work on all cell types in a tumour (Gatenby (2009); Paczkowski et al. (2021))). It has also long been clear that cancer cells mutate a lot (Nowell, 1976). Indeed, high genomic instability and mutation rate are considered to be an 'enabling characteristic' of cancer (Hanahan and Weinberg, 2011). Seen properly, this high mutation provides the raw material for rapid exploration of the fitness landscape and local adaptation to the tissue environment (Scott and Marusyk, 2017). Thus, it should come as no great surprise that we should see intratumoral heterogeneity arise as local populations of cells adapt to their particular environment.

When we take the perspective of invasion dynamics, it seems very clear that there are two obvious environments that emerge: the core, and the invasion front. With this core–front framework in place, and with advances in medical imaging (Zahir et al., 2020), it has rapidly become apparent that there are often clear phenotypic or genetic differences between cells on the front versus the core of a growing tumour. Lloyd et al. (2016), for example, showed clear differences in cell phenotypes between core and peripheral cell populations in breast cancer. Here, when compared to the tumour core, invasion front cells had relatively higher rates of proliferation, and had higher scores for biomarkers associated with (bizarrely) hypoxia and anaerobic glycolosis. These last two are surprising and interesting, because the edge of the tumour is relatively well oxygenated, but this 'pseudohypoxia' produces acid as a by-product, which can damage surrounding tissue and hamper the immune system. The invasion front cells in Lloyd et al.'s study also showed markers associated with resistance to this acidic environment. At around the same time as Lloyd's study, Hoefflin et al. (2016) were discovering a similar core–front pattern in renal cancers: cells on the invasion front showed higher rates of proliferation and signalling than did cells in the core of the tumour

(Figure 9.1). Similar patterns have also been found in laryngeal squamous cell carcinoma, with a notably higher abundance of malignant proliferative cells on the invasion front of tumours (Song et al., 2020). Core–front heterogeneity has also been observed in rates of glucose consumption in breast and lung cancers, suggesting new diagnostic tools and metrics (Jiménez-Sánchez et al., 2021). In recent experiments involving competition between two breast cancer cell lines, the more proliferative and motile cell line rapidly came to dominate the expanding edge of the cell population (Freischel et al., 2021).

In these studies and their associated modelling, we can see a recapitulation of many of the ideas discussed in Chapter 3 of this book. Essentially, spatial sorting and/or r-selection playing out upon the invasion front. We see people building cancer-specific models (e.g., Lloyd et al., 2016; Gallaher et al., 2019; Jiménez-Sánchez et al., 2021), observing the phenomena of increasing rD on invasion fronts, and giving it field-specific language and interpretation. Interestingly, much of the language and commentary around these models implies that the dominant paradigm in the cancer field is that cell populations are adapting to different microenvironments (e.g., Maley et al., 2017; Zahir et al., 2020; Fiandaca et al., 2021). This is fine, but there is less appreciation, I think, that the evolutionary shifts being observed might be a natural consequence of the dynamics of invasion. That is, that we expect quite particular evolutionary shifts purely

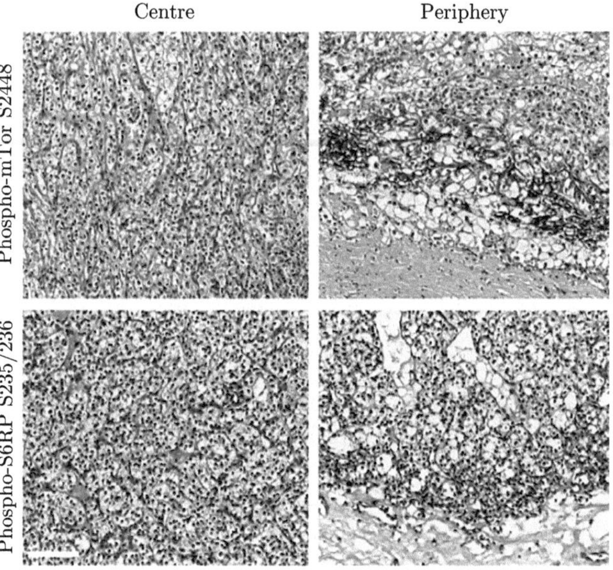

Figure 9.1 *Sections of renal cancer tissue from the tumour core versus the invading edge. Rows show staining for two different biomarkers indicating cells with high rates of proliferation. Columns show representative sections from the core and the periphery of a renal tumour. Figure from Hoefflin et al. (2016).*

because we have an expanding population. This broader view is apparent in the modelling and the data, but appears not to have been fully appreciated by many of the researchers in the field (Gidoin and Peischl, 2019). This broader view is important, because it improves the generality of our predictions: it is not just adaptation to some arbitrary environment, it is an evolutionary response to the process of invasion itself. In short, as tumours grow, the base expectation should be that a core and edge environment will emerge, but that the edge environment has spatial sorting and serial foundering as fundamental features. Under this view, it is clear that (depending upon trade-offs), tumours are likely to become more malignant, and more likely to metastasize as they grow; this expectation flows directly from the ecological and evolutionary dynamics of invasion.

Let's pause from surveying the literature for a moment and set up a basic verbal model of a tumour. It is worth doing, because tumours have a few particular features that are worth noting. The model is simple, but it describes in words an amalgam of theoretical models of cancer (e.g., Orlando et al., 2013; Lloyd et al., 2016; Gidoin and Peischl, 2019; Gallaher et al., 2019; Fiandaca et al., 2021), and is consistent with observations (Lloyd et al., 2016; Freischel et al., 2021; Zhao et al., 2021). First, we have a somatic mutation occur that causes a cell to break the rules of its assigned role in the body and to start proliferating. This cell rapidly becomes a ball of cells, all clones of the original. This goes on for some time until we have many millions of cells in the ball; enough that we start to see diversity arising from further somatic mutations. We can expect some of these mutations to occur close enough to the edge that we get sectoring driven by spatial drift and mutation surfing (Chapter 5). At the same time, we have a growing geometrical problem: cells on the surface of the ball have access to oxygen and nutrients and those towards the centre of the ball are getting very little. This resource gradient (particularly that of oxygen) sets up quite distinct environments, to which our cells begin to adapt. If you are a cell with very little oxygen, you spend more effort on your anaerobic pathway: you chew up glucose and produce acid as a by-product. In the core of the tumour, cells that are better at this will outcompete those that aren't. But of course this makes your microenvironment quite acidic, so cells that are better at dealing with an acidic environment also tend to do better (Zhao et al., 2021). Interestingly, one of the responses cells can have to being bathed in acid is to become more mobile (Gillies, 2022). This can be a plastic response, but it also might arise via spatial sorting in a metapopulation (the metapopulation effect; Section 7.4.1). In this case, the extinction-recolonization dynamic of a metapopulation is caused by higher death rates of cells in the core, favouring a colonizer phenotype (Labi and Erlacher, 2015; Gallaher et al., 2019). So now we have higher dispersal rates emerging in the core of the tumour. This causes a net flow of genes from the core to the edge; genes that are adapted to hypoxia and acidity, and that are more motile. Once these genotypes emerge on the invasion front, they rapidly take it over simply because they have a more dispersive phenotype. In the process, of course, we see the acid-tolerant hypoxic-adapted genes carried along with them. These genes turn out to be very handy in this new environment; sure, they now have abundant oxygen and resources, but the acid protects them from the immune system and destroys nearby (normal) cells that are in the way of further expansion. The stage is now set for things to get nasty as we see these invasive cell genotypes subject to ongoing spatial sorting and r-selection.

It is a basic model and it comes with a raft of assumptions and will clearly not apply as neatly to more complex tumours, but it is an interesting starting point, and we can see the eco-evolutionary dynamics at play: spatial sorting, r-selection, spatial drift. We didn't drill down on the implications of spatial drift here, but we might expect this, on average, to slow the growth of a tumour down. We know that the vast majority of mutations are detrimental, and we know that mutations can accumulate on invasion fronts despite being detrimental (Chapter 5). This expansion load can slow down invasions or even cause them to temporarily halt, and there is evidence from both modelling and empirical work that this does indeed happen in tumours (McFarland et al., 2013). While expansion load might well slow an invasion down, we also know that these slow parts of the invasion front will be transient; they will be invaded from the side, by fitter genotypes (Section 7.1.2); again, this is a theoretical expectation that we now see recapitulated in tumour growth (Fu et al., 2022).

In addition to these basic expectations, we also see fascinating wrinkles in this work on cancer that echo other aspects we have touched upon in this book. The experimental work of Freischel et al. (2021), for example, examined experimental 'tumours' composed of a mix of two cell lines: a 'pioneering' phenotype and an 'engineering' phenotype. While the pioneering phenotype rapidly came to dominate the growing edge of the tumour, it was tumours containing a mix of the two cell types that spread most rapidly. Clearly there is some kind of facilitation occurring here, but it may also be possible that they observed anomalous invasion speeds (Section 7.3). We also often see conditions that might lead to pushed waves. Certainly if motile cells emerge in the core of young tumours, we would transiently see a pushed dynamic as genes flowed from core to edge. The model of Orlando et al., for example, included an interesting wrinkle: that cells would move preferentially in a direction that would improve their fitness. This equates, in their model, to cells moving towards lower density areas at a rate determined by the spatial gradient in fitness. Although there is no formal analysis of it, it seems likely that their model would result in a pushed wave (Section 6.2.2).

9.1.1 Eco-evolutionary management of tumours?

It is clear that tumours are—just like cane toads—invasive populations. They are subject to the same eco-evolutionary dynamics that have been discovered in invasive species. There has, in the last 10 years, been an explosion of research on the ecological and evolutionary dynamics of cancer, with researchers porting many ideas from ecology and evolution across to cancer biology (e.g., Korolev et al. (2014)). This has culminated, quite recently, in calls for the eco-evolutionary management of cancer (Maley et al., 2017; Zahir et al., 2020). It is an exciting idea. Our capacity to treat cancer has seen astonishing improvements within my lifetime, but there is still a long way to go. Does an eco-evolutionary perspective open up new management ideas?

We know from the theory developed in previous chapters that increasing rD is a basic expectation on the invasion front of a pulled wave (Chapter 3). We also know that these pulled invasions tend to move fast and can be very unpredictable (Chapters 5 and 6). In short, pulled dynamics seem very likely to result in bad outcomes for a cancer patient. Pushed waves, by contrast, tend to have slower rates of evolution, and be more predictable. On balance, a tumour with a pushed dynamic—particularly if that pushed dynamic is caused by an Allee effect—seems a much less dangerous beast. And of course we also know that there are many different ways to transform a pulled wave into a pushed wave, including species interactions or a coarse-grained environment (Chapter 6). Predation, for example, might be one way of generating a pushed wave: 'predation' by the immune system, for example, is thought to not only prevent many tumours establishing, but also to drive selection for immune evasion in established tumours (Angelova et al., 2018). Perhaps we should be building on the idea of an eco-evolutionary approach to cancer diagnosis (Maley et al., 2017) and extending it to classifying the dynamics at play: is our tumour pushed, or pulled? How quickly might it evolve from pushed to pulled? Perhaps we should be looking for treatments that target density dependence: that force cells to cooperate, for example, at once both controlling the growth of a tumour but also radically reducing the chance of new tumours establishing? The possibility of Allee effects in cancer and how we might manipulate them has, however, barely been investigated (Korolev et al., 2014). From the perspective of invasion dynamics, this seems like a big omission.

Another avenue worth pondering is trade-offs. What are the fundamental trade-offs that cancer cells must operate within? These trade-offs, operating between r, D, and K must exist and must place fundamental constraints upon the evolutionary pathways available. If we knew these constraints, we would be in a much better position to understand the likely evolutionary divergence between edge and core. Such fundamental trade-offs are already being used to develop drug combinations that exploit rapid evolution: evolution in response to one drug induces susceptibility to a second (Zhao et al., 2016). But can we generalize this idea to predict and/or manipulate the dynamics of invasion? Are there fundamental trade-offs between r and D in cancer cells that limit the maximum speed at which a tumour can grow? Does the shape of these trade-offs determine whether a tumour is more likely to become metastatic or simply grow faster? How quickly might these trade-offs evolve? Amy Boddy et al. (2018) raised this issue of trade-offs and how they might affect cancer evolution. A clear case was made that basic life history trade-offs in ecology—for example, survival versus reproduction; survival versus movement; movement versus reproduction—should (and often do) occur in the life history of cancer cells. Delineating the form of such trade-offs and how the trade-offs themselves might evolve would have clear implications for predicting the evolutionary trajectory of tumour populations (Section 7.2). Here, strong progress has already been made by Hausser and Alon (2020), who have developed a new modelling approach to identify, from gene expression data, the key trade-offs faced by tumours and where cells in a tumour sit along the various trade-off curves. Mapping these trade-off spaces to the fundamental parameters of invasion, r, K, and D, would be a valuable exercise. That these curves will affect the eco-evolutionary dynamics of tumour growth has recently been noticed by cancer researchers, who developed a model of tumour growth involving trade-offs between r and D (Gallaher et al. (2019); Figure 9.2); their results were largely in line with what we would expect from earlier models (e.g., Olivieri et al., 1995; Burton et al., 2010), albeit with a few anomalies that might be explained by how their modelled traits map to invasion speed.

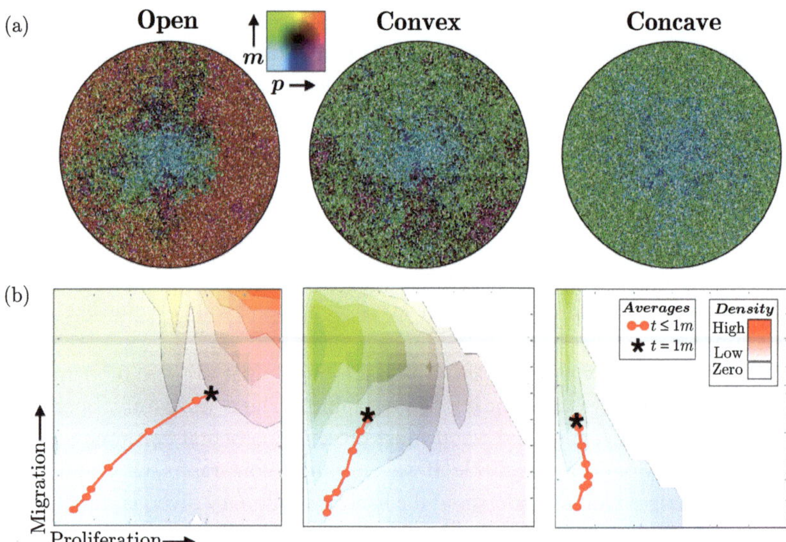

Figure 9.2 *A model examining evolution of cell migration (m) versus cell proliferation (p) in a tumour under different trade-off shapes. In all cases, there is a negative trade-off between migration and proliferation, but the shape of this trade-off is changed to be either convex, concave (when viewed from above), or straight line. The simulation imagines a small population of founder cells setting off an invasion that takes on the shape of a growing circle. The top row of circles shows an example of the model space at the end of a simulation, with colours representing the life history strategies evolved in each location. The bottom row of figures shows the density of trait values in the entire population at the end of the simulation, with the red line showing the evolutionary trajectory of the system (mean trait value every five time steps, starting in the bottom left and ending on the black asterisk). From Gallaher et al. (2019).*

Finally, another interesting avenue to explore is the mutation rate of tumour cells (Korolev et al., 2014; Gidoin and Peischl, 2019). It is a surprisingly long journey from a well-behaved somatic cell to an utterly selfish one, and cancer biologists talk about 'driver mutations' that provide the necessary steps. But for every one of these driver mutations that might occur (and be favoured by selection), there are likely hundreds of thousands of other mutations that will occur first: mutation is a random process, and the chances of a random mutation being beneficial is very small indeed. Cancer biologists tend to talk about these other mutations as 'passenger mutations' (because they are not drivers of tumorogenesis), but it would be a mistake to assume that all these 'passengers' are benign and have no effect on cell fitness. We know that deleterious mutations can accumulate on invasion fronts, and can lead to stalled invasions caused by 'mutational meltdown' on the invasion front (Section 5.6). This effect is quite sensitive to the balance between mutation rate and movement; thus either reducing the movement rate of cancer cells, or increasing their mutation rate, should move us towards a dynamic in which mutational meltdown is more likely, either globally or on the invasion front. Of course, given that we can expect 'invasion from the side' of fitter genotypes on the invasion front, it is likely that this manipulation would work best very early in the establishment of tumour: before cell population size and invasion front size give us a reasonable chance of a beneficial mutation occurring somewhere in the frontal population. Then perhaps this timing might be extended if there are ways to generate spatially enhanced drift on the invasion front.

It is clear that the growth and development of a tumour is an eco-evolutionary process, and when we think of it in these terms, new strategies emerge for management. How feasible it is to enact these strategies is another matter entirely, and is a question for ongoing research. But simply being able to conceive new strategies is a useful advance; it opens up new parts of idea space; parts that might be filled as technology advances.

9.2 Cane toads

Invasive species are a fairly major problem nowadays. The vast flow of people and goods across the globe has eroded biogeographic barriers everywhere, and caused species to be introduced to places they never would have reached unassisted. Many of these introductions, of course, fail in the early stages, but the sheer number of introductions means eventually something will stick. When something does stick, the results can be spectacular: wholesale invasion across entire continents and dramatic changes in the invaded communities (Mack et al., 2000). The roll call of troublesome invasive species really is too long to list, but cane toads (*Rhinella marina* [née *Bufo marinus*]) would have to be one of the most well-known invasive species. The IUCN declared them one of the world's 100 worst invasive species in 2001 (Lowe et al., 2001) and, at least in Australia, they have also happily invaded popular culture. They have also proved very amenable to study and (no doubt in part because of the public's fascination with them) have been subject to a massive research effort over many years (Shine, 2018).

Toads were introduced to north-eastern Australia in 1935 by an overzealous sugar industry (Section 1.2.2), and they have been spreading across the country ever since (Figure 9.3). At the time of writing (2022), they have spread westwards more than 2,300 km across northern Australia from their original release points. They now occupy around 1.5 million square kilometres of the country, and they still have a long way left to spread in Western Australia (WA; Kearney et al., 2008). Toads are highly toxic, and their toxins—bufadienolides and other nasties—are unlike any that Australian predators have ever dealt with (Phillips et al., 2003). As a consequence, many native predators—ranging from fish to reptiles and mammals—are badly impacted when toads arrive (Shine, 2010). For many of these predator species, particularly the large generalist predators—goannas (metre long lizards); snakes; and quolls (carnivorous marsupials)—the arrival of toads causes rapid local extinction. The removal of these large generalist predators then causes indirect effects to ramify down the food chain as smaller species are released from predation pressure and increase in number (Doody et al., 2015). In short, it's a mess.

One silver lining to have come from the poorly thought out decision to introduce toads to Australia is that we have an astonishing natural laboratory in which to study an invasion of continental scale. In 1974, Jeanette Covacevich and

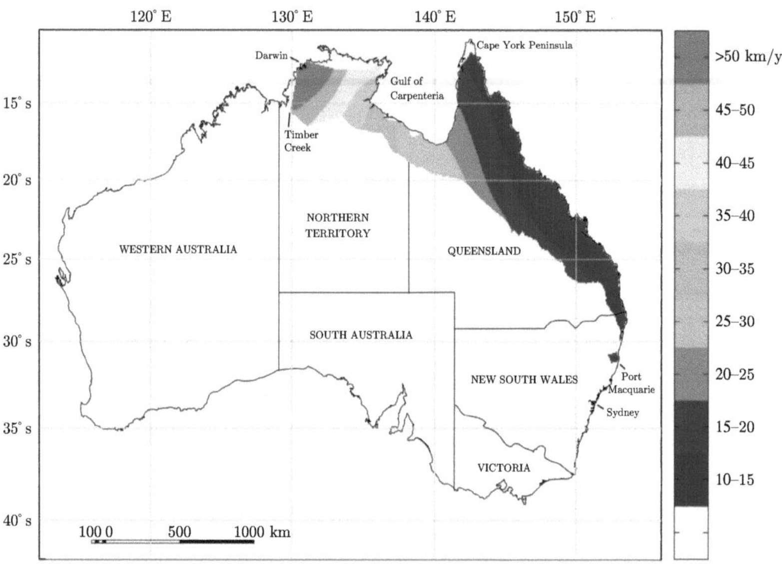

Figure 9.3 *The speed of the toad invasion front as it moved across northern Australia. Reproduced from Urban et al. (2008).*

Michael Archer at the Queensland Museum realized that the public's fascination with toads would allow them to reconstruct the toads' current distribution and invasion history (Covacevich and Archer, 1975). They did citizen science before citizen science was even a thing: they wrote to every school in north-eastern Australia asking whether toads were in town or not, and if so, when they had arrived. The result was an astonishing dataset. Not only could it be used to map the toads' current distribution, but it could also be used to reconstruct the invasion, and could be built upon over time (Sabath et al., 1981; Easteal et al., 1985). The data were indeed built upon; in fact, they became part of a much larger dataset tracking the global distribution of toads, pulled together by Simon Easteal and colleagues (Easteal, 1981). Happily for this story, these researchers well understood the value of this dataset and decided to make it available in a public data repository (Floyd et al., 1981). The only sensible public data repository that existed in the 1980s was of course the university library, where I was absolutely delighted to find the data—pages and pages of neatly typed out numbers in a little book—when I went looking for it near the end of my PhD in 2004.

By updating the historical dataset with recent records, we were now able to reconstruct the toads' invasion from 1935 to 2006. The result very clearly showed a steady acceleration of the invasion front during this time: an invasion that once progressed at 10 km/y was now progressing at closer to 55 km/y (Phillips et al., 2006, 2007; Urban et al., 2008). This observation (with support from the Australian Research Council) set me off on a set of experiments all aimed at looking for evolutionary shifts in traits related to rates of dispersal and reproduction. Sure enough, toads on the invasion front moved substantially further than their conspecifics in the range core (Phillips et al., 2008), and this shift looks to have a genetic basis (dispersal rate is heritable and lab-reared animals show a similar divergence in dispersal rate; Phillips et al. (2010c)); clear evidence for an increase in D. Additionally, tadpoles and metamorph toads from the invasion front grew substantially faster than their conspecifics from the core (Phillips, 2009), suggesting that the invasion front might manifest a higher population growth rate, r.

In the years since, there has been a tremendous amount of work looking at shifts in toad phenotypes between invasion front and core. I cannot do justice to the extent of this work here, but divergence has now been observed in the toads' immune systems (Brown et al., 2018); in their rates of spinal disease (Brown et al., 2007) and parasitism (Phillips et al., 2010b); in their fecundity (frontal toads producing fewer eggs, Hudson et al., 2015); in their levels of phenotypic plasticity (Ducatez et al., 2016); and in their limb lengths (Clarke et al., 2019), among many other traits. There have also been some clear trade-offs that emerged, notably a clear trade-off between dispersal and fecundity (Hudson et al., 2015; Kelehear and Shine, 2020). There is even work showing that the life history of one of the toads' parasites (a nematode lungworm) has diverged between the invasion front and core populations (Kelehear et al., 2012). It is a boggling list of changes, most (but not all) of which align with the basic ideas of Chapter 3.

But, of course, running through all this work is the question of how we might use what we have learned to control toads. Are there lessons from theory that we can bring to bear on the control of cane toads?

9.2.1   Eco-evolutionary management of cane toads

First, it is worth noting that in 2005, 70 years after they were introduced, the only tool available for controlling toads was to go outside with a torch and catch as many as you could. Faced with a species capable of moving 50 km in a year and producing 30,000 eggs in a sitting, it is fairly clear that any such efforts are about as futile as trying to empty the ocean with a bucket. In 2022, after 17 years of intensive research into toads and their biology, are we in a better position?

The answer is yes. We definitely have some new tools (most notably some chemical attractants that can be used to very effectively trap and remove tadpoles; see Crossland and Shine (2011a)), but we also have some new ideas that look good in theory and might one day deliver important containment and control options (Tingley et al., 2017). But before we embark on these, a quick tale of an idea that didn't work. In 2008, it was revealed that the lungworm parasite found in many toad populations was not a native species but was, in fact, a parasite that toads had brought with them from South America (Dubey and Shine, 2008). There was a brief flurry of excitement about this: excellent empirical work by Crystal Kelehear showed that the parasite reduced growth, survivorship, and movement of young toads (Kelehear et al., 2009) as well as the growth rates of adult toads (Kelehear et al., 2011); perhaps we could use this parasite in some way to effect control? This idea rapidly fell apart, however: not only was this parasite already highly prevalent across the toads' range, but it was also missing from the invasion front. The former implied that its impact must be minimal, or at least is already being felt; the latter observation implied that some combination of spatial sorting, stochastic processes, and transmission dynamics was causing it to be lost from the invasion front (Phillips et al., 2010b). The big focus at the time was on how to stop the invasion front (the main impact of toads happens soon after they arrive); it was clear that even if the lungworm had some nasty effect on toads, we couldn't use it to slow the invasion down because the invasion front would simply leave the parasite behind. The idea was dead on theoretical grounds, but the final nail in this idea's coffin came when it was discovered that the parasite would infect endemic frog species and cause increased mortality (Pizzatto and Shine, 2011).

9.2.1.1   *A waterless barrier and a genetic backburn*

In 2010, a creative ecologist called Mike Letnic pulled me aside at a workshop to point out a striking observation: the toad invasion would one day have to skirt around the edge of the Great Sandy Desert. This meant they would have to move through a very narrow corridor of suitable habitat (Figure 9.4). Moreover,

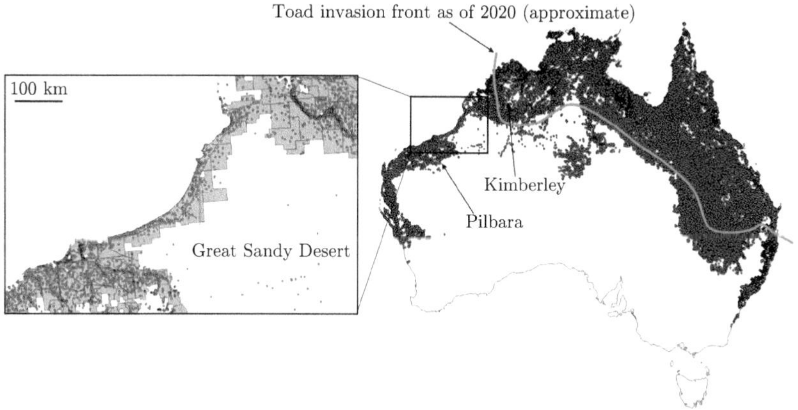

Figure 9.4 *Map of Australia showing artificial waterbodies within the toads' potential distribution. The inset map shows the narrow strip along the coast (between the Kimberley and Pilbara regions) through which toads will need to pass if they are to invade the Pilbara. Polygons in the inset map are pastoral property boundaries.*

Mike argued, that narrow corridor was only suitable because of the groundwater being brought to the surface to provide water to cattle. Remove these water sources, Mike argued, and the toad invasion would stop. The reason for this is the extreme wet–dry seasonality of northern Australia: for six months of the year, it is wet (very wet); for the other six months, it is hot and very dry. Mike's point was that during these dry periods, toads relied critically on these artificial waterpoints. Mike and his team followed up this idea with a set of experiments showing that when toads were prevented from accessing these waterpoints in the dry season, they rapidly perished (Florance et al., 2011; Letnic et al., 2014). They also made some estimates of connectivity between waterpoints and showed that it was indeed likely that toads could move between these waterpoints and so successfully skirt around the Great Sandy Desert (Figure 9.4).

It just so happened that I had been working on a model of spread across a landscape of points in collaboration with Stuart Baird (he of the wrinkles in Chapter 7). We had the model sketched out but had never really done anything with it. Together with Reid Tingley, we adapted this model to describe the spread of toads across this particular landscape of points. The model confirmed that toads would indeed be able to spread down this corridor, but also that we could halt that spread if we established a 'waterless barrier' about 70-km wide. That is, if we prevent toads from accessing waterpoints across a 70-km stretch of country within this corridor, the invasion would be stopped (Tingley et al., 2013). Not only that, but later modelling work done by Darren Southwell, who incorporated cost, suggested that if we chose our location carefully, this could be achieved for a surprisingly small amount of money (Southwell et al., 2017). This was a pretty exciting idea: we manage a stretch of country such that $r = 0$ in that part of space, and the invasion stops. Achieve this, and we keep toads out of about 270,000 km^2 of their potential range. Here was a containment effort worth trying.

In most parts of the world, 70 km is a lot of country. Most of Paris, for example, sits within a circle 40 km in diameter. But in this part of the world, you might drive 70 km and still be on the same cattle property. Reid, Darren, and myself spent some time there in 2015. We wanted to discuss this idea with the locals: the Nyangumarta and Karajarri landowners, the pastoral landholders, and representatives of government and non-government organizations. We ran a workshop in the nearest town to discuss the idea, and we drove from property to property to talk with the pastoralists: were our maps of their infrastructure correct? Was our assumption about no other water being around in the dry season reasonable? By and large, we found the locals supportive of the idea; a few tweaks to our ideas were suggested, but there were no obvious stop signs that we hadn't seen. Still, we came to appreciate that 70 km really is a lot of country.

We had been driving around for about a week when we washed up at Anna Plains Station. We'd been directed to a campsite near the homestead, and we tumbled out of the car at dusk. We were tired and hungry, and no one could be bothered putting up tents. We made a fire, ate food out of cans, and slowly revived our spirits with a box of cheap red wine, passed at intervals between us to top up our tin cups. We reviewed all the country we had driven past, and all the people we had met. There were plenty of places where this barrier could work if it only needed to be 50-km wide, but 70 km really narrowed down our options. There always seemed to be something—an irrigated orchard, a township, a natural waterbody—that made a 70-km stretch tricky to find. We gazed onto our drinks and into the fire. Someone pointed out that the barrier only needed to be 70-km wide because we needed to stop these crazy, highly dispersive toads that had evolved on the invasion front. We gazed some more. Half-jokingly, I pointed out that we should just introduce toads from the core of the range to the north side of the barrier. That way, we would only need to stop toads that move about 10 km per year. Only after it was out of my mouth did it occur to all of us that this wasn't such a bad idea. Within minutes, we had given this notion a moniker: the genetic backburn (Figure 9.5).

Backburning is a practice well understood by those who fight bush fires. If you want to stop a bush fire, you establish a firebreak downwind of your oncoming fire front. You then start a new bush fire on the upwind side of this firebreak such that it burns against the wind, back towards the oncoming fire front. If all goes well, the main fire front is driven by the wind onto this burnt country and simply goes out for want of fuel. The genetic backburn is a very similar idea: we establish a waterless barrier (the 'firebreak'), and then we set off a new invasion of (low dispersing) toads that spread back towards the oncoming invasion front of (highly dispersive) toads. What happens when the two invasion fronts meet? Well, space has already been occupied by the less dispersive phenotype, so the only way for the more dispersive phenotype to successfully invade into that space is to outcompete (i.e., to be fitter than) the less dispersive phenotype. But we

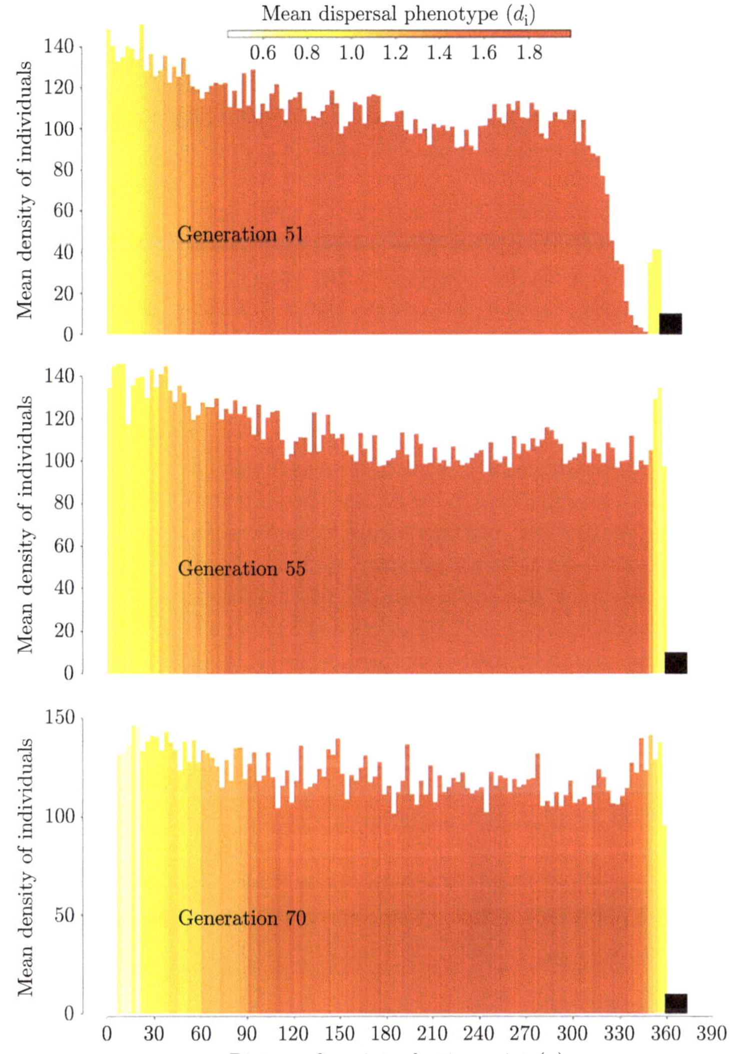

Figure 9.5 *A genetic backburn to prevent highly dispersive phenotypes reaching a dispersal barrier. The three panels show three time points in a simulation model with the black square at x = 360 indicating the dispersal barrier. At generation 51, a population has been invading from left to right and evolving increased dispersal ability (yellow to red colour ramp), while trading off competitive ability (seen in lower density of individuals as we move from left to right). At generation 50, a small number of (low dispersal, high competition) individuals is transplanted from the core of the range. At generation 55 (second panel), we see the two invasion fronts meeting; by generation 70, we can see the less dispersive phenotypes invading into the bulk of the population. From Phillips et al. (2016).*

know that the conditions on the invasion front select for traits that maximize rD, whereas in the core of the range, traits that maximize K (i.e., competitive ability) should be favoured (Burton et al., 2010). We know that invasion front toads pay a price for their speed: they are less fecund, and they suffer higher rates of disease (Brown et al., 2007; Hudson et al., 2015; Kelehear and Shine, 2020). We also know that after the invasion front has passed, dispersal rates in the toad population drop (Lindstrom et al., 2013), and this is a strong sign that natural selection is favouring less dispersive phenotypes behind the invasion front (Perkins et al., 2016). So, all up, I would be willing to bet that when those invasion fronts meet, it will be the highly dispersive phenotypes that retreat. If we time this correctly, a simple model of the idea suggests that genetic backburning can dramatically improve the strength of the barrier (Phillips et al., 2016). If we time this correctly, it seems very likely that a 50-km barrier could be sufficient to stop the invasion front.

It is a fun idea, but before we leave it, it is worth pointing out that the barrier (the firebreak) is essential here. It has been proposed that we might slow an invasion down by moving the core phenotypes to the invasion front, but without a barrier, this would be a dangerous strategy. Without a barrier, we set off an invasion in both directions: that is, we move the invasion forward. Not only that, but we risk the possibility that a few long-dispersing individuals do move through the backburn and establish themselves on the new invasion front. The result would be an invasion that gets a jump forward, picks up some extra genetic diversity, and then very rapidly assumes the same ghastly speed as it was achieving before the intervention (or possibly faster given it has now picked up some extra genetic variation). With the barrier in place, however, we have a situation in which those few long-dispersing individuals are killed as they move into the barrier region. Indeed, any individual moving into that region is killed, which sets up a kind of reverse spatial sorting (Section 7.4.3) and should drive dispersal rates down in the part of the population adjacent to the barrier. The barrier is as essential to a genetic backburn as the firebreak is to a standard backburn.

9.2.1.2 *Gene drives*

The advent of CRISPR-Cas9 technology means we now have capacity to edit genomes with pinpoint accuracy (Doudna and Charpentier, 2014). This has opened the possibility that we might introduce selfish genetic elements into a genome. Such elements either copy themselves onto alternative alleles, or they simply kill competing alleles at the gamete stage. Either way they exhibit 'super-mendelian' inheritance, and this allows them to increase in frequency against the force of natural selection (Burt, 2003; Noble et al., 2017). This allows us to 'drive' novel genes into a population, even if those novel genes carry a fitness cost. It is astonishing technology, and opens up a suite of new options for managing wild populations, including invasives (Esvelt et al., 2014; Champer et al., 2016; Rode et al., 2019). Gene drives will be discussed in their own section below, but here I provide a small entrée, using toads as a vehicle.

The first proposal for the use of gene drives in toads is that we might knock out a gene necessary for the production of the most highly toxic bufadienolides (toad toxins) to produce 'detoxified toads' (Tingley et al., 2017). This is, apparently, a fairly simple modification: we just need to break a single gene (and let us assume that this gene has no other function other than toxin production). We can create a genetic element whose sole purpose is to break that gene, and we can use a simple drive system to drive that element through the population. If successful, the result would be a population of toads that is dramatically less toxic. Given that the main impact of toads is through poisoning naive native predators, low-toxin toads should be low-impact toads. Or so it is hoped, anyway.

It is an attractive idea, but let us remember that the main impact of toads happens as they arrive in a new area; naive predators die in droves. This suggests that we really need low-toxicity toads to be the first to arrive; the invasion front needs to be made up of low-toxicity toads. This won't happen without careful thought. The considerations in Section 3.9 suggest that the invasion speed of the gene drive is unlikely to be as fast as the invasion speed of the toad population itself (recall that toads can have 30,000 eggs in a sitting, so $r(0)$ for toads is likely a large number). So unless the drive allele is brought to a high frequency on the invasion front, such that it surfs, it is likely to be left behind. Achieving such a high frequency would require releasing gene drive toads well in front of the invasion front, essentially setting off a new invasion in front of the oncoming one.

9.3 Gene drives

Gene drives are an astonishing piece of technology that allow us to drive genetic variants into a population (Section 9.2.1.2). Gene drives are engineered selfish genetic elements, and they can push quite hard against the force of natural selection, so it is feasible to drive variants into a population that reduce that population's fitness. Of course, when we introduce a gene drive into a population, we set off a new invasion; we set off the spread of an advantaged disadvantageous allele.

As we saw above (Section 9.2.1.2), drives might be used to spread a 'desirable' trait through a population. If that trait has only minor fitness effects—that is, it doesn't really affect the density of the population—then the invasion of the drive allele would essentially follow the basic model set out in the Fisher–KPP equations (Section 2.1). It should spread with a constant speed determined by the dispersal rate of drive carriers, and the growth rate of the drive allele (when the drive allele is rare).

A more aggressive use of gene drives is to use them to send unwanted populations extinct. The idea here is that we use a drive to skew sex ratios to the point that the population collapses (Hickey and Craig, 1966; Hamilton, 1967). If sex is genetically determined, then we simply drive a gene (or chromosome) that determines maleness (Galizi et al., 2014; Prowse et al., 2017; Rode et al., 2019). As a consequence, the population sex ratio rapidly runs towards being dominated by 'daughterless males', and eventually the population collapses through a lack of reproductive females (Schliekelman et al., 2005). It is a clever idea and has broad use, potentially applicable to any diploid, sexually reproducing organism. Gene drives for population suppression are not just a theoretical idea: they have been demonstrated in laboratory populations of mosquitoes and fruit flies (e.g., Kyrou et al., 2018; Meccariello et al., 2021; Hammond et al., 2021).

While the idea of using gene drives for population suppression is simple in principle, when we consider a population in space (as opposed to an aspatial laboratory population), the mathematics quickly becomes a lot harder than the Fisher–KPP model. The reason for this is that, as the wildtype population goes extinct in one part of space, it sets up a density gradient between areas yet to be invaded by the drive and areas where the population has gone extinct. This gradient causes asymmetric gene flow: the parts of space in which the drive is present receive a lot of wildtype immigrants, which work to reduce the frequency of the drive alleles in that part of space. Thus, with suppression drives, the drive needs to push not just against the force of natural selection, but also against the force of the gene flow that emerges from suppression (Girardin and Débarre, 2021). A thorough mathematical model of suppression gene drives needs to account for this process, and this requires a model with an advection term (similar to those used in Sections 7.4.2 and 8.3.2); advection makes things complicated.

Once space is part of the picture, we have two species of wave to watch: the population wave, and the gene drive wave (Figure 9.6; Girardin and Débarre (2021)). If we start with a population at carrying capacity across all space and drop a handful of drive carriers into the middle, we can watch a gene drive wave

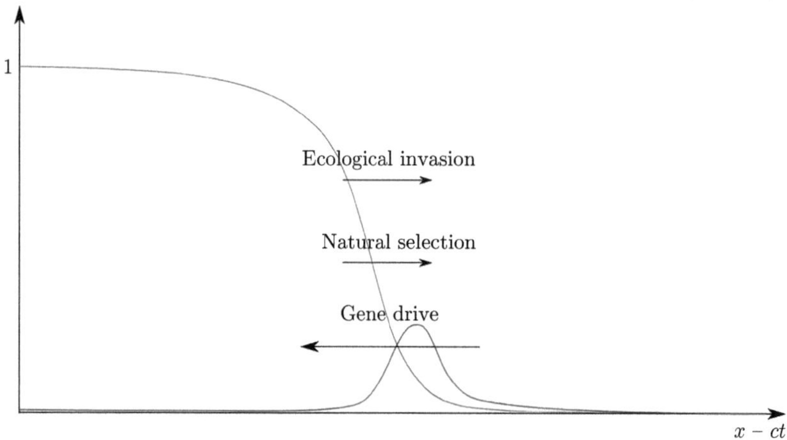

Figure 9.6 *The speed of a suppression gene drive is determined by the balance of three 'forces': the growth rate of the drive; the force of natural selection; and the force of the emergent ecological invasion. After Girardin and Débarre (2021).*

emerge as the drive increases in frequency and spreads across space, similar to Fisher's 'advantageous allele' (Section 2.1). We also see population waves emerge. This happens because, as mentioned above, the drive causes local extinction, so the population wave we hope to see is a retreating wave, strongly coupled to the drive wave. But it is still a population wave, so potentially all the ecological and evolutionary dynamics of invasive populations come into play. In addition to the issue of asymmetrical gene flow, we also have spatial drift that arises (Chapter 5): by chance, the vacant space behind the drive wave might be colonized by a handful of wildtype individuals who don't carry the drive gene. This then decouples the population wave from the drive wave and sets off a new invasion front that does not carry the gene drive: the wildtype allele happily 'surfs' across space leaving the drive to play catchup. What can emerge across space is a 'chasing' dynamic, in which wildtypes repeatedly penetrate through the drive wave and set off new population invasions (Eckhoff et al. (2017); Champer et al. (2021); Figure 9.7). Essentially anything can happen after that. Depending upon the parameter settings, we can see the drive eventually winning, the drive going extinct, or a kind of stochastic coexistence of the drive and the wildtype (Champer et al., 2021; Paril and Phillips, 2022). It is clear that space changes the outcomes here in very non-trivial ways.

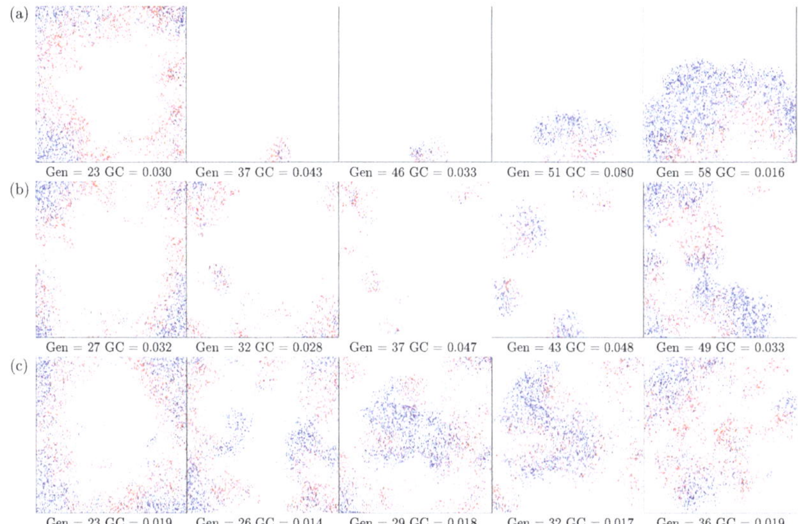

Figure 9.7 *Three examples of the chasing dynamic that can emerge in spatially explicit stochastic simulation of gene drives. Each row is a different simulation, with columns representing time points ('Gen=') in the simulation. In each case, gene drive-carrying individuals (coloured red) have been introduced into the centre of a population of wildtype individuals (coloured blue). In each case, the drive wave is penetrated (sometimes multiple times) by wildtypes, which founder into the vacant space generated by the local extinction of a drive-carrying population. From Champer et al. (2021).*

The spatial dynamics of suppression gene drives are an active area of research. It is tricky (Girardin and Débarre, 2021), and things are still being worked out, but we do at least have some patterns that have emerged from models. First, the result is sensitive to the scale of dispersal relative to the scale of local dynamics (Champer et al., 2021). As dispersal distances increase, the chance of drive failure decreases: the reason for this is not clear, but it appears to be that as dispersal increases, the drive wave gets wider faster than the population wave does (presumably because of advection), so penetration by wildtypes becomes less likely (Paril and Phillips, 2022). A second observation is that 'faster' drive constructs (drives that have very high fitness relative to wildtypes) are more likely to manifest chasing. The reason here is that these drive waves are relatively narrow, so are more easily penetrated by a wildtype; they also tend to move quite quickly such that the escaped wildtype population might never be recolonized by the drive (Paril and Phillips, 2022).

9.3.1   Local dynamics matter

It is possible to create drives that exhibit a range of local dynamics. The simplest drive—a homing drive—contains a genetic element that creates a double stranded break in the wildtype allele on the homologous chromosome (e.g., Windbichler et al., 2011). Cellular machinery then repairs the cut using the drive allele as a template. In this way, the drive replaces the wildtype allele, and an individual inheriting only one copy of the drive allele will produce offspring that all inherit the drive. When we think of this system, it is clear that the per-capita growth rate of the drive is frequency-dependent: when most of the population is wildtype, a drive allele has a good chance of meeting a susceptible target; when most of the population already carries the drive, an individual drive allele has a much lower chance of being paired with a susceptible target. Such a drive essentially exhibits logistic growth. As a consequence, we can expect a simple homing drive (in the absence of demographic fitness effects) to have a pulled wave.

This basic logic also applies to another class of drives: 'shredder drives'. These drives are placed either on a sex chromosome or an autosome and are designed to 'shred' (i.e., render dysfunctional) sex chromosomes associated with females. In mammals (that have an XY sex-determination system), we might have an 'X-shredder' that shreds the X-chromosome; in reptiles, frogs, and birds (that use a ZW system), by contrast, we might have a 'W-shredder' that shreds the W-chromosome. Both drives would result in a strong skew towards (drive-carrying) males and can, in principle, drive a population to extinction (e.g., Holman, 2019). These drives were initially designed to step around the fact that homing drives tend to produce drive-resistant alleles (Noble et al., 2017; KaramiNejadRanjbar et al., 2018). Similar to homing drives, however, it is clear that the shredders have the highest per-capita growth when they are rare. Once again, we would expect such drives (in the absence of demographic fitness effects) to exhibit a pulled wave.

It is also possible to design drives that exhibit local bistability (Section 6.2.1.2): drives that will tend to extinction unless brought above a threshold frequency. One way to do this is to design drives with a relatively high selective disadvantage in heterozygotes. If we do this, we have a situation akin to hybrid zones (Section 8.4): when an allele is at low frequency, it is mostly found in a heterozygous state, and if it is selected against in this state, then it will have a negative growth rate at low frequency. If frequencies are increased above a threshold, however, the allele is increasingly found in a homozygous state; so the effect of low heterozygote fitness becomes less relevant. One drive system that can cook heterozygote disadvantage is a toxin-antidote drive. These drives involve a 'toxin' and an 'antidote' and often mimic in some ways naturally occurring drives. The toxin-antidote-dominant-embryo drive (TADE) is one such system (Champer et al., 2020). Under the TADE system, the drive consists of two parts: a pair of CRISPR/cas9 scissors that cut (and so disrupt) an essential gene, and a recoded copy of that gene that is immune to the CRISPR/Cas9 scissors. Thus, the scissors act as a 'toxin' by destroying an essential gene, and the recoded copy is the 'antidote'. Under the TADE system, the essential gene is haplolethal: unless an embryo has two working copies of the gene (i.e., one working copy on each chromosome), the embryo is non-viable. In this way, drive-carrying females that mate with a wildtype male will have around half of their offspring rendered inviable (Figure. 9.8). All of their viable offspring carry the drive (so there is skewed inheritance), but there is a large fitness cost to this heterozygous mating: the overall cost depends upon the frequency of the drive in the population. When the drive is rare, the fitness cost is at a maximum and may be sufficient to cause negative growth of the drive; growth may only become positive once the drive is brought above a threshold frequency. Because this drive (or any other with strong heterozygote disadvantage) no longer has the highest per-capita growth when it is rare, we can expect a pushed dynamic (Chapter 6) to emerge.

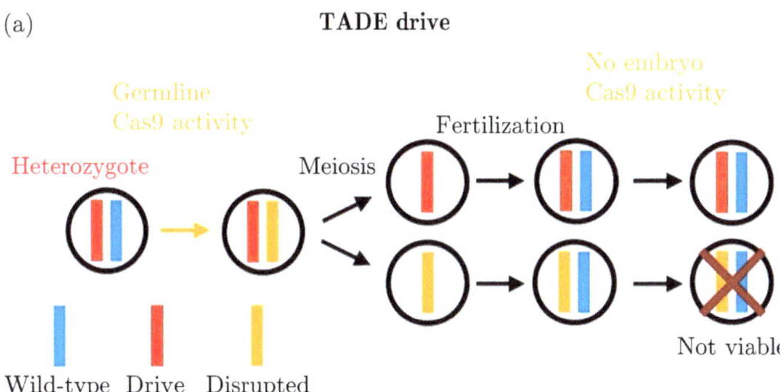

Figure 9.8 *Skewed inheritance under the TADE gene drive system. Here the essential gene is haplolethal, so embryos are only viable when they have a working copy (either the wildtype or the drive allele) on both chromosomes. Reproduced from Champer et al. (2020).*

These local drive dynamics matter, because they change the fundamental invasion dynamics of the drive. If we ignore demographic fitness effects for a moment, it is clear that we can expect the pulled drives to spread rapidly, and be very difficult to stop (Girardin et al., 2019). From a biosecurity perspective, a single drive carrier transported to a new location could set off an invasion of the drive, with all its attendant consequences (Esvelt and Gemmell, 2017). It might be one thing to want to eradicate cane toads in Australia, but it would be an outrage if the biotechnology we create for doing that ends up driving them extinct in their native range. Because of this, drives that generate a pushed dynamic would be preferable (Tanaka et al., 2017). Here, such drives would naturally be subject to range-pinning (Section 6.3.3): a deliberate effort would be required to bring a target population up above the threshold drive frequency beyond which the drive will spread. Without that effort, the drive will not spread, so a few accidental introductions here or there are of little consequence.

But of course pushed drive waves will be harder to establish, slower to spread, and may face particular difficulties when pushing up against clines in population density (Figure 9.6): either natural clines or those resulting from the action of the drive itself (Tanaka et al., 2017). They are reminiscent, in many ways, of the situation of invading *Wolbachia* (Section 6.2.1.2). Additionally, in situations where the target species falls naturally into discrete populations within the target range; range-pinning—the very feature that makes these drives preferable—reduces the capacity of the drive to spread of its own accord.

Interestingly, however, many toxin-antidote drives may be less subject to the problem of 'chasing'. Because toxin-antidote drives tend to have slower growth rates than homing or shredder drives, they tend to exhibit relatively wide invasion waves. These wide drive waves will have a lower chance of being penetrated by wildtypes, and so will have a lower chance of manifesting the 'chasing' dynamic that can lead to failure in faster drives (Paril and Phillips (2022); Figure 9.7).

In addition to the local dynamics of the drive, the local dynamics of the population is also important when we consider gene drives for population suppression. While chasing looks like a difficulty to overcome in many situations, it should also be the case that populations subject to Allee effects are much less likely to exhibit chasing than ones that are free of Allee effects (Beaghton and Burt, 2021). The reason for this is obvious when we consider that the Allee effect causes a higher effective population size (and so lower drift) on the invasion front (Chapter 6). Thus, the chances of a wildtype population emerging on the other side of the drive wave are greatly reduced. As a consequence, another way to manage the risk of chasing would be—prior to the introduction of a suppression drive—to manipulate the target population in a way that causes an Allee effect to arise. Doing so might then be expected to make the drive substantially less prone to chasing.

9.4   Spreading pathogens

Given this book was largely written during the coronavirus pandemic of 2020–2021, it was probably inevitable that there would be a section on the spread of pathogens. When a pathogen jumps ship and finds itself in a novel host population, it is pretty clear that we have an invasion afoot. It is the invasion of a novel (and tiny) predator across a naive (and relatively large) prey population. Such events happen all the time. Among many examples, we watched Rinderpest sweep through the ungulates of Africa; chestnut blight decimate North American chestnut trees; white-nosed syndrome sweep through bat populations across North America, and chytrid fungus decimate frog species worldwide. In humans, among others, we had the Black Death sweeping through Europe in the fourteenth century; and the 'Spanish' flu, the spread of HIV, various Ebola outbreaks, Zika, SARS, MERS, and of course SARS-CoV-2 in the twentieth and twenty-first centuries. In all cases, these were invasions, starting from a few individuals in a single location and spreading to infect many individuals over a much larger area. Because these are all invasions, they are subject to all the same basic eco-evolutionary processes as toads or tumours.

I was outrageously fortunate enough to be situated in Western Australia (WA) during the coronavirus pandemic of 2020–2021. While much of the rest of the world struggled through the pandemic, life in WA was largely unaffected. SARS-CoV-2 was not able to invade WA; instead, it was deliberately let in in early 2022, when vaccination rates were about as high as we could expect them to get. This astonishing outcome was achieved through a combination of natural isolation (WA is a long way from the rest of the world) and strong public health measures. Although the public health officials didn't think about it in these terms, the public health measures were aimed at effecting 'range-pinning' (*sensu* Lewis and Kareiva (1993); Section 6.3.3), thus preventing spread. The first and most critical move was to 'close' the borders: no one was allowed in except under extraordinary circumstances, and then under strict quarantine guidelines. This did not stop every infected case from arriving in WA, but it dramatically reduced the rate at which SARS-CoV-2 cases arrived. The second important measure was contact tracing. Whenever a case arrived in the community, every contact of that case was found, went into isolation, and had their tonsils swabbed for a PCR test every few days. This had the effect of reducing the growth rate of cases. When there are only a few cases, it is possible to trace and isolate every contact, so when there are only a few cases, the growth rate can be brought below replacement. When there are a lot of cases, however, contact tracers rapidly become overwhelmed and growth rates climb into positive territory. Thus, contact tracing imposes a strong Allee effect (Section 6.2.1.1); if arriving case numbers are small enough (thanks to closed borders), we can keep the population in this negative growth part of state space; the invasion becomes pinned.

This idea of pinning an invasion is purely ecological manipulation, but it is clearly effective. I feel quite strongly that this idea is underappreciated by those who manage invasions. It is clearly well appreciated by health officials and epidemic modellers in Australia, but there are many other settings (e.g., tumour growth, invasive species) where the idea has received little to no attention. This is a shame, because it can achieve dramatic outcomes.

But beyond coronavirus and range-pinning, let us consider the eco-evolutionary dynamics of pathogen spread. There are a few peculiarities worth flagging: let's dispense with those now. The first peculiarity is that we need to be very clear about the difference between transmission and dispersal. They are related: transmission is the movement of pathogens between hosts; dispersal is the movement of pathogens through space (often effected by a host). This is quite an obvious distinction, but one worth being clear about, because the vast majority of work on the ecology and evolution of pathogens concerns transmission rather than dispersal and if we're not careful, we might get them confused. Transmission is very closely tied to the (temporal) fitness of a pathogen (Anderson and May, 1992); dispersal is closely tied to the spatial fitness of a pathogen (*sensu* Chapter 3). A second peculiarity worth mentioning is that we typically think about pathogens from the perspective of the host. Most ecological and evolutionary models of pathogens track the number of hosts infected with, say, the flu rather than the number of flu particles within each host. Again, an obvious distinction, but useful to keep in mind, because this makes a lot of host–pathogen models a kind of metapopulation model, with hosts as the habitat patch and transmission playing the role of dispersal between patches (hence why it is useful to keep transmission/dispersal separate in our minds; they are actually both a kind of dispersal). It is also clear that pathogens evolve within the host, and, just as within metapopulations, we may see conflict between traits maximizing within-host fitness versus those maximizing between-host fitness (transmission) (Ewald, 1994). A final peculiarity worth mentioning is that we are often looking at newly emerged pathogens; some parasite that has successfully switched host, or been transported to a new, genetically distinct, host population. Such pathogens are often likely to be poorly adapted to their new host, so there is plenty of room for improvement in their new role (Visher et al., 2021). Given time, we can expect quite large evolutionary shifts to occur.

With these caveats in mind, let us turn to a very simple, but powerful, model of disease dynamics: the *SI* model (Kermack et al., 1927). The model takes a host population and divides it into two partitions: those who are susceptible to infection (S) and those who are infected (I):

$$\frac{dS}{dt} = -\beta SI \tag{9.1}$$

$$\frac{dI}{dt} = \beta SI - \gamma I, \tag{9.2}$$

where β is the transmission rate, and γ is the rate at which infectious individuals are lost from the population (by death or through recovering and becoming resistant). It is a disarmingly simple model, and yet it does a decent job of recapitulating the dynamics of epidemics. From it flows the idea of herd immunity—the idea that if we vaccinate a proportion of the population, we can drop the growth rate below zero: it is clear that the growth rate of infected is: $\beta S - \gamma$, which equals zero when $S = \frac{\gamma}{\beta}$. Thus, if the number of susceptibles S drops below $\frac{\gamma}{\beta}$, the outbreak is heading towards deterministic extinction. This threshold value, $S_t = \frac{\gamma}{\beta}$, is often written as $S_t = \frac{1}{R_0}$, where $R_0 = \frac{\beta}{\gamma}$ is the number of new infections likely to arise per infected individual very early in the epidemic (when I is tiny compared with S).

This model has been extended in all sorts of interesting directions, but it was the basic model kicking around in the back of my head when I argued that invasions might select for more virulent pathogens (Phillips and Puschendorf, 2013). My basic argument was that, on a pathogen's invasion front, I is always tiny compared to S; there are always a lot of susceptible individuals around. Thus, anything that increases the pathogen's β should be favoured. It is the same basic argument for why we expect growth rates to increase on invasion fronts (Section 3.2). But in this case, I argued, higher β might very often be associated with higher virulence: essentially, it doesn't matter to the pathogen how quickly it damages its host; the only thing that matters is rapid transmission.

It was a reasonable verbal model. It helped explain patterns we were seeing in the impact of chytrid fungus in frog populations over time (Phillips and Puschendorf, 2013). But it lacked generality. Happily (and once again), several groups were independently working on similar ideas. These two groups independently developed spatially explicit models of pathogen evolution. Surprisingly, they came to nearly complete opposite conclusions: on one hand, we were seeing the process of invasion increase virulence (Griette et al., 2015); on the other, we were seeing virulence decrease on the invasion front (Osnas et al., 2015). The reason for this discrepancy came down to their basic assumptions about trade-offs. The Osnas et al. model assumed a strong trade-off between infectiousness and movement: highly infectious pathogen strains damaged their host and so experienced lower rates of movement than less infectious strains. By contrast, the Griette et al. model ignored the effect of infectiousness on movement and instead focused on a trade-off between epidemic versus endemic fitness. Clearly, the evolution of virulence on invasion fronts is, like other life history traits, going to be very sensitive to the trade-offs involved (Section 7.2).

The basic reason that trade-offs will matter can be seen if we extend the SI model (equation 9.1, 9.2) into space. If we simply add a diffusion term to both of the SI equations and assume a single rate of diffusion, it turns out that the speed at which the invasion progresses is given as: $c = 2\sqrt{D(\beta S_0 - \gamma)}$, where S_0 is the density of susceptibles on the leading tip of the invasion (or the carrying capacity of disease-free hosts, if you prefer; Shigesada and Kawasaki (1997)).

If we take the basic idea that evolution on invasion fronts works to maximize the speed (Section 3.3), we can see that it has three parameters to play with here: D, β, and γ. The Griette model essentially imagined a trade-off between β and γ. The Osnas model assumed a trade-off between D and β. The truth is that we have three parameters in play and they all likely have some trade-off structure between them. For how many diseases do we know what these trade-offs look like?

9.5 Conclusions

What themes might we draw from this haphazard sample of case studies? The first, I think, is obvious: that invasion dynamics applies to a broad set of situations, from the spread of an allele to the growth of a tumour or the spread of an invasive species. The ideas explored in this book give us thought tools with which to approach each of these apparently different problems. Sometimes we might find purely ecological strategies for management (e.g., range-pinning SARS-CoV-2, chemotherapy for tumours, a waterless barrier for toads). But even when we execute a purely ecological manipulation, an evolutionary perspective matters: either because it gives us a sense of how the population might evolve in response to our management strategy (e.g., resistance to chemotherapeutants), or because it gives us additional tools with which we can strengthen the management outcome (e.g., the genetic backburn). Sometimes, an evolutionary perspective gives us ideas for management that we simply would never have dreamed of otherwise (e.g., mutational meltdown for stopping tumours).

A second theme that comes through is that density-dependent responses are really important. Pulled waves move like wildfire: they are hard to stop, they are unpredictable, and (entirely unlike wildfire) they evolve so as to accelerate over time. Pushed waves, by contrast, are much more manageable beasts. Perhaps the very first question we should be asking when confronted with an invasion is, 'what class of invasion is this: is it pushed, or pulled?' We should look very hard at how the rates of dispersal and population growth change with density, and whether we can manipulate that relationship to shift our invasion into a preferable class. WA was essentially free of SARS-CoV-2 for the first two years of that pandemic, and this was achieved by limiting dispersal and imposing an Allee effect. This combination of manipulations caused range-pinning, which is clearly a powerful method for containment. The basic principles behind range-pinning apply to any invasion. Where else might we be able to apply that idea? One obvious place is in the design of gene drives, but what about other invasions?

A third theme that emerges, I think, is that trade-offs matter. Will pathogens become more virulent as they spread? The answer depends critically on trade-offs between transmission, dispersal, and clearance rates. Will a tumour evolve to become metastatic? Once again, the answer depends to a large extent on the trade-offs that the cells in that tumour need to make. Will a genetic backburn work? Again, the answer depends to a large extent on trade-offs. We often know perilously little about trade-offs between r, K, and D in the organisms we seek to manage. Perhaps we should step the acquisition of this knowledge up a little higher on our list of priorities. Knowing this information gives us a good insight into what might happen under various management scenarios (including the do-nothing scenario).

A final theme I see is that invasion fronts are tricky things, and they often confound our intuition. Let's introduce a directly transmitted parasite to the invasion front to slow it down; that will work, right? Well, no, it won't. Let's introduce a gene drive onto the invasion front to make our invader less dangerous; that will work, right? Well, maybe, but probably not under any scenario you are going to like. Let's stop a pulled invasion by concentrating all our management resources on the leading tip of the invasion; that will work, right? Well, probably not (you'd probably be better off working out how to impose an Allee effect). If we develop a gene drive that produces males and rapidly runs to fixation in a well-mixed population, it will spread across space and eradicate the target population, right? Well, maybe; it depends on how much spatial drift you have to deal with. Faced with such counterintuitive outcomes, it is clear that models are a valuable tool. We don't all need to be mathematical modellers, however, to think intelligently about how an invasion may behave under various manipulations. We just need to understand the basic processes at play. It is my hope that this book has set those processes out in an approachable way; if we can improve 'process literacy' among managers, we will improve communication and management decisions.

I also hope that this book has imparted some sense of the very human process of scientific discovery. We are all individually fallible: mistakes are made; ideas rise and fall. But through iteration, and a community commitment to testing ideas against observation and experiment, clarity eventually emerges. We have seen this play out across the history of invasion biology, but quite intensely in the last few decades as an evolutionary perspective on invasions has rapidly emerged.

This evolutionary perspective has greatly deepened our understanding of invasions. Invasions are a fascinating, complex, and beautiful phenomenon, and they are worthy objects of study. But they also matter to us in many important ways. It is my hope that this book not only serves as a useful introduction, but that it also stimulates new ideas and new strategies for management. After all, invasions are everywhere, and they often carry consequences.

Appendix A
Diffusion confusion

$$\frac{\partial N(t,x)}{\partial t} = D\frac{\partial^2 N(t,x)}{\partial x^2}$$

This is Fisher's equation, with the growth part dropped off. It's just straight diffusion. Even with half of the equation missing, it's intimidating.

I used to stare at the diffusion term for what felt like hours, willing it to make sense to me. It never did. It just sat there, stubbornly defiant. I just couldn't get an intuitive grasp as to why a second derivative was needed. I remembered from calculus that a second derivative described the change in slope: but how on earth did the change in slope of N against x describe movement? I also read various derivations of diffusion, but most of them very quickly lost me in mathematical complexity: I wished I had done a maths or physics degree and had misspent my youth with equations, rather than with snakes.

A combination of two things eventually made it click. First, a colleague attempted to explain it to me at a conference. He got completely messed up with the algebra, but he explained the gist of it well, which was the main thing. And then I found an appendix entitled 'Diffusion for ecologists' in Peter Turchin's excellent book (1998), which presented the same argument, but with the algebra neatly laid out. The good news is that the diffusion term does make sense and it really does have a very intuitive meaning: it really is just the balance of comings and goings at a location.

A.1 Diffusion explained

We can build up an intuitive understanding of this as follows.

First, let's go back to our nice simple one-dimensional space, populated with one-dimensional cats. The density of cats along the string at any given time is given by $N(t,x)$. We might want to plot this density against x. We might get something like Figure A.1, or not. The shape of this function is irrelevant; I am just trying to give you a picture to work with.

Now, let's imagine that we admit a very small period of time (τ) to elapse in which our cats move. A proportion, m, of the cats are going to leave their current location and move either left or right along our string. Because we are considering only a very small instant of time, cats that are a long way from x are irrelevant: they can't possibly get to x in the tiny amount of time we afford. We only need to worry about the coming and going of cats from nearby locations on either side of x: at $x + \lambda$ and $x - \lambda$. We will now do a quick bit of accounting to work out how

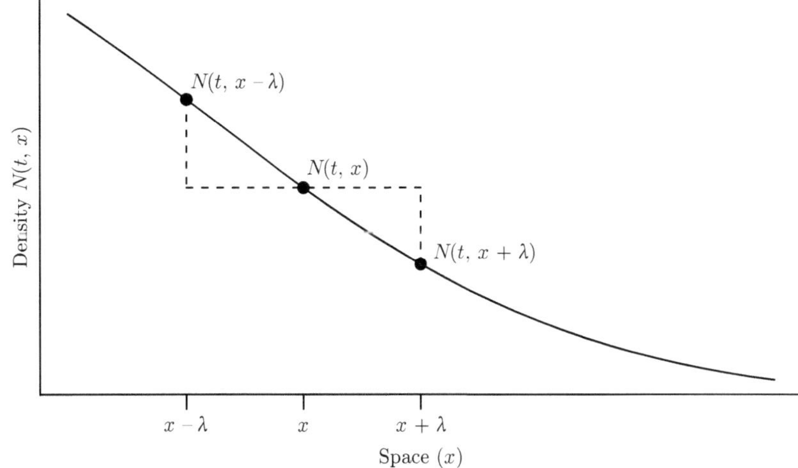

Figure A.1 *The density of cats along our string, and a few useful landmarks.*

the density of cats changes at an arbitrary location. We will describe the change in N at location x: $N(t + \tau, x) - N(t, x)$.

$N(t + \tau, x)$ is equal to the density already at x plus the arrivals minus the departures. The density already at x is $N(t, x)$; the density of cats arriving at x is $\frac{m}{2}N(t, x - \lambda) + \frac{m}{2}N(t, x + \lambda)$[1]; and the density of cats departing x is $mN(t, x)$. Putting this together, we get:

$$N(t + \tau, x) = N(t, x) + \frac{m}{2}N(t, x - \lambda) + \frac{m}{2}N(t, x + \lambda) - mN(t, x)$$

the number that were there initially, plus arrivals, minus departures. To express as a change in N, we subtract $N(t, x)$ from both sides, and rearranging we get:

$$N(t + \tau, x) - N(t, x) = \frac{m}{2}\left[\left(N(t, x + \lambda) - N(t, x)\right) - \left(N(t, x) - N(t, x - \lambda)\right)\right]$$

I have arranged this in a very particular way, so you can see what is coming. The parenthetical terms inside the square brackets are both differences in N across space. They are the vertical dashed lines in Figure A.1. Now we do a little mathematical sleight of hand. We multiply the right-hand side of our equation by $\frac{\lambda^2}{\lambda^2}$[2], and we divide both sides of our equation by τ.

$$\frac{N(t + \tau, x) - N(t, x)}{\tau} = \frac{m\lambda^2}{2\tau}\left[\frac{\frac{N(t, x + \lambda) - N(t, x)}{\lambda} - \frac{N(t, x) - N(t, x - \lambda)}{\lambda}}{\lambda}\right]. \quad \text{(A.1)}$$

The result is quite an eyeful, but it will all make sense in a moment. First, the left-hand side of equation (A.1) is simply a change in N per time. If τ is sufficiently small, then:

$$\frac{N(t + \tau, x) - N(t, x)}{\tau} \approx \frac{\partial N}{\partial t}.$$

[1] The proportion arriving at x from $x - \lambda$ is half of m, because half of the m cats at $x - \lambda$ went in the other direction, to $x - 2\lambda$.

[2] Which is just multiplying by one, so we haven't broken the truth of our equation.

On the right-hand side of equation (A.1), let's tackle the stuff in the big square brackets. The first fraction in the numerator is a change in N per unit of x. It is the slope of $N(t, x)$ just to the right of x. The second term is the slope just to the left of x. So the numerator on the right-hand side is a change in slope, and the denominator is expressing that change per unit of x. In other words, if we assume λ is sufficiently small, we have our mysterious second derivative:

$$\left[\frac{\frac{N(t, x+\lambda) - N(t, x)}{\lambda} - \frac{N(t, x) - N(t, x-\lambda)}{\lambda}}{\lambda} \right] \approx \frac{\partial^2 N}{\partial x^2}.$$

Of equation (A.1), this leaves only that innocent little term outside the square brackets, $\frac{m\lambda^2}{2\tau}$. This is a constant that is usually just denoted D, the 'diffusion coefficient'.

By substituting these various simplifications into equation (A.1), we have

$$\frac{\partial N(t, x)}{\partial t} = D \frac{\partial^2 N(t, x)}{\partial x^2}, \tag{A.2}$$

which brings us back to where we started: the first part of Fisher's reaction–diffusion model. Our derivation is a little fast and loose towards the end, but the aim is to develop an intuitive feeling for what diffusion is, rather than go deep on theory. Hopefully, it is a little less mysterious now.

A.2 Implications of diffusion

In 1889, Francis Galton designed a strange contraption, the Quincunx,[3] which consisted of a board with many pegs set into it, in offset rows (see Figure A.2).

[3] Also known as the "Galton Box", or "Bean Machine".

The purpose of this device was to demonstrate a simple principle: that order can emerge from randomness. If a small ball is dropped from the top into the centre of the pegs, it is very difficult to know where that ball will end up. The ball will hit the top row of pegs and bounce either left or right before falling through to the next row; when it hits the next row, it will bounce left or right before falling through to the next row; and so on. It is very hard to know into which slot it will finally fall: the ball follows a random path and the slot into which the ball finally comes to rest at the bottom of the Quincunx is a random outcome.

At this stage, you might have said to Galton, 'Frank, that seems rather obvious, and surely it didn't justify all the effort you put into building this funny thing.' To which he would have replied, 'Drop a thousand balls into it.' And that's when you would have seen the magic. Each of those balls follows its own haphazard trajectory, but at the bottom of the box, as the balls one-by-one fall into place, a beautiful bell-shaped curve emerges. The central slots always contain the most balls, and the number of balls in each slot falls away in a shapely arc as we move outwards from the centre.

The distribution of balls at the bottom of the Quincunx is an approximation of a normal distribution.[4] The central limit theorem tells us that the normal distribution emerges whenever we add lots of random events together, and we can

[4] It is an approximation, because we are dealing with discrete rows of pegs and discrete slots into which the balls could fall, but if we had many balls, many rows, and many slots the result would become indistinguishable from Normal

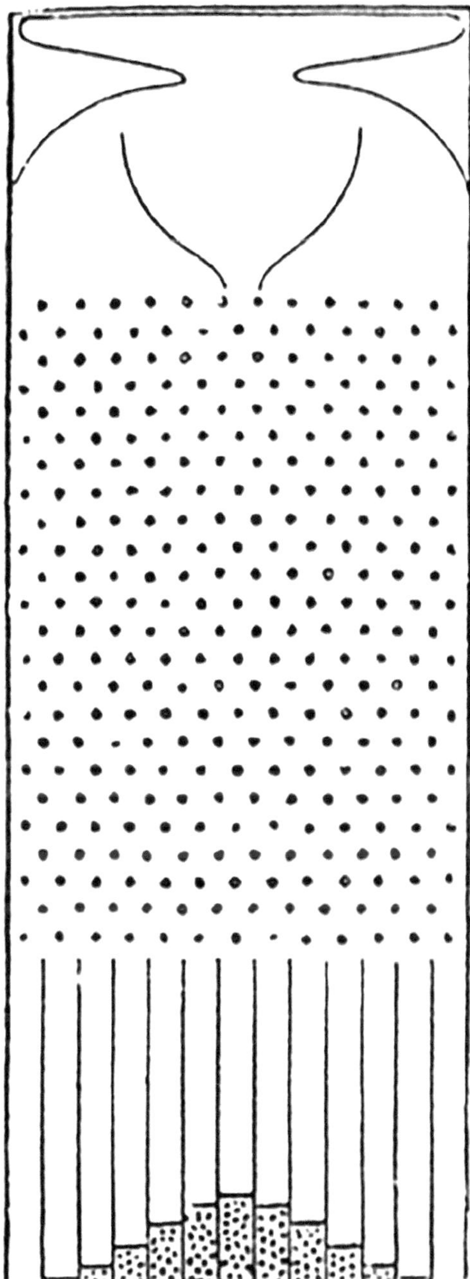

Figure A.2 *The original design of the Quincuncx. Small balls are dropped into the top, and they then bounce through all the pegs before falling into one of the slots at the bottom.*

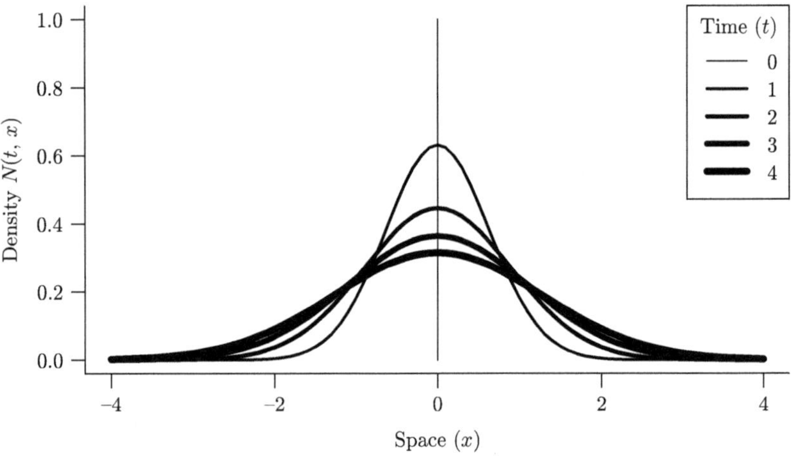

Figure A.3 *The diffusion of cats along a string as time progresses, with all cats initially being placed at x = 0.*

think of the Quincuncx as a kind of adding machine, in that each row of pegs adds another random event to a ball's location.

Why are we talking about Galton's strange box, and normal distributions? Well, diffusion, as we have seen, is a description of random movements in an infinitesimal of time. In our small interval of time, τ, some of our cats moved left; some of them right. When a ball encounters a row of pegs in Galton's box, it bounces either left or right. Each row of pegs in the Galton box can be thought of as an interval of time in our diffusion model. In Galton's box, all the balls start in the centre. If we started our diffusion model at $t = 0$ with all of our cats piled on top of each other at $x = 0$, we may well expect a normal distribution of cats to emerge as time progresses. This is, in fact, precisely what happens.

It transpires that if our initial density at $t = 0$ is all piled on $x = 0$, the solution to the diffusion equation (A.2) is

$$N(t, x) = \frac{N_0}{\sqrt{4\pi Dt}} e^{-\frac{x^2}{4Dt}},$$

which is the equation for a normal distribution (with mean $\mu = 0$ and variance $\sigma^2 = 2Dt$) multiplied by our initial density, N_0[5]. We can now watch how the density of our cats might spread out over time (Figure A.3).

[5] This little nugget was worked out by Albert Einstein in his annus mirabilis of 1905. It followed from observations by Robert Brown that small particles shed from pollen would move erratically in the water even though the particles were not actually alive. This became known as Brownian motion and was conjectured to be due to water molecules bashing the particles about in a relentless and utterly haphazard way. Most of the maths had already been worked out, but Einstein's important contribution was the insight that a complete physical description of Brownian motion could be used to infer the existence and mass of atoms. He was right, of course.

Appendix B
Probability distributions, random numbers, and simulation

B.1 Probability distributions

Probability theory imagines a world in which, for a given process, there is a complete set of possible outcomes called the sample space, Ω. For example, if you toss a coin (and assume it can't land on its edge), the sample space is heads or tails: $\Omega = \{H, T\}$. That is a complete list of all the possibilities that might eventuate. If you enter the running of the bulls at Pamploma, the proportion of your body that will sustain bruising is somewhere between 0 and 1: $\Omega = [0, 1]$; again, no other outcomes are possible. The number of photons hitting your retina right now will be a positive integer: $\Omega = \{0, 1, 2, \ldots\}$ (negative and partial photons are not possible). And so on.

B.1.1 Discrete probability distributions

A probability distribution describes how the probability of a random outcome, X, is distributed across the sample space. Probability distributions are subtle and astonishing things, but let's start with the simplest. For a fair coin, the probability distribution is $P(X = H) = 0.5; P(X = T) = 0.5$. That's it. The coin example is very straightforward, but note that (a) there is a probability associated with every possible outcome in the sample space (heads, or tails), and (b) the sum of probabilities across the sample space is 1. These are very firm rules: a probability distribution must account for the whole space, and sum to one.

We might generalize the idea of a coin toss to a 'Bernoulli trial', which is essentially a coin toss with an arbitrary probability, p, of turning up heads. Bernoulli trials can be used to describe the many situations in which the outcome is binary: you will be hit by space junk in the next 10 minutes, or not; the fish you just caught is female, or not—that kind of thing. The Bernoulli distribution can be described with a function as

$$f(p) = P(X = k \mid p) = \begin{cases} 1 - p, & k = 0. \\ p, & k = 1. \end{cases} \tag{B.1}$$

This should be read as 'the probability that the outcome is k given the parameter, p, is either p or $1 - p$'. Note that all possible outcomes are accounted, and $p + 1 - p = 1$. The coin example (and its generalization) is easy to imagine, because the outcomes are few, and discrete (there are only heads/tails, success/fail, 0/1, possible). Things get more complicated from here.

The next simplest distribution to imagine is the binomial distribution. This is another important one, and describes the number of successes you might see from a given number, n, of Bernoulli trials. So the number of heads we might see if we tossed a coin $n = 5$ times; the number of people in Australia that will be hit by space junk in the next 10 minutes; the number of female fish you caught when you caught three fish; and so on.

What is the probability that we get four heads from five coin tosses? Let's take the five coin tosses ($n = 5$), assume we have a fair coin (so $p = 0.5$), and look at all the events that will lead us to score four heads ($k = 4$). The events giving $k = 4$ are:

$$\text{A:HHHHT} \quad \text{B:HHHTH} \quad \text{C:HHTHH} \quad \text{D:HTHHH} \quad \text{E:THHHH}$$

There are five of them, but all of them involve getting four heads, and one tail. The probability of getting the first combination $P(A) = p^4(1-p)$; the probability of getting the second combination $P(B) = p^4(1-p)$; the third $P(C) = p^4(1-p)$; and so on. Each of these combinations has the same probability of happening. So the probability of seeing one of these combinations is the probability of A or B or C or D or E, which is simply the sum of probabilities over all five events: $5 \times p^4(1-p)$. In general (arbitrary values of n and k), we get the binomial distribution:

$$f(p, n) = P(X = k \mid p, n) = \binom{n}{k} p^k (1 - p)^{n-k}. \tag{B.2}$$

The binomial coefficient, $\binom{n}{k}$, is a combinatorial function that yields the number of combinations possible when we choose k things from a set of n things (disregarding the order, each thing is drawn): $\binom{n}{k} = \frac{n!}{k!(n-k)!}$. The binomial distribution given in equation (B.2) is quite an elegant thing to look at when it is plotted across the sample space (Figure B.1). Each of those points in Figure B.1 describes the probability of that number of 'successes' happening given 10 trials and a probability p of success in each trial. As with the Bernoulli distribution, if you sum those probabilities over all possible k, the answer is 1.

Hopefully, this starts to give you the idea of how probability can be distributed across the sample space. To bring it back to coin tosses again, the series in Figure B.1 with $p = 0.5$ is equivalent to tossing a fair coin 10 times. We can see immediately that five heads is the most likely outcome, but there is also a pretty good chance of getting either four or six heads. By contrast, getting either 0 or 10 heads from 10 coin tosses has a very low probability of occurring. These results mesh perfectly with intuition, but we have built the function up from scratch using basic probability. It is important to appreciate that probability distributions are often process models: they describe how probability is distributed across sample space under some process or another. People often make the mistake (and I have made it in the past) of using a particular distribution because it seems to fit the data;

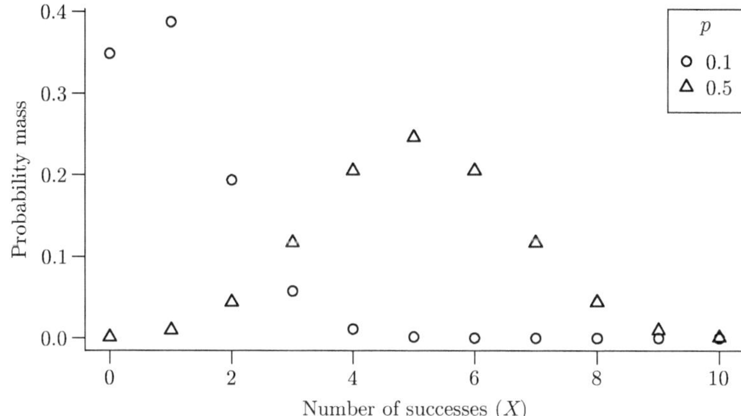

Figure B.1 *The binomial distribution with n = 10, compared across two values of p.*

instead, we should use distributions that fit the process by which we imagine the data came into being.

There are several other important discrete probability distributions, each describing a probability distribution arising from a different process. The functions that describe discrete probability distributions are called 'probability mass functions', because all of the probability falls as a point mass onto each integer. The height of the points on a probability mass function (PMF) equates perfectly with the probability of that outcome occurring.

B.1.2 Continuous probability distributions

We will not generate an exhaustive list of discrete probability distributions. Let us turn instead to continuous sample spaces. Here, an outcome could land anywhere on the number line, not just on an integer. These distributions are a little harder to get our heads around, but they obey the same rules: they describe probability across all of the sample space, and when we sum probabilities across all sample spaces, we get 1. The simplest of these is the uniform distribution, and the most famous is the normal distribution. Let's get the boggling bit out of the way, and start with the uniform.

The uniform distribution has a sample space encompassing the entire number line, but is defined entirely by the upper u and lower l bounds of an interval on the number line. The uniform distribution is:

$$f(l, u) = P(X = x \mid l, u) = \begin{cases} \frac{1}{u-l} & \text{for} \quad l \leq x \leq u, \\ 0 & \text{otherwise} \end{cases} \tag{B.3}$$

which is a fancy way of saying that it is a perfectly flat line at a height of $\frac{1}{u-l}$ running between l and u. It looks really boring. Figure B.2 shows an example of the uniform distribution with $l = 1$ and $u = 3$. It shows us visually that there

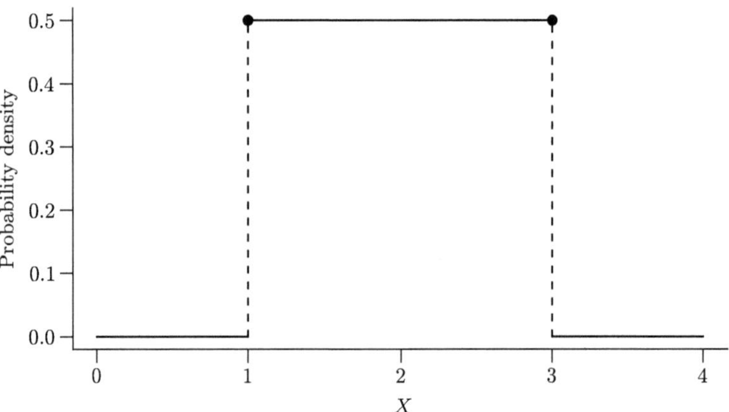

Figure B.2 *The uniform distribution between 1 and 3 on the number line.*

is absolutely no chance of an event happening outside of the interval between l and u. We will never see the event $x = 0$, for example.

But what about inside that interval? What is the probability that we will see the $x = 2.35$? If we were dealing with a PMF, we could answer this question by going to the number in question on the x-axis and reading off the corresponding value for the function on the y-axis. In this case, if we go to $X = 2.35$, we find that the 'probability density' is 0.5. But that can't be the correct probability that $x = 2.35$. It implies that half of our observations would be 2.35, which is clearly incorrect. So what's going on? You will notice that between Figures B.1 and B.2, I have changed the vertical axis label from 'Probability mass' to 'Probability density'. There is a very good reason for that, though it is slightly boggling.

There are actually an infinite number of possible numbers in the interval between 1 and 3 on the number line. In fact, pick any interval on the number line and you will be able to find an infinite number of numbers between its upper and lower bounds. The simplest way to see that this is true is that were I to write it out to full precision, the number 2.35 would actually need an infinite number of trailing zeros: 2.350000. ... Change one of those zeros to a 1, and we have a different number, albeit one that may be infinitesimally different from 2.35. Given that $x = 2.35$ is only one of an infinite number of possibilities within the interval, the probability of getting $x = 2.35$ (or any other number) is, in fact, zero.

To get around this awkward issue of dividing by infinity, we need to stop thinking about precise numbers (like we did with discrete distributions) and think about intervals, and about the probability of seeing an outcome falling within a given interval. When we do this, we see that the term 'Probability density' (used on the y-axis of Figure B.2) makes sense: it is a probability per unit of space along the number line; a probability within an interval.

The uniform distribution, being perfectly flat, says that an event can happen anywhere between l and u with equal probability. Looking at Figure B.2, we can see that the probability of an event occurring in the interval [1, 2] should be the same as the probability of an event occurring in the interval [2, 3], because these sub-intervals are the same size. These sub-intervals also make up the entire interval of the uniform distribution, so $P(X \in [1, 2]) + P(X \in [2, 3]) = \frac{1}{2} + \frac{1}{2} = 1$. Now, let us look at the area under the curve, within each of these sub-intervals. For each sub-interval, we have a simple rectangle of width 1 and height $\frac{1}{2}$. So the area under the curve within each sub-interval is $\frac{1}{2}$, which is the same as the probability that an event will occur in that sub-interval.

The fact that the area under the curve and the probability are the same is not a coincidence. We have a probability density (probability per unit of space, the height of the curve), and we have multiplied it by a unit of space (the length of our sub-interval), leaving us with a probability. Continuous sample spaces require 'probability density functions' rather than 'probability mass functions', and the probability actually lives in the area under the curve, rather than the height of the curve.

We could, in fact, look at a smaller division of the interval: [1, 1.5], [1.5, 2], [2, 2.5], [2.5, 3]. Again, these probabilities ($4 \times \frac{1}{4}$) and areas would be the same and would sum to 1. We could, in fact, keep making the intervals smaller and smaller. The number of rectangles would get larger and larger, but their area (and probabilities) would get smaller, but always sum to 1. Taken to the limit of an infinite number of infinitely small sub-intervals, all added together, we end up with the integral. The integral is just a sum of many tiny parts;[1] it represents the area under a curve, and the area under the curve of a continuous probability distribution must sum to 1.

So, for a PMF, the sum over the sample space must equal 1.

$$\sum_{k} f(X = k) = 1.$$

For a probability density function (PDF), the integral over the sample space must equal 1:

$$\int f(X = x)\,dx = 1.$$

The above equations describe the result when we sum/integrate across the entire sample space. But it is also very useful, as we will see shortly, for us to sum over subsets of sample space. If we start at the minimum value of our sample space and sum forward from there, the resulting functions are called cumulative distribution functions, because they describe how probability accumulates as we move through the sample space. We can generate cumulative mass functions (CMFs), and cumulative density functions (CDFs), $F()$, for the discrete and continuous cases, respectively. A CMF/CDF reports the probability that an outcome happens below some value: $F(x) = P(X \leq x)$.[2]

So a CMF is

$$F(k) = \sum_{\min(X)}^{k} f(X = k),$$

----

[1] It is very often very hepful to remember that the integral is just a sum.

[2] You will note that at this stage I am using the convention of k to denote an outcome in a discrete probability space, and x to denote an outcome in a continuous space

and a CDF is

$$F(x) = \int_{\min(X)}^{x} f(X = x)\,dx.$$

The cumulative distribution functions have a minimum value of 0 and a maximum (after summing/integrating over the entire sample space) of 1. They become very handy when we go to generate random samples from a given distribution. But before we can do that, a quick aside on random numbers.

B.2 Random numbers

Random numbers are very useful things. Without random numbers, there would be no secure transactions on the internet; we wouldn't be able to do robust experimental design; many statistical analyses would be impossible; and stochastic simulation wouldn't be a thing. Random numbers are useful, but it is very difficult to generate truly random numbers: most procedures that people think up end up having some bias in them and so, in the long run, allow patterns to develop. True randomness is an absence of pattern; it is the whitest of white noise. Happily, unless you are very serious about cryptography, the 'pseudo-random' numbers that computers can generate are, for most intents and purposes, close enough to random.[3] Certainly, for our modelling needs, pseudorandom numbers are perfectly excellent.

In most modelling applications, we will want to draw random numbers from specified probability distributions. We are using random numbers to sample, in an unbiased way, these probability distributions. It turns out that if you can generate random numbers in the interval between 0 and 1 on the number line, you can sample any probability distribution that you care to dream up.

B.3 Sampling from specified distributions

Let's assume that we have some process that gives us a random number between 0 and 1. Technically, this is a sample from the uniform distribution (equation (B.3)). Let's call this sample, y, so, $y \sim \text{Unif}(l = 0, u = 1)$.[4] We know that y must fall somewhere between 0 and 1, which is also the domain within which probability makes sense. The domain $[0, 1]$ is also the domain across which cumulative distribution functions are defined. These functions describe how probability accumulates as we move across the sample space, so they are naturally bounded within $[0, 1]$.

We want this random number, y, to actually be a sample from some other probability distribution, $f(x)$. We want our random number, y, to turn up in our sample space, $X \in \Omega$. To have this happen, we simply project y through our cumulative distribution function, $F(x)$. To do this, we simply set $F(x) = y$ and then solve for x to find the value of x that would yield the cumulative probability given by y. This is a little bit of magic: you can turn a uniform distribution into any other distribution in this way. The results are best seen graphically. In Figure B.3,

[3] Pseudorandom number generators on computers are deterministic algorithms: if you start them with the same number – the "seed" – you will get the same string of random numbers. For low-security applications, computers will, at the time a random number is needed, usually get the highest precision time on the computer's clock, and use this as the seed.

[4] This notation, $\sim$, says that "y is drawn from a uniform distribution between 0 and 1".

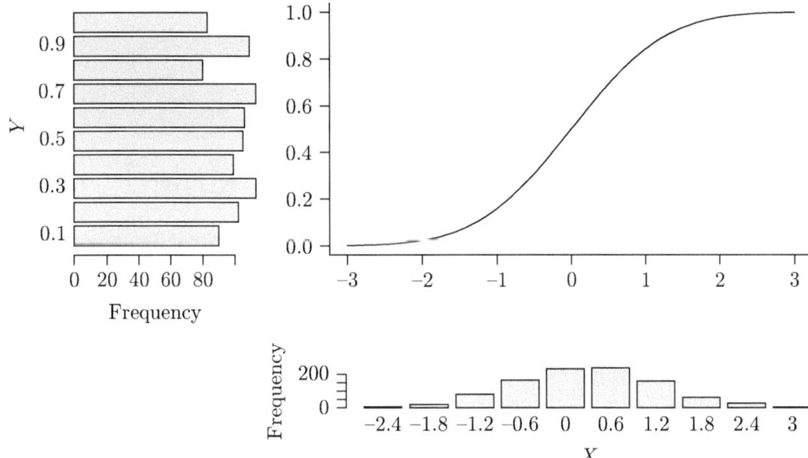

Figure B.3 *An example of how to take samples from a specified probability distribution. Here, we generate 1,000 random numbers, y, uniformly distributed between 0 and 1. We then project these numbers through a cumulative distribution function (in this case, the CDF of the standard normal distribution), to have them arrive as x values in the sample space of the distribution function.*

we take 100 random numbers between 0 and 1, and we project them through the CDF for the standard normal distribution. Our sample of ys goes from looking like a flat distribution to the characteristic bell shape of a normal distribution. You can sample from any conceivable probability distribution in this way.

Appendix C
Natural invasions with documented trait shifts

Invasive populations are exceedingly common, and there is a long ecological tradition of studying invasive species in the field. More recently, as concerns have grown around the impact of climate change, ecologists have also begun documenting invasions driven by climatic shift; species shifting their ranges. Strangely, very few of these studies had looked for evidence of evolution along the invasion axis. The striking results from the toad invasion threw a light onto the possibility of evolution in invasive populations, and there was a rapid effort, across the world, to document trait shifts between core and invasion front populations. Table C.1 documents natural invasions by plants and animals in which trait shifts have been looked for and (mostly) observed. This table builds off the table put together by Angela Chuang and Chris Peterson (2016) with updates to about 2023, and some minor additions and amendments.

Table C.1 *Natural invasions in which trait shifts have been observed between core and invasion front populations.*

Species name	Ecological driver	Trait category	Trait change in edge populations	Reference
Fungi				
Seiridum cardinale	Invasion	Life history	Increased growth rate, smaller spore size	Garbelotto (2015)
Plants				
Bitter vine (*Mikania micrantha*)	Invasion	Morphology	Decreased plume loading	Huang (2015)
			Decreased seed mass	
South African ragwort (*Senecio inaequidens*)	Invasion	Morphology	Larger pappus in diaspores	Monty (2010)
Madagascar ragwort (*Senecio madagascariensis*)	Invasion	Morphology	Greater pappus volume	Bartle (2013)
		Metabolism	Higher germination rates	
Japanese honeysuckle (*Lonicera japonica*)	Invasion	Morphology	More branches	Kilkenny (2013)
			Greater biomass	
		Life history	Higher survivorship	
Purple loosestrife (*Lythrum salicaria*)	Invasion	Reproduction	Smaller flowers	Colautti (2013)
			Earlier flowering	
Lodgepole pine (*Pinus contorta*)	Natural colonization	Morphology	Decreased seed wing loading	Cwynar (1987)
Vertebrate animals				
Fish				
Vendace (*Coregonus albula*)	Invasion	Morphology	Decreased body length	Amundsen (2012)
			Decreased female gonad length	Bohn (2004)
		Metabolism	Faster growth rate	
		Reproduction	Lower age of sexual maturity	Amundsen (2012)
			Decreased fecundity	

continued

Table C.1 *continued*

Species name	Ecological driver	Trait category	Trait change in edge populations	Reference
African jewelfish (*Hemichromis letourneuxi*)	Invasion	Forage	Greater gut fullness	Lopez (2012)
		Morphology	Better body condition	
			Higher gonadosomatic index	
Round goby (*Neogobius melanostomus*)	Invasion	Forage	Greater gut fullness	Raby (2010)
		Morphology	Increased male length	Gutowsky (2011)
			Increased female length	
		Metabolism	Higher resting metabolic rate	Myles-Gonzalez (2015)
		Reproduction	Lower age to sexual maturity	Macinnis (2000)
				Gutowsky (2012)
		Behaviour	Higher gonadosomatic index	Gutowsky (2012)
			More aggressive	Groen (2012)
			More dispersive	Myles-Gonzalez (2015)
			Bolder	
European catfish (*Siluris glanis*)	Invasion	Morphology	Decreased body length	Carolbruguera (2009)
			Better body condition	
			Faster growth rate	
Amphibians				
Cane toad (*Rhinella marina*)	Invasion	Behaviour	Faster range expansion rate	Phillips (2006)
			Higher movement rate	Brown (2007)
			Increased tendency to move	Alford (2009)

Table C.1 *continued*

			Greater movement endurance	Llewelyn (2010)
			Sickness behaviour	Brown (2018)
		Morphology	Increased female and male tibia length	Phillips (2006)
			Increased body length	Brown (2013)
		Metabolism	Faster tadpole growth rate	Phillips (2009)
		Physiology	More neutrophils	Brown (2015)
			Higher phagocytosis rates	
			Higher spinal arthritis frequency	Brown (2007)
Asian common toad (*Duttaphrynus melanostictus*)	Invasion	Morphology	No difference after 10 generations of spread	Licata (2021)
African clawed frog (*Xenopus laevis*)	Invasion	Morphology	Smaller body size; larger leg length	Louppe (2017)
		Behaviour	Increased stamina and movement distance	
Green treefrog (*Hyla cinerea*)	Climate change	Morphology	Increased relative leg length	Edwards (2023)
Birds				
Common waxbill (*Estrilda astrild*)	Invasion	Morphology	Redder breast pigmentation saturation in males	Cardoso (2014)
			Reduced red face mask area in females	
Icelandic black-tailed godwit (*Limosa limosa islandica*)	Natural colonization	Morphology	Longer male bill length	Gunnarsson (2012)
			Higher male bill/wing ratio	

continued

Table C.1 *continued*

Species name	Ecological driver	Trait category	Trait change in edge populations	Reference
House sparrow (*Passer domesticus*)	Invasion	Behaviour	More exploratory of novel habitats	Liebl (2012)
			Faster consumption of novel foods	
		Physiology	Greater corticosterone response to stressors during breeding season	Liebl (2014)
			Reduced expression of stress hormone receptors	Liebl (2014)
Western bluebird (*Sialia mexicana*)	Natural colonization	Behaviour	Higher aggression score	Duckworth (2006)
			Lower social tendency	
Common myna (*Acridotheres tristis*)	Invasion	Behaviour	Higher movement rates	Burstal (2020)
	Invasion	Morphology	Larger wing length and higher wing loading	Berthouly-Salazar (2012)
Mammalia				
Eurasian beaver (*Castor fiber*)	Natural colonization	Life history	Younger age at dispersal	Hartman (1997)
North American red squirrel (*Tamiasciurus hudsonicus*)	Natural colonization	Morphology	Larger cranial features	Goheen (2003)
Human (Laryngeal tumour)	Invasion	Physiology	Malignant proliferative cells on expanding edge	Song (2020)
Human (Breast tumour)	Invasion	Physiology	Proliferative cells on expanding edge	Lewis (2016)
Human (Renal tumour)	Invasion	Physiology	Proliferative cells on expanding edge	Hoefflin (2016)
Human (Renal tumour)	Invasion	Physiology	Highly dispersive clones originating in tumour centre	Zhao (2021)
Humans	Invasion	Life history	Larger family sizes	Moreau (2011)

Table C.1 *continued*

Voles (*Microtus agrestis*)	Metapopulation	Morphology	Larger body size and relative leg length	Forsman (2011)
Bank voles (*Myodes glareolus*)	Invasion	Behaviour	Greater risk aversion and faster habituation	(Eccard 2023; Mazza 2023)
Reptilia				
Brown anole (*Anolis sagrei*)	Invasion	Physiology	Decreased critical thermal minimum	Kolbe (2014)
Invertebrate animals				
Crustacea				
Signal crayfish (*Pacifastacus leniusculus*)	Invasion	Morphology	Decreased male length	Hudina (2012)
			Better body condition	Rebrina (2015)
		Physiology	Better female hepatopancreatic state	
		Reproduction	Better female gonadal state	
Mud fiddler crab (*Uca pugnax*)	Climate change	Metabolism	Faster developmental rate	Sanford (2006)
Echinoidea				
Longspine sea urchin (*Centrostephanus rodgersii*)	Climate change	Reproduction	Greater gonadosomatic index	Ling (2008)
Gastropoda				
Angular unicorn (*Acanthinucella spirata*)	Climate change	Morphology	Differences in relative height of shell spire	Hellberg (2001)
Insecta				
Hemlock woody adelgid (*Adelges tsugae*)	Climate change	Physiology	Greater cold resistance	Butin (2005)
Banded demoiselle damselfly (*Calopteryx splendens*)	Climate change	Morphology	Higher wing aspect ratio	Hassall (2009)

continued

Table C.1 *continued*

Species name	Ecological driver	Trait category	Trait change in edge populations	Reference
			Greater flight ability in males	
Azure damselfly (*Coenagrion puella*)	Climate change	Morphology	Longer postnodal section of the wing	Hassall (2008)
Dainty damselfly (*Coenagrion scitulum*)	Climate change	Behaviour	Increased juvenile activity	Therry (2014)
		Morphology	Increased flight muscle ratio	Therry (2014a)
			Decreased wing loading	Therry (2014b)
			Lower body fat	
			Lower body mass	
		Metabolism	Faster growth rate	
			Decreased female larval development rate	
			Decreased male larval development rate	
		Physiology	Stronger encapsulation (immune) response	
Red-eyed damselfly (*Erythromma najas*)	Climate change	Morphology	Higher wing aspect ratio in males	Hassall (2009)
Multicolored Asian lady beetle (*Harmonia axyridis*)	Invasion	Metabolism	Faster flight speed	Lombaert (2014)
Colorado potato beetle (*Leptinotarsa decemlineata*)	Invasion	Metabolism	Faster developmental rate	Lyytinen (2009)
Argentine ant (*Linepithema humile*)	Invasion	Morphology	Lower queen/worker thorax volume ratio	Abril (2013)
Ground beetle (*Merizodus soledadinus*)	Invasion	Morphology	Greater thorax length	Laparie (2013)
			Greater thorax width	
			Greater femur length	

Table C.1 *continued*

			Greater interocular width	
			Greater elytron length	
			Greater body mass	
Glanville fritillary butterfly (*Melitaea cinxia*)	Natural colonization Metabolism		Higher flight metabolic rate	Haag (2005)
Speckled wood butterfly (*Pararge aegeria aegeria*)	Climate change	Morphology	Greater female and male thorax size	(Hill1999; Hughes 2003)
			Greater female and male mass	
			Greater female and male abdomen	
		Reproduction	Larger clutch sizes	
Comma (*Polygonia c-album*)	Climate change	Morphology	Increased male thorax breadth	Braschler (2007)
Monarch butterfly (*Danaus plexippus*)	Colonization/ migration	Morphology	Larger relative wings in natural colonizers and migrating populations	Freedman (2020)
Kudzu bug (*Megacopta cribraria*)	Invasion	Morphology	Larger wings and body size on invasion front	Lovejoy (2021)
Northern tamarisk beetle (*Diorhabda carinulata*)	Invasion	Life history	Higher reproductive rate	Clark (2022)
		Dispersal	Increased dispersal of unmated beetles at low density	
Bush crickets (*Conocephalus* spp. and *Metrioptera* spp.)	Climate change	Morphology	Higher proportions of long-winged individuals	Simmons (2004)
Pygmy grasshopper (*Tetrix subulata*)	Metapopulation	Morphology	Higher proportion of long-winged individuals in disturbed habitats	Berggren (2012)

continued

Table C.1 *continued*

Species name	Ecological driver	Trait category	Trait change in edge populations	Reference
Red-shouldered soapberry bug (*Jadera haematoloma*)	Metapopulation	Morphology	Higher proportion of long-winged individuals in disturbed habitats	Comerford (2023)
Arachnids				
Wasp spider (*Argiope bruennichi*)	Climate change	Reproduction	Higher reproduction on expanding edge	Wolz (2020)
		Behaviour	No evidence of increased ballooning behaviour	
Orb spider (*Cyrtophora citricola*)	Invasion	Behaviour	Higher dispersal at one invasion front, but not at another	Chuang (2022)
Nematoda				
Nematode lungworm (*Rhabdias pseudosphaerocephala*)	Natural colonization	Life history	Lower age of sexual maturity	Kelehear (2012)
		Morphology	Larger female body size	
			Larger larvae body size	
		Life history	Higher survival of larva	Mayer (2021)

Appendix D
Experimental invasions

Invasions in nature are subject to all sorts of chance and contingency. To get rigorous, then, scientists around the world have developed systems in which we can run replicate invasions in a laboratory setting (Table D.1). These experimental invasions have allowed rigorous tests of theory, from basic spread theory through to more complex models incorporating stochasticity and evolution.

Table D.1 *Experimental invasions developed to test key theoretical predictions.*

Organism	Species	Treatments	Space	Recombination	Notes	Reference
Bacteria	*Escherichia coli*	None	2D	N	Two fluorescently labelled strains on a Petri dish show sectoring	Hallatschek et al. (2007)
Flour beetles	*Tribolium castaneum*	None	1D	Y	Replicate invasions to test predictions about expansion variability	Melbourne and Hastings (2009)
Yeast	*Saccharomyces cerevisiae*	None	2D	Y	Two fluorescently labelled strains on a Petri dish show sectoring	Hallatschek and Nelson (2010)
Bacteria	*Pseudomonas aeruginosa*	None	2D	N	Invasions on a Petri dish, demonstration of a trade-off between competition and expansion	van Ditmarsch et al. (2013)
Protist	*Tetrahymena* sp	None	1D	Y	To test a model that estimates demographic variance	Giometto et al. (2014)
Protist	*Tetrahymena* sp	None	1D	Y	Experimental evolution showing increased dispersal	Fronhofer and Altermatt (2015)
Yeast	*Saccharomyces cerevisiae*	Glucose: sucrose ratio	1D	Y	Used sucrose to impose an Allee effect and demonstrate a pushed wave and other theoretical predictions	Gandhi (2016)
Plant	*Arabidopsis thaliana*	Evolving vs. non-evolving; continuous vs. fragmented habitat	1D	Y	Tested evolution of dispersal and effect of habitat fragmentation	Williams et al. (2016a)
Wasps	*Trichogramma chilonis*	Varying patch connectivity	2D	Y	Showed that too much connectivity could cause invasions to fail through Allee effects.	Morel-Journel et al. (2016)
Bacteria	*Escherichia coli*	Well-mixed vs. colonies on plates	2D	N	Mutants arriving early or on the invasion front were favoured	Fusco (2016)

Table D.1 *continued*

Bacteria	*Escherichia coli*	High mutation vs. low mutation; spatial vs. well-mixed	1D/2D	N	Showed accumulation of deleterious mutations on invasion fronts	Bosshard et al. (2017)
Bean beetles	*Callosobruchus maculatus*	Control vs. spatially re-assorted	1D	Y	Tested spatial sorting, showed it increased dispersal and variability of invasions	Ochocki and Miller (2017)
Flour beetles	*Tribolium castaneum*	Evolving vs. non-evolving; continuous vs. fragmented habitat	1D	Y	Tested adaptation to novel conditions and effect on spread rate	Szűcs et al., (2017)
Flour beetles	*Tribolium castaneum*	Control vs. spatially re-assorted	1D	Y	Tested spatial sorting, showed it increased dispersal and variability of invasions	Weiss-Lehman et al. (2017)
Protist	*Tetrahymena sp.*	None	1D	Y	Found increased dispersal and shift to negative density-dependent dispersal on invasion front	Fronhofer et al. (2017)
Bacteria	*Escherichia coli*	None	2D	N	Increased dispersal; strong trade-off between dispersal and growth rate	Ni et al. (2017)
Bacteria	*Escherichia coli*	High/low nutrient media	2D	N	Evolution along a growth-dispersal trade-off; trajectory depended on environment	Fraebel et al. (2017)
Spider mite	*Tetranychus urticae*	Control vs. shuffled and high vs. low genetic variance	1D	Y	Faster range expansions in control replicates; no difference between high and low genetic variance, which they assign to condition-dependent dispersal (kin competition)	Van Petegem et al. (2018)

continued

Table D.1 *continued*

Organism	Species	Treatments	Space	Recombination	Notes	Reference
Wasps	*Trichogramma chilonis*	High and low carrying capacity	1D	Y	Faster range expansion in higher carrying capacity populations	Haond et al. (2018)
Plant	*Arabidopsis thaliana*	Varied connectivity and density dependence	1D	Y	Density-dependent growth and landscape patchiness both slow dispersal speed	Williams and Levine (2018)
Plant	*Arabidopsis thaliana*	Stressed vs not	1D	Y	Larger seed size and plant height evolved, as did heat tolerance and silique production	Lustenhouwer et al. (2018)
Bacteria	*Escherichia coli*	Random sample of front vs. furthest forward	2D	N	Evolution of dispersal and expansion load: random sampling favoured expansion load	Bosshard et al. (2019)
Bacteria	*Pseudomonas aeruginosa*	Transplant experiments	2D	N	Showed that low velocity mutants could not take over an invasion front and the reverse was true only if the higher velocity mutant arises on or near the invasion front	Deforet et al. (2019)
Yeast	*Saccharomyces cerevisiae*	Galactose vs. glucose vs. sucrose	1D	Y	Demonstrate diversity is preserved on invasion fronts with pushed dynamics (sucrose)	Gandhi et al. (2019)
Bacteria	*Escherichia coli*	Rough agar vs. smooth agar	2D	N	Spatially enhanced drift can overwhelm selection	Gralka and Hallatschek (2019)
Bean beetles	*Callosobruchus maculatus*	Control vs. spatially re-assorted	1D	Y	Showed that dispersal-demography trait correlations affected invasion dynamics	Ochocki et al. (2019)
Bacteria	*Escherichia coli*	Genotyping through invasion history	2D	N	Showed that beneficial mutations primarily accumulate early; deleterious mutations slow down the invasion	Bosshard et al. (2020)

Table D.1 *continued*

Virus	*Bacteriophage T7*	Phages invading susceptible bacteria	2D	N	Bacteria represent a heterogeneous environment that can cause phages to spread with a pushed wave	Hunter et al. (2021)
Bacteria	*Escherichia coli*	Multiple gene deletion lines	2D	N	Demographic noise has a large genetic component and so can evolve	Yu et al. (2021)
Human	Breast cancer cells	Two cell lines that vary in competitive ability	3D	N	Pioneering cell line dominated the invasion front under normal conditions; reversed under high pH	Freischel et al. (2021)
Human	Prostate cancer lines	Mixed populations grown as a spheroid	3D	N	Differentiation of lines between invasion front and core	Paczkowski et al. (2021)
Bacteria	*Raoultella planticola*	Two strains in competition	2D	N	Dented fronts emerge as a result of a trade-off between competition and invasion speed	Lee et al. (2022)
Wasps	*Trichogramma chilonis*	High and low connectivity	1D	Y	Pushed dynamics resulting from reducing the connectivity of patches	Dahirel et al. (2021)
Protist	*Tetrahymena* sp	Uniform environment vs. gradient in pH	1D	Y	Identify genes under selection during invasion; genetic swamping slows adaptation	Moerman et al. (2022)
Wasps	*Trichogramma chilonis*	High and low carrying capacity; landscape heterogeneity	1D	Y	Range pinning can occur as a consequence of density-dependent dispersal	Morel-Journel et al. (2022)
Beetles	*Tribolium castaneum*	Spatially sorted (disperser and non-disperser) and control	1D	Y	Higher dispersal rates in disperser lines; sex bias in phenotypic change	Arnold (2023)
Wasps	*Trichogramma chilonis*	High and low connectivity	1D	y	Evolution of density-dependent dispersal following range expansion	Dahirel et al. (2023)

Acknowledgments

What drives someone to write a book? We need motives, of course, because writing a book is a lot of work. We need good motives because a book always requires more than just the author's time: there is the reader's time to think of, but also the time of all those people who support the author in one way or another. Here's my thanks to those people.

For encouragement to start, I thank Craig Moritz, who also challenged me to be clear about my argument from the outset. (For the record, I wasn't clear at the outset, but various lines of argument have nonetheless emerged.) For encouragement to finish, I thank Stuart Baird, who rather heroically read the entire first draft, kindly pointed out (at least some of) my errors, and whose capacity to go rapidly between whimsy and deep theoretical insight is an endless source of both amusement and inspiration. I always learn a lot from conversations with Stuart, despite the fun.

On the logistics, fellowships from the Australian Research Council and the Western Australian government gave me the time needed to pull this together. In 2020, my bosses at the University of Melbourne gave me their blessing to move to Western Australia 'for a year, to write a book'. So, to Ute Roessner and Steve Swearer, thanks for helping me carve out the time, and sorry I didn't come back.

I have had all manner of wonderful discussions and correspondences about invasions with various other excellent people over the years. For these influences on the book, thanks to Rick Shine, Greg Brown, Michael Kearney, Justin Travis, Olivia Burton, Alex Perkins, Mark Urban, Crystal Kelehear, Matt Greenlees, Georgia Ward-Fear, Reid Tingley, Tom Lindstrom, Lee-Anne Rollins, Stephan Peischl, Justin Welbergen, Megan Higgie, Conrad Hoskin, Jeremy Vanderwal, Rob Puschendorf, Ross Alford, Vincent Calvez, Elodie Vercken, Flo Débarre, Léo Girardin, Greg Clarke, Phil Erm, Sally Drapes, Matt Hall, Louise Nørgaard, Andrew Coates, Adam Smart, Jeff Paril, Charlie Robin, and Nick Golding. Also thanks to Tim Dempster for his endless magnanimity and enthusiasm, and for his robust disagreement about the Minjerribah story (for the record, he may be correct that it was water, not wind, that cleaved the island, but historical sources vary and I'm sticking to my version for perhaps less than rational reasons).

Several research assistants have, in their spare work time, assisted with tables and indexes and chasing copyright permissions. So thanks to Ingrid Crossing, Erin Campbell-Hooper, and Haylee D'Agui for their efforts here. You are heroes each.

To my whip-smart, funny, and big-hearted wife, Taegan. Thanks for your support with this project at a time when you had your own big projects on the boil. We moved interstate, went through a pandemic, and moved house five times while this book was being written. Sometimes you had to move me as well as all the furniture. And finally, to my sons, Archer and Griffin, thanks for reminding me, daily, that life is an astonishing and wonder-filled journey.

Bibliography

Alfaro, M., Girardin, L., Hamel, F., and Roques, L. (2020). When the Allee threshold is an evolutionary trait: Persistence vs. extinction. *arXiv:2009.09657 [math]*.

Allee, W. C. (1938). The social life of animals.

Allgayer, R. L., Scarpa, A., Fernandes, P. G., Wright, P. J., Lancaster, L., Bocedi, G., and Travis, J. M. J. (2021). Dispersal evolution in currents: Spatial sorting promotes philopatry in upstream patches. *Ecography*, 44(2):231–241.

Altwegg, R., Collingham, Y. C., Erni, B., and Huntley, B. (2013). Density-dependent dispersal and the speed of range expansions. *Diversity and Distributions*, 19(1):60–68.

Anderson, R. M., and May, R. M. (1992). *Infectious Diseases of Humans: Dynamics and Control*. Oxford University Press.

Angelova, M., Mlecnik, B., Vasaturo, A., Bindea, G., Fredriksen, T., Lafontaine, L., Buttard, B., Morgand, E., Bruni, D., Jouret-Mourin, A., Hubert, C., Kartheuser, A., Humblet, Y., Ceccarelli, M., Syed, N., Marincola, F. M., Bedognetti, D., Van den Eynde, M., and Galon, J. (2018). Evolution of Metastases in Space and Time under Immune Selection. *Cell*, 175(3):751–765.e16.

Aronson, D. G. and Weinberger, H. F. (1975). Nonlinear diffusion in population genetics, combustion, and nerve pulse propagation. In Goldstein, J. A., editor, *Partial Differential Equations and Related Topics*, Lecture Notes in Mathematics, pages 5–49, Berlin, Heidelberg. Springer.

Austerlitz, F., Jung-Muller, B., Godelle, B., and Gouyon, P.-H. (1997). Evolution of Coalescence Times, Genetic Diversity and Structure during Colonization. *Theoretical Population Biology*, 51(2):148–164.

Baker, H. G. (1955). Self-compatibility and establishment after 'long-distance' dispersal. *Evolution; International Journal of Organic Evolution*, 9(3):347–349.

Baker, H. G. and Stebbins, GL., editors (1965). *The Genetics of Colonizing Species: Proceedings of the First International Union of Biological Sciences Symposia on General Biology*. Academic Press, New York.

Barnett, L. K., Phillips, B. L., and Hoskin, C. J. (2017). Going feral: Time and propagule pressure determine range expansion of Asian house geckos into natural environments. *Austral Ecology*, 42(2):165–175.

Barrett, S. C. H. (2002). The evolution of plant sexual diversity. *Nature Reviews Genetics*, 3(4):274–284.

Barrett, S. C. H., Colautti, R. I., Dlugosch, K. M., and Rieseberg, L. H. (2016). *Invasion Genetics: The Baker and Stebbins Legacy*. John Wiley & Sons.

Bartoń, K. A., Hovestadt, T., Phillips, B. L., and Travis, J. M. J. (2012). Risky movement increases the rate of range expansion. *Proceedings of the Royal Society B: Biological Sciences*, 279(1731):1194–1202.

Barton, N. H. (2001). Adaptation on the edge of a species' range. In Silvertown, J. and Antonovics, J., editors, *Integrating Ecology and Evolution in a Spatial Context*, pages 365–392. Blackwell Science, Oxford.

Barton, N. H. and Charlesworth, B. (1984). Genetic Revolutions, Founder Effects, and Speciation. *Annual Review of Ecology & Systematics*, 15:133–164.

Barton, N. H., and Turelli, M. (2011). Spatial waves of advance with bistable dynamics: Cytoplasmic and genetic analogues of Allee effects. *The American Naturalist*, 178(3):E48–75.

Barton, NH. (1979). Gene flow past a cline. *Heredity*, 43(3):333–339.

Baxter-Gilbert, J., Riley, J. L., Wagener, C., Mohanty, N. P., and Measey, J. (2020). Shrinking before our isles: The rapid expression of insular dwarfism in two invasive populations of guttural toad (Sclerophrys gutturalis). *Biology Letters*, 16(11):20200651.

Beaghton, P. J. and Burt, A. (2021). Gene drives and population persistence vs elimination: The impact of spatial structure and inbreeding at low density.

Benichou, O., Calvez, V., Meunier, N., and Voituriez, R. (2012). Front acceleration by dynamic selection in Fisher population waves. *Physical Review E*, 86(4):041908.

Berec, L., Angulo, E., and Courchamp, F. (2007). Multiple Allee effects and population management. *Trends in Ecology & Evolution*, 22(4): 185–191.

Berestycki, H., Matano, H., and Hamel, F. (2009). Bistable traveling waves around an obstacle. *Communications on Pure and Applied Mathematics*, 62(6):729–788.

Berggren, H., Tinnert, J., and Forsman, A. (2012). Spatial sorting may explain evolutionary dynamics of wing polymorphism in pygmy grasshoppers. *Journal of Evolutionary Biology*, 25(10):2126–2138.

Beverton, RJH. and Holt, SJ. (1957). *On the Dynamics of Exploited Fish Populations*, volume 19 of *Fisheries Investigation Series 2*. UK Ministry of Agriculture.

Bîrzu, G., Hallatschek, O., and Korolev, K. S. (2018). Fluctuations uncover a distinct class of traveling waves. *Proceedings of the National Academy of Sciences*, 115(16):E3645–E3654.

Bîrzu, G., Matin, S., Hallatschek, O., and Korolev, K. S. (2019). Genetic drift in range expansions is very sensitive to density dependence in dispersal and growth. *Ecology Letters*, 22(11):1817–1827.

Boddy, A. M., Huang, W., and Aktipis, A. (2018). Life History Trade-Offs in Tumors. *Current Pathobiology Reports*, 6(4):201–207.

Bosshard, L., Dupanloup, I., Tenaillon, O., Bruggmann, R., Ackermann, M., Peischl, S., and Excoffier, L. (2017). Accumulation of Deleterious Mutations During Bacterial Range Expansions. *Genetics*, page genetics.300144.2017.

Bosshard, L., Peischl, S., Ackermann, M., and Excoffier, L. (2019). Mutational and Selective Processes Involved in Evolution during Bacterial Range Expansions. *Molecular Biology and Evolution*, 36(10):2313–2327.

Bouin, E., Calvez, V., Meunier, N., Mirrahimi, S., Perthame, B., Raoul, G., and Voituriez, R. (2012). Invasion fronts with variable motility: Phenotype selection, spatial sorting and wave acceleration. *Comptes Rendus Mathematique*, 350(15):761–766.

Boukal, D. S. and Berec, L. (2002). Single-species Models of the Allee Effect: Extinction Boundaries, Sex Ratios and Mate Encounters. *Journal of Theoretical Biology*, 218(3):375–394.

Bourret, S. L., Kovach, R. P., Cline, T. J., Strait, J. T., and Muhlfeld, C. C. (2022). High dispersal rates in hybrids drive expansion of maladaptive hybridization. *Proceedings of the Royal Society B: Biological Sciences*, 289(1986):20221813.

Bowler, D. E. and Benton, T. G. (2005). Causes and consequences of animal dispersal strategies: Relating individual behaviour to spatial dynamics. *Biological Reviews*, 80(2):205–225.

Braude, R., Clarke, P. M., and Mitchell, K. G. (1954). Analysis of the breeding records of a herd of pigs. *The Journal of Agricultural Science*, 45(1):19–27.

Bridle, J. R., Polechová, J., Kawata, M., and Butlin, R. K. (2010). Why is adaptation prevented at ecological margins? New insights from individual-based simulations. *Ecology Letters*, 13(4):485–494.

Brown, G., Phillips, B. L., Dubey, S., and Shine, R. (2015). Invader immunology: Invasion history alters immune-system function in cane toads (Rhinella marina) in tropical Australia. *Ecology Letters*, 18:57–65.

Brown, G. P., Holden, D., Shine, R., and Phillips, B. L. (2018). Invasion history alters the behavioural cost of immune system activation in cane toads. *Journal of Animal Ecology*, 87:716–726.

Brown, G. P., Shilton, C., Phillips, B. L., and Shine, R. (2007). Invasion stress and spinal arthritis in cane toads. *Proceedings of the National Academy of Sciences of the United States of America*, 104:17698–17700.

Brown, W. L. J. and Wilson, E. O. (1956). Character Displacement. *Systematic Zoology*, 5(2):49–64.

Burt, A. (2003). Site-specific selfish genes as tools for the control and genetic engineering of natural populations. *Proceedings of the Royal Society of London. Series B: Biological Sciences*, 270(1518):921–928.

Burton, O. J., Travis, J. M. J., and Phillips, B. L. (2010). Trade-offs and the evolution of life-histories during range expansion. *Ecology Letters*, 13:1210–1220.

Case, T. J. and Taper, M. L. (2000). Interspecific competition, environmental gradients, gene flow, and the coevolution of species' borders. *American Naturalist*, 155(5):583–605.

Champer, J., Buchman, A., and Akbari, O. S. (2016). Cheating evolution: Engineering gene drives to manipulate the fate of wild populations. *Nature Reviews Genetics*, 17(3):146–159.

Champer, J., Kim, I. K., Champer, S. E., Clark, A. G., and Messer, P. W. (2020). Performance analysis of novel toxin-antidote CRISPR gene drive systems. *BMC Biology*, 18(1):27.

Champer, J., Kim, I. K., Champer, S. E., Clark, A. G., and Messer, P. W. (2021). Suppression gene drive in continuous space can result in unstable persistence of both drive and wild-type alleles. *Molecular Ecology*, 30(4):1086–1101.

Chuang, A. and Peterson, C. R. (2016). Expanding population edges: Theories, traits, and trade-offs. *Global Change Biology*, 22(2):494–512.

Clark, J. S. (1998). Why trees migrate so fast: Confronting theory with dispersal biology and the paleorecord. *American Naturalist*, 152(2): 204–224.

Clarke, G. S., Shine, R., and Phillips, B. L. (2019). May the (selective) force be with you: Spatial sorting and natural selection exert opposing forces on limb length in an invasive amphibian. *Journal of Evolutionary Biology*, 32(9):994–1001.

Cole, L. C. (1954). The population consequences of life-history phenomena. *The Quarterly Review of Biology*, 29:103–137.

Comerford, M. S., La, T. M., Carroll, S., and Egan, S. P. (2023). Spatial sorting promotes rapid (mal)adaptation in the red-shouldered soapberry bug after hurricane-driven local extinctions. *Nature Ecology & Evolution*, 7(11):1856–1868.

Comins, H. N., Hamilton, W. D., and May, R. M. (1980). Evolutionarily stable dispersal strategies. *Journal of Theoretical Biology*, 82(2):205–30.

Courchamp, F., Clutton-Brock, T., and Grenfell, B. (1999). Inverse density dependence and the Allee effect. *Trends in Ecology & Evolution*, 14(10):405–410.

Covacevich, J. and Archer, M. (1975). The distribution of the cane toad, Bufo marinus, in Australia and its effects on indigenous vertebrates. *Memoirs of the Queensland Museum*, 17:305–310.

Cox, G. W. (2013). *Alien Species and Evolution: The Evolutionary Ecology of Exotic Plants, Animals, Microbes, and Interacting Native Species*. Island Press.

Crossland, M. R. and Shine, R. (2011a). Cues for cannibalism: Cane toad tadpoles use chemical signals to locate and consume conspecific eggs. *Oikos*, 120(3):327–332.

Crossland, M. R., and Shine, R. (2011b). Embryonic exposure to conspecific chemicals suppresses cane toad growth and survival. *Biology Letters*, page rsbl20110794.

Crow, J. F. and Kimura, M. (1970). *An Introduction to Population Genetics Theory*. Burgess Publishing Company, Minneapolis, Minnesota.

Cwynar, L. C. and MacDonald, G. M. (1987). Geographical variation of lodgepole pine in relation to population history. *The American Naturalist*, 129(3):463.

Dahirel, M., Bertin, A., Calcagno, V., Duraj, C., Fellous, S., Groussier, G., Lombaert, E., Mailleret, L., Marchand, A., and Vercken, E. (2023). Landscape connectivity alters the evolution of density-dependent dispersal during pushed range expansions. *Peer Community Journal*, 3: e114.

Dahirel, M., Bertin, A., Haond, M., Blin, A., Lombaert, E., Calcagno, V., Fellous, S., Mailleret, L., Malausa, T., and Vercken, E. (2021). Shifts from pulled to pushed range expansions caused by reduction of landscape connectivity. *Oikos*, 130(5):708–724.

Darwin, C. (1859). *On the Origin of Species by Means of Natural Selection, or the Preservation of Favoured Races in the Struggle for Life*. John Murray, London.

Dayan, T. and Simberloff, D. (2005). Ecological and community-wide character displacement: The next generation. *Ecology Letters*, 8(8):875–894.

Deforet, M., Carmona-Fontaine, C., Korolev, K. S., and Xavier, J. B. (2019). Evolution at the Edge of Expanding Populations. *The American Naturalist*, 194(3):291–305.

Dobzhansky, T. and Wright, S. (1943). Genetics of natural poulations X. Dispersion rates in Drosophila pseudoobscura. *Genetics*, 28(4):304–340.

Doody, J. S., Soanes, R., Castellano, C. M., Rhind, D., Green, B., McHenry, C. R., and Clulow, S. (2015). Invasive toads shift predator–prey densities in animal communities by removing top predators. *Ecology*, 96(9):2544–2554.

Doudna, J. A. and Charpentier, E. (2014). The new frontier of genome engineering with CRISPR-Cas9. *Science*, 346(6213):1258096.

Dubey, S. and Shine, R. (2008). Origin of the parasites of an invading species, the Australian cane toad (Bufo marinus): Are the lungworms Australian or American? *Molecular Ecology*, 17:4418–4424.

Ducatez, S., Crossland, M., and Shine, R. (2016). Differences in developmental strategies between long-settled and invasion-front populations of the cane toad in Australia. *Journal of Evolutionary Biology*, 29(2):335–343.

Dunbar, SR. (1983). Travelling wave solutions of diffusive Lotka-Volterra equations. *Journal of Mathematical Biology*, 17:11–32.

Easteal, S. (1981). The history of introductions of Bufo marinus (Amphibia: Anura); a natural experiment in evolution. *Biological Journal of the Linnean Society*, 16:93–113.

Easteal, S., Van Beurden, E. K., Floyd, R. B., and Sabath, M. D. (1985). Continuing geographical spread of Bufo marinus in Australia: A range expansion between 1974 and 1980. *Journal of Herpetology*, 19:185–188.

Eckhoff, P. A., Wenger, E. A., Godfray, H. C. J., and Burt, A. (2017). Impact of mosquito gene drive on malaria elimination in a computational model with explicit spatial and temporal dynamics. *Proceedings of the National Academy of Sciences*, 114(2):E255–E264.

Edmonds, C. A., Lillie, A. S., and Cavalli-Sforza, L. L. (2004). Mutations arising in the wave front of an expanding population. *Proceedings of the National Academy of Sciences*, 101(4):975–979.

Elliott, E. C., and Cornell, S. J. (2012). Dispersal polymorphism and the speed of biological invasions. *PLOS ONE*, 7(7):e40496.

Elliott, E. C., and Cornell, S. J. (2013). Are anomalous invasion speeds robust to demographic stochasticity? *PloS One*, 8(7):e67871–e67871.

Elton, C. S. (1958). *The Ecology of Invasions by Animals and Plants*. Methuen, London.

Erm, P. and Phillips, B. L. (2020). Evolution transforms pushed waves into pulled waves. *The American Naturalist*, 195(3):E87–E99.

Estoup, A., Ravigné, V., Hufbauer, R., Vitalis, R., Gautier, M., and Facon, B. (2016). Is there a genetic paradox of biological invasion? *Annual Review of Ecology, Evolution, and Systematics*, 47(1):51–72.

Esvelt, K. M. and Gemmell, N. J. (2017). Conservation demands safe gene drive. *PLOS Biology*, 15(11):e2003850.

Esvelt, K. M., Smidler, A. L., Catteruccia, F., and Church, G. M. (2014). Concerning RNA-guided gene drives for the alteration of wild populations. *eLife*, 3:e03401.

Ewald, P. W. (1994). *Evolution of Infectious Disease*. Oxford University Press, New York.

Felsenstein, J. (1988). Sex and the evolution of recombination. In Michod, R. E. and Levin, B. R., editors, *The Evolution of Sex: An Examination of Current Ideas*, pages 74–86. Sinauer, Sunderland, MA.

Fiandaca, G., Delitala, M., and Lorenzi, T. (2021). A mathematical study of the influence of hypoxia and acidity on the evolutionary dynamics of cancer. *Bulletin of Mathematical Biology*, 83(7):83.

Fife, P. C. (1979). *Mathematical Aspects of Reacting and Diffusing Systems*. Springer, New York.

Fisher, R. A. (1918). The correlation between relatives on the supposition of mendelian inheritance. *Transactions of the Royal Society of Edinburgh*, 52:399–433.

Fisher, R. A. (1930). *The Genetical Theory of Natural Selection*. Dover, New York.

Fisher, R. A. (1937). The wave advance of advantageous genes. *Annals of Eugenics*, 7(355-369):355–369.

Florance, D., Webb, J. K., Dempster, T., Kearney, M. R., Worthing, A., and Letnic, M. (2011). Excluding access to invasion hubs can contain the spread of an invasive vertebrate. *Proceedings of the Royal Society B: Biological Sciences*, 278:2900–2908.

Floyd, R. B., Boughton, W. C., Easteal, S., Sabath, M. D., and van Beurden, E. K. (1981). The distribution records of the marine toad (Bufo marinus). 1. Australia. *AES Working Paper 3/81*. Brisbane: Griffith University.

Forsman, A., Merilä, J., and Ebenhard, T. (2011). Phenotypic evolution of dispersal-enhancing traits in insular voles. *Proceedings of the Royal Society B: Biological Sciences*, 278(1703):225–232.

Fraebel, D. T., Mickalide, H., Schnitkey, D., Merritt, J., Kuhlman, T. E., and Kuehn, S. (2017). Environment determines evolutionary trajectory in a constrained phenotypic space. *eLife*, 6:e24669.

Frank-Kamenetskii, DA. and Zeldovich, Y. (1938). A theory of thermal propagation of flame. *Acta Physicochim USSR*, 9:341–350.

Freischel, A. R., Damaghi, M., Cunningham, J. J., Ibrahim-Hashim, A., Gillies, R. J., Gatenby, R. A., and Brown, J. S. (2021). Frequency-dependent interactions determine outcome of competition between two breast cancer cell lines. *Scientific Reports*, 11(1):1–18.

Fronhofer, E. A. and Altermatt, F. (2015). Eco-evolutionary feedbacks during experimental range expansions. *Nature Communications*, 6:6844.

Fronhofer, E. A., Gut, S., and Altermatt, F. (2017). Evolution of density-dependent movement during experimental range expansions. *Journal of Evolutionary Biology*, 30(12):2165–2176.

Fu, X., Zhao, Y., Lopez, J. I., Rowan, A., Au, L., Fendler, A., Hazell, S., Xu, H., Horswell, S., Shepherd, S. T. C., Spencer, C. E., Spain, L., Byrne, F., Stamp, G., O'Brien, T., Nicol, D., Augustine, M., Chandra, A., Rudman, S., Toncheva, A., Furness, A. J. S., Pickering, L., Kumar, S., Koh, D.-M., Messiou, C., ap Dafydd, D., Orton, M. R., Doran, S. J., Larkin, J., Swanton, C., Sahai, E., Litchfield, K., Turajlic, S., and Bates, P. A. (2022). Spatial patterns of tumour growth impact clonal diversification in a computational model and the TRACERx Renal study. *Nature Ecology & Evolution*, 6(1):88–102.

Galizi, R., Doyle, L. A., Menichelli, M., Bernardini, F., Deredec, A., Burt, A., Stoddard, B. L., Windbichler, N., and Crisanti, A. (2014). A synthetic sex ratio distortion system for the control of the human malaria mosquito. *Nature Communications*, 5(1):3977.

Gallaher, J. A., Brown, J. S., and Anderson, A. R. A. (2019). The impact of proliferation-migration tradeoffs on phenotypic evolution in cancer. *Scientific Reports*, 9(1):2425.

Gandon, S. (1999). Kin competition, the cost of inbreeding and the evolution of dispersal. *Journal of Theoretical Biology*, 200(4):345–364.

Garnier, J., and Lewis, M. A. (2016). Expansion under climate change: The genetic consequences. *Bulletin of Mathematical Biology*, 78(11): 2165–2185.

Gatenby, R. A. (2009). A change of strategy in the war on cancer. *Nature*, 459(7246):508–509.

Gidoin, C., and Peischl, S. (2019). Range expansion theories could shed light on the spatial structure of intra-tumour heterogeneity. *Bulletin of Mathematical Biology*, 81(11):4761–4777.

Gillies, R. J. (2022). Cancer heterogeneity and metastasis: Life at the edge. *Clinical & Experimental Metastasis*, 39(1):15–19.

Gilpin, M. E. (1991). The genetic effective size of a metapopulation. *Biological Journal of the Linnean Society*, 42(1-2):165–175.

Giometto, A., Rinaldo, A., Carrara, F., and Altermatt, F. (2014). Emerging predictable features of replicated biological invasion fronts. *Proceedings of the National Academy of Sciences*, 111(1):297–301.

Girardin, L., Calvez, V., and Débarre, F. (2019). Catch me if you can: A spatial model for a brake-driven gene drive reversal. *Bulletin of Mathematical Biology*, 81(12):5054–5088.

Girardin, L. and Débarre, F. (2021). Demographic feedbacks can hamper the spatial spread of a gene drive. *Journal of Mathematical Biology*, 83(6):67.

Gomulkiewicz, R., and Holt, R. D. (1995). When does evolution by natural-selection prevent extinction. *Evolution*, 49(1):201–207.

Gomulkiewicz, R., Holt, R. D., and Barfield, M. (1999). The effects of density dependence and immigration on local adaptation in a black-hole sink environment. *Theoretical Population Biology*, 55:283–296.

Goodsman, D. W. and Lewis, M. A. (2016). The minimum founding population in dispersing organisms subject to strong Allee effects. *Methods in Ecology and Evolution*, 7(9):1100–1109.

Gralka, M. and Hallatschek, O. (2019). Environmental heterogeneity can tip the population genetics of range expansions. https://elifesciences.org/articles/44359.

Griette, Q., Raoul, G., and Gandon, S. (2015). Virulence evolution at the front line of spreading epidemics: VIRULENCE EVOLUTION AT THE FRONT LINE. *Evolution*, 69(11):2810–2819.

Guo, J.-S. and Lin, Y.-C. (2013). The sign of the wave speed for the Lotka-Volterra competition-diffusion system. *Communications on Pure and Applied Analysis*, 12(5):2083–2090.

Haldane, J. B. S. (1924). A mathematical theory of natural and artificial selection—I. *Transactions of the Cambridge Philosophical Society*, 23:19–41.

Haldane, J. B. S. (1956). The relation between density regulation and natural selection. *Proceedings of the Royal Society of London. Series B, Biological Sciences*, 145(920):306–308.

Hallatschek, O., Hersen, P., Ramanathan, S., and Nelson, D. R. (2007). Genetic drift at expanding frontiers promotes gene segregation. *Proceedings of the National Academy of Sciences*, 104(50):19926–19930.

Hallatschek, O. and Nelson, D. R. (2008). Gene surfing in expanding populations. *Theoretical Population Biology*, 73(1):158–170.

Hallatschek, O. and Nelson, D. R. (2010). Life at the front of an expanding population. *Evolution*, 64(1):193–206.

Hamilton, W. D. (1967). Extraordinary sex ratios. *Science*, 156(3774):477–488.

Hamilton, W. D. and May, R. M. (1977). Dispersal in stable habitats. *Nature*, 269:578–581.

Hammond, A., Pollegioni, P., Persampieri, T., North, A., Minuz, R., Trusso, A., Bucci, A., Kyrou, K., Morianou, I., Simoni, A., Nolan, T., Müller, R., and Crisanti, A. (2021). Gene-drive suppression of mosquito populations in large cages as a bridge between lab and field. *Nature Communications*, 12(1):4589.

Hanahan, D., and Weinberg, R. A. (2011). Hallmarks of cancer: The next generation. *Cell*, 144(5):646–674.

Hanski, I., Breuker, C. J., Schöps, K., Setchfield, R., and Nieminen, M. (2002). Population history and life history influence the migration rate of female Glanville fritillary butterflies. *Oikos*, 98:87–97.

Hanski, I. and Gaggiotti, O. E., editors (2004). *Ecology, Genetics, and Evolution of Metapopulations*. Elsevier, London.

Haond, M., Morel-Journel, T., Lombaert, E., Vercken, E., Mailleret, L., and Roques, L. (2018). When higher carrying capacities lead to faster propagation. *bioRxiv*, page 307322.

Harrison, S. and Hastings, A. (1996). Genetic and evolutionary consequences of metapopulation structure. *Trends in Ecology & Evolution*, 11(4):180–183.

Hassell, M. and Comins, H. (1976). Discrete time models for two-species competition. *Theoretical Population Biology*, 9(2):202–221.

Hastings, A. (1983). Can spatial variation alone lead to selection for dispersal? *Theoretical Population Biology*, 1:244–251.

Hausser, J. and Alon, U. (2020). Tumour heterogeneity and the evolutionary trade-offs of cancer. *Nature Reviews Cancer*, 20(4):247–257.

Hawley, D. M., Osnas, E. E., Dobson, A. P., Hochachka, W. M., Ley, D. H., and Dhondt, A. A. (2013). Parallel patterns of increased virulence in a recently emerged wildlife pathogen. *PLOS Biology*, 11(5):e1001570.

Hickey, W. A., and Craig, G. B. (1966). Genetic distortion of sex ratio in a mosquito, AEDES AEGYPTI. *Genetics*, 53(6):1177–1196.

Higgie, M., Chenoweth, S., and Blows, M. W. (2000). Natural selection and the reinforcement of mate recognition. *Science*, 290:519–521.

Hoefflin, R., Lahrmann, B., Warsow, G., Hübschmann, D., Spath, C., Walter, B., Chen, X., Hofer, L., Macher-Goeppinger, S., Tolstov, Y., Korzeniewski, N., Duensing, A., Grüllich, C., Jäger, D., Perner, S., Schönberg, G., Nyarangi-Dix, J., Isaac, S., Hatiboglu, G., Teber, D., Hadaschik, B., Pahernik, S., Roth, W., Eils, R., Schlesner, M., Sültmann, H., Hohenfellner, M., Grabe, N., and Duensing, S. (2016). Spatial niche formation but not malignant progression is a driving force for intratumoural heterogeneity. *Nature Communications*, 7(1):ncomms11845.

Hoffmann, AA., Iturbe-Ormaetxe, I., Callahan, AG., Phillips, BL., Billington, K., Axford, JK., Montgomery, B., Turley, AP., and O'Neill, SL. (2014). Stability of the wMel Wolbachia Infection following Invasion into Aedes aegypti Populations. *PLoS neglected tropical diseases*, 8(9):e3115.

Holman, L. (2019). Evolutionary simulations of Z-linked suppression gene drives. *Proceedings of the Royal Society B: Biological Sciences*, 286(1912):20191070.

Holt, R. D., and Gomulkiewicz, R. (1997). How does immigration influence local adaptation? A reexamination of a familiar paradigm. *The American Naturalist*, 149(3):563–572.

Holt, R. D., Gomulkiewicz, R., and Barfield, M. (2003). The phenomenology of niche evolution via quantitative traits in a 'black-hole' sink. *Proceedings of the Royal Society B-Biological Sciences*, 270:215–224.

Hosono, Y. (1998). The minimal speed of traveling fronts for a diffusive Lotka–Volterra competition model. *Bulletin of Mathematical Biology*, 60(3): 435–448.

Hudson, C. M., Phillips, B. L., Brown, G. P., and Shine, R. (2015). Virgins in the vanguard: Low reproductive frequency in invasion front toads. *Biological Journal of the Linnean Society*, 116(4):743–747.

Hughes, C. L., Dytham, C., and Hill, J. K. (2003). Evolutionary trade-offs between reproduction and dispersal in populations at expanding range boundaries. *Proceedings of the Royal Society Biological Sciences Series B*, 270:S147–S150.

Hughes, C. L., Dytham, C., and Hill, J. K. (2007). Modelling and analysing evolution of dispersal in populations at expanding range boundaries. *Ecological Entomology*, 32:437–445.

Hui, C. and Richardson, D. M. (2017). *Invasion Dynamics*. Oxford University Press.

Hunter, M., Krishnan, N., Liu, T., Möbius, W., and Fusco, D. (2021). Virus-host interactions shape viral dispersal giving rise to distinct classes of traveling waves in spatial expansions. *Physical Review X*, 11(2):021066.

Hutchinson, G. (1951). Copepodology for the ornithologist. *Ecology*, 32:571–577.

Jessop, T. S., Ariefiandy Achmad, Purwandana Deni, Claudio, C., Imansyah Jeri, Benu Yunias Jackson, Fordham Damien A., Forsyth David M., Mulder Raoul A., and Phillips Benjamin L. (2018). Exploring mechanisms and origins of reduced dispersal in island Komodo dragons. *Proceedings of the Royal Society B: Biological Sciences*, 285(1891):20181829.

Jiménez-Sánchez, J., Bosque, J. J., Jiménez Londoño, G. A., Molina-García, D., Martínez, Á., Pérez-Beteta, J., Ortega-Sabater, C., Honguero Martínez, A. F., García Vicente, A. M., Calvo, G. F., and Pérez-García, V. M. (2021). Evolutionary dynamics at the tumor edge reveal metabolic imaging biomarkers. *Proceedings of the National Academy of Sciences*, 118(6):e2018110118.

KaramiNejadRanjbar, M., Eckermann, K. N., Ahmed, H. M. M., Sánchez C. H. M., Dippel S., Marshall, J. M., and Wimmer, E. A. (2018). Consequences of resistance evolution in a Cas9-based sex conversion-suppression gene drive for insect pest management. *Proceedings of the National Academy of Sciences*, 115(24):6189–6194.

Karlin, S. and Taylor, A. (1998). *An Introduction to Stochastic Modeling*. Academic Press, San Diego.

Kawasaki, K., Shigesada, N., and Iinuma, M. (2017). Effects of long-range taxis and population pressure on the range expansion of invasive species in heterogeneous environments. *Theoretical Ecology*, 10(3):269–286.

Keane, R. M. and Crawley, M. J. (2002). Exotic plant invasions and the enemy release hypothesis. *Trends in Ecology & Evolution*, 17(4):164–170.

Kearney, M. R., Phillips, B. L., Tracy, C. R., Christian, K. A., Betts, G., and Porter, W. P. (2008). Modelling species distributions without using species distributions: The cane toad in Australia under current and future climates. *Ecography*, 31:423–434.

Keenan, V. A. and Cornell, S. J. (2021). Anomalous invasion dynamics due to dispersal polymorphism and dispersal–reproduction trade-offs. *Proceedings of the Royal Society B: Biological Sciences*, 288(1942):20202825.

Keitt, T. H., Lewis, M. A., and Holt, R. D. (2001). Allee effects, invasion pinning, and species' borders. page 14.

Kelehear, C., Brown, G. P., and Shine, R. (2011). Influence of lung parasites on the growth rates of free-ranging and captive adult cane toads. *Oecologia*, 165(3):585–592.

Kelehear, C., Brown, G. P., and Shine, R. (2012). Rapid evolution of parasite life history traits on an expanding range-edge. *Ecology Letters*, pages 329–337.

Kelehear, C. and Shine, R. (2020). Tradeoffs between dispersal and reproduction at an invasion front of cane toads in tropical Australia. *Scientific Reports*, 10(1):486.

Kelehear, C., Webb, J. K., and Shine, R. (2009). Rhabdias pseudosphaerocephala infection in Bufo marinus: Lung nematodes reduce viability of metamorph cane toads. *Parasitology*, 136(8):919–927.

Kermack, W. O., McKendrick, A. G., and Walker, G. T. (1927). A contribution to the mathematical theory of epidemics. *Proceedings of the Royal Society of London. Series A, Containing Papers of a Mathematical and Physical Character*, 115(772):700–721.

Kimura, M. (1985). *The Neutral Theory of Molecular Evolution*. Cambridge University Press.

Kirkpatrick, M. and Barton, N. H. (1997). Evolution of a species' range. *American Naturalist*, 150:1–23.

Klopfstein, S., Currat, M., and Excoffier, L. (2006). The fate of mutations surfing on the wave of range expansion. *Molecular Biology and Evolution*, 23(3):482–490.

Kolmogorov, AN., Petrovskii, IG., and Piskunov, NS. (1991). *Selected Works of A. N. Kolmogorov I*, : 248– 270. Tikhomirov, V. M. editor A study of the diffusion equation with increase in the amount of substance, and its application to a biological problem Kluwer Translated by Volosov, V. M. from Bull. Moscow Univ., Math. Mech., 1, 1–26, 1937

Korolev, K. S. (2015). Evolution arrests invasions of cooperative populations. *Physical Review Letters*, 115(20):208104.

Korolev, K. S., Xavier, J. B., and Gore, J. (2014). Turning ecology and evolution against cancer. *Nature Reviews Cancer*, 14(5):371–380.

Kot, M., Lewis, M. A., and vandenDriessche, P. (1996). Dispersal data and the spread of invading organisms. *Ecology*, 77(7):2027–2042.

Kottler, E. J., Dickman, E. E., Sexton, J. P., Emery, N. C., and Franks, S. J. (2021). Draining the swamping hypothesis: Little evidence that gene flow reduces fitness at range edges. *Trends in Ecology & Evolution*, 36(6): 533–544.

Kramer, A. M., Dennis, B., Liebhold, A. M., and Drake, J. M. (2009). The evidence for Allee effects. *Population Ecology*, 51(3):341– 354.

Kubisch, A., Fronhofer, E. A., Poethke, H. J., and Hovestadt, T. (2013). Kin competition as a major driving force for invasions. *The American Naturalist*, 181(5):700–706.

Kyrou, K., Hammond, A. M., Galizi, R., Kranjc, N., Burt, A., Beaghton, A. K., Nolan, T., and Crisanti, A. (2018). A CRISPR–Cas9 gene drive targeting doublesex causes complete population suppression in caged Anopheles gambiae mosquitoes. *Nature Biotechnology*, 36(11):1062–1066.

Labi, V. and Erlacher, M. (2015). How cell death shapes cancer. *Cell Death & Disease*, 6(3):e1675–e1675.

Le Roux, J. (2022). *The Evolutionary Ecology of Invasive Species*. Academic Press, London.

Lee, H., Gore, J., and Korolev, K. S. (2022). Slow expanders invade by forming dented fronts in microbial colonies. *Proceedings of the National Academy of Sciences*, 119(1):e2108653119.

Letnic, M., Webb, J. K., Jessop, T. S., Florance, D., and Dempster, T. (2014). Artificial water points facilitate the spread of an invasive vertebrate in arid Australia. *Journal of Applied Ecology*, 51:795–803.

Levine, J. M., Adler, P. B., and Yelenik, S. G. (2004). A meta-analysis of biotic resistance to exotic plant invasions. *Ecology Letters*, 7(10):975–989.

Levins, R. and Culver, D. (1971). Regional Coexistence of Species and Competition between Rare Species. *Proceedings of the National Academy of Sciences*, 68(6):1246–1248.

Lewis, M. A., Petrovskii, S. V., and Potts, J. R. (2016). *The Mathematics Behind Biological Invasions*, volume 44 of *Interdisciplinary Applied Mathematics*. Springer International Publishing, Cham.

Lewis, MA. and Kareiva, P. (1993). Allee dynamics and the spread of invading organisms. *Theoretical Population Biology*, 43(2):141–158.

Lewontin, R. C. (1965). Selection for colonizing ability. In Baker, HG. and Stebbins, GL., editors, *The Genetics of Colonizing Species*, pages 79–94. Academic Press, New York.

Lewontin, R. C. and Cohen, D. (1969). On population growth in a randomly varying environment. *Proceedings of the National Academy of sciences*, 62(4):1056–1060.

Lindstrom, T., Brown, G. P., Sisson, SA., Phillips, B. L., and Shine, R. (2013). Rapid shifts in dispersal behaviour on an expanding range edge. *Proceedings of the National Academy of Sciences USA*, 110(33):13452–13456.

Lloyd, M. C., Cunningham, J. J., Bui, M. M., Gillies, R. J., Brown, J. S., and Gatenby, R. A. (2016). Darwinian dynamics of intratumoral heterogeneity: Not solely random mutations but also variable environmental selection forces. *Cancer Research*, 76(11):3136–3144.

Lockwood, J. L., Hoopes, M. F., and Marchetti, M. P. (2007). *Invasion Ecology*. Blackwell, Maiden, MA.

López-Fernández, E., and López, J. I. (2018). The impact of tumor eco-evolution in renal cell carcinoma sampling. *Cancers*, 10(12):485.

Lotka, AJ. (1932). The growth of mixed populations: Two species competing for a common food supply. *Journal of the Washington Academy of Sciences*, 22:461–469.

Lowe, S., Browne, M., Boudjelas, S., and De Poorter, M. (2001). 100 of the world's worst invasive species: A selection from the global invasive species database. 2002(2/4/02).

Luque, G. M., Vayssade, C., Facon, B., Guillemaud, T., Courchamp, F., and Fauvergue, X. (2016). The genetic Allee effect: A unified framework for the genetics and demography of small populations. *Ecosphere*, 7(7): e01413.

Lutscher, F., Popovic, L., and Shaw, A. K. (2023). How mutation shapes the rate of population spread in the presence of a mate-finding Allee effect. *Theoretical Ecology*, 16(4):255–269.

Lynch, M., Conery, J., and Burger, R. (1995). Mutation accumulation and the extinction of small populations. *The American Naturalist*, 146(4): 489–518.

Macarthur, R., and Levins, R. (1967). The limiting similarity, convergence, and divergence of coexisting species. *The American Naturalist*, 101(921):377–385.

MacArthur, R. H. and Wilson, E. O. (1967). *The Theory of Island Biogeography*. Princeton University Press, Princeton, NJ.

Mack, R. N., Simberloff, D., Lonsdale, W. M., Evans, H., Clout, M., and Bazzaz, F. (2000). Biotic invasions: Causes, epidemiology, global consequences and control. *Ecological Applications*, 10:689–710.

Macnair, M. (1987). Heavy metal tolerance in plants: A model evolutionary system. *Trends in Ecology & Evolution*, 2:354–359.

Maley, C. C., Aktipis, A., Graham, T. A., Sottoriva, A., Boddy, A. M., Janiszewska, M., Silva, A. S., Gerlinger, M., Yuan, Y., Pienta, K. J., Anderson, K. S., Gatenby, R., Swanton, C., Posada, D., Wu, C.-I., Schiffman, J. D., Hwang, E. S., Polyak, K., Anderson, A. R. A., Brown, J. S., Greaves, M., and Shibata, D. (2017). Classifying the evolutionary and ecological features of neoplasms. *Nature Reviews Cancer*, 17(10):605–619.

Mayr, E. (1999). *Systematics and the Origin of the Species, from the Viewpoint of a Zoologist*. Harvard Univeristy Press, Cambridge, MA.

McFarland, C. D., Korolev, K. S., Kryukov, G. V., Sunyaev, S. R., and Mirny, L. A. (2013). Impact of deleterious passenger mutations on cancer progression. *Proceedings of the National Academy of Sciences*, 110(8):2910–2915.

Meccariello, A., Krsticevic, F., Colonna, R., Del Corsano, G., Fasulo, B., Papathanos, P. A., and Windbichler, N. (2021). Engineered sex ratio distortion by X-shredding in the global agricultural pest Ceratitis capitata. *BMC Biology*, 19(1):78.

Melbourne, B. A. and Hastings, A. (2008). Extinction risk depends strongly on factors contributing to stochasticity. *Nature*, 454(7200):100–103.

Melbourne, B. A., and Hastings, A. (2009). Highly variable spread rates in replicated biological invasions: Fundamental limits to predictability. *Science*, 325(5947):1536–1539.

Méndez, V., Llopis, I., Campos, D., and Horsthemke, W. (2011). Effect of environmental fluctuations on invasion fronts. *Journal of Theoretical Biology*, 281(1):31–38.

Mitchell, C. E. and Power, A. G. (2003). Release of invasive plants from fungal and viral pathogens. *Nature*, 421:625–627.

Möbius, W., Murray, A. W., and Nelson, D. R. (2015). How obstacles perturb population fronts and alter their genetic structure. *PLOS Computational Biology*, 11(12):e1004615.

Möbius, W., Tesser, F., Alards, K. M. J., Benzi, R., Nelson, D. R., and Toschi, F. (2019). The collective effect of finite-sized inhomogeneities on the spatial spread of populations in two dimensions. *arXiv:1910.05332 [cond-mat, q-bio]*.

Moerman, F., Fronhofer, E. A., Altermatt, F., and Wagner, A. (2022). Selection on growth rate and local adaptation drive genomic adaptation during experimental range expansions in the protist Tetrahymena thermophila. *Journal of Animal Ecology*, 91(6):1088–1103.

Mollison, D. (1972a). Possible velocities for a simple epidemic. *Advances in Applied Probability*, 4(2):233–257.

Mollison, D. (1972b). The rate of spatial propagation of simple epidemics. In *Proceedings of the Sixth Berkeley Symposium on Mathematical Statistics and Probability*, volume 3, pages 579–614. Berkeley and Los Angeles, University of California Edinburgh, UK.

Moreau, C., Bhérer, C., Vézina, H., Jomphe, M., Labuda, D., and Excoffier, L. (2011). Deep human genealogies reveal a selective advantage to be on an expanding wave front. *Science (New York, N.Y.)*, 334(6059):1148–50.

Morel-Journel, T., Haond, M., Lamy, L., Muru, D., Roques, L., Mailleret, L., and Vercken, E. (2022). When expansion stalls: An extension to the concept of range pinning in ecology. *Ecography*, 2022(2).

Morel-Journel, T., Piponiot, C., Vercken, E., and Mailleret, L. (2016). Evidence for an optimal level of connectivity for establishment and colonization. *Biology Letters*, 12(11):20160704.

Nei, M., Maruyama, T., and Chakraborty, R. (1975). The bottleneck effect and genetic variability in populations. *Evolution*, 29:1–10.

Neubert, M. G., Kot, M., and Lewis, M. A. (1995). Dispersal and pattern formation in a discrete-time predator-prey model. *Theoretical population biology*, 48(1):7–43.

Neubert, M. G., Kot, M., and Lewis, M. A. (2000). Invasion speeds in fluctuating environments. *Proceedings of the Royal Society of London. Series B: Biological Sciences*, 267(1453):1603–1610.

Newman, W. I. (1980). Some exact solutions to a non-linear diffusion problem in population genetics and combustion. *Journal of Theoretical Biology*, 85(2):325–334.

Nichols, R. A. and Hewitt, G. M. (1994). The genetic consequences of long distance dispersal during colonization. *Heredity*, 72(3):312–317.

Noble, C., Olejarz, J., Esvelt, K. M., Church, G. M., and Nowak, M. A. (2017). Evolutionary dynamics of CRISPR gene drives. *Science Advances*, page 8.

Nowell, P. C. (1976). The clonal evolution of tumor cell populations. *Science*, 194(4260):23–28.

Ochocki, B. M. and Miller, T. E. X. (2017). Rapid evolution of dispersal ability makes biological invasions faster and more variable. *Nature Communications*, 8:ncomms14315.

Ochocki, B. M., Saltz, J. B., and Miller, T. E. X. (2019). Demography-dispersal trait correlations modify the eco-evolutionary dynamics of range expansion. *The American Naturalist*, 195(2):231–246.

Okubo, A., Maini, P. K., Williamson, M. H., and Murray, J. D. (1989). On the spatial spread of the grey squirrel in Britain. *Proceedings of the Royal Society of London. B. Biological Sciences*, 238(1291):113–125.

Olivieri, I., and Gouyon, P.-H. (1997). 13 - Evolution of migration rate and other traits: The metapopulation effect. In Hanski, I., and Gilpin, M. E., editors, *Metapopulation Biology*, pages 293–323. Academic Press, San Diego.

Olivieri, I., Michalakis, Y., and Gouyon, P.-H. (1995). Metapopulation genetics and the evolution of dispersal. *The American Naturalist*, 146(2):202–228.

Orlando, P. A., a. Gatenby, R., and Brown, J. S. (2013). Tumor evolution in space: The effects of competition colonization tradeoffs on tumor invasion dynamics. *Frontiers in Oncology*, 3:1–12.

Osnas, E. E., Hurtado, P. J., and Dobson, A. P. (2015). Evolution of pathogen virulence across space during an epidemic. *The American Naturalist*, 185(3):332–342.

Otto, S. P. and Day, T. (2007). *A Biologist's Guide to Mathematical Modeling in Ecology and Evolution*. Princeton University Press.

Owen, M. R., and Lewis, M. A. (2001). How predation can slow, stop or reverse a prey invasion. *Bulletin of Mathematical Biology*, 63(4):655–684.

Pachepsky, E., and Levine, J. M. (2011). Density dependence slows invader spread in fragmented landscapes. *The American Naturalist*, 177(1):18–28.

Paczkowski, M., Kretzschmar, W. W., Markelc, B., Liu, S. K., Kunz-Schughart, L. A., Harris, A. L., Partridge, M., Byrne, H. M., and Kannan, P. (2021). Reciprocal interactions between tumour cell populations enhance growth and reduce radiation sensitivity in prostate cancer. *Communications Biology*, 4(1):1–13.

Pannell, J. R. (2015). Evolution of the mating system in colonizing plants. *Molecular Ecology*, 24(9):2018–2037.

Paril, J. F. and Phillips, B. L. (2022). Slow and steady wins the race: Spatial and stochastic processes and the failure of suppression gene drives. *Molecular Ecology*, 31(17):4451–4464.

Peck, J. R., Yearsley, J. M., and Waxman, D. (1998). Explaining the geographic distributions of sexual and asexual populations. *Nature*, 391(6670):889–892.

Peischl, S., Dupanloup, I., Kirkpatrick, M., and Excoffier, L. (2013). On the accumulation of deleterious mutations during range expansions. *Molecular Ecology*, 22(24):5972–5982.

Peischl, S., and Gilbert, K. J. (2019). Evolution of dispersal can rescue populations from expansion load. *The American Naturalist*, 195(2):349–360.

Peischl, S., Kirkpatrick, M., and Excoffier, L. (2015). Expansion load and the evolutionary dynamics of a species range. *The American Naturalist*, 185(4):E81–E93.

Perkins, S. E., Boettiger, C., and Phillips, B. L. (2016). After the games are over: Life-history trade-offs drive dispersal attenuation following range expansion. *Ecology and Evolution*, 6(18):6425–6434.

Perkins, T. A. (2012). Evolutionarily labile species interactions and spatial spread of invasive species. *The American Naturalist*, 179(2):E37–E54.

Perkins, T. A., Phillips, B. L., Baskett, M. L., and Hastings, A. (2013). Evolution of dispersal and life-history interact to drive accelerating spread of an invasive species. *Ecology Letters*, 16(8):1079–1087.

Petrovskii, S. and Morozov, A. (2009). Dispersal in a statistically structured population: Fat tails revisited. *The American Naturalist*, 173(2): 278–289.

Phillips, B. L. (2009). The evolution of growth rates on an expanding range edge. *Biology Letters*, 5(6):802–804.

Phillips, B. L. (2012). Range shift promotes the formation of stable range edges. *Journal of Biogeography*, 39(1):153–161.

Phillips, B. L. (2015). Evolutionary processes make invasion speed difficult to predict. *Biological Invasions*, 17:1949–1960.

Phillips, B. L. and Baird, S. J. E. (2015). Spatial sorting unlikely to promote maladaptive hybridization: Response to Lowe, Muhlfeld, and Allendorf. *Trends in Ecology & Evolution*, 30(10):564–565.

Phillips, B. L., Brown, G. P., Greenlees, M., Webb, J. K., and Shine, R. (2007). Rapid expansion of the cane toad (Bufo marinus) invasion front in tropical Australia. *Austral Ecology*, 32:169–176.

Phillips, B. L., Brown, G. P., and Shine, R. (2003). Assessing the potential impact of cane toads on Australian snakes. *Conservation Biology*, 17(6): 1738–1747.

Phillips, B. L., Brown, G. P., and Shine, R. (2010a). Evolutionarily accelerated invasions: The rate of dispersal evolves upwards during range advance of cane toads. *Journal of Evolutionary Biology*, 23:2595–2601.

Phillips, B. L., Brown, G. P., and Shine, R. (2010b). Life-history evolution in range-shifting populations. *Ecology*, 91(6):1617–1627.

Phillips, B. L., Brown, G. P., Travis, J. M. J., and Shine, R. (2008). Reid's paradox revisited: The evolution of dispersal in range-shifting populations. *The American Naturalist*, 172(s1):S34–S48.

Phillips, B. L., Brown, G. P., Webb, J. K., and Shine, R. (2006). Invasion and the evolution of speed in toads. *Nature*, 439:803.

Phillips, B. L., Kelehear, C., Pizzatto, L., Brown, G. P., Barton, D., and Shine, R. (2010c). Parasites and pathogens lag behind their host during periods of host range-advance. *Ecology*, 91:872–881.

Phillips, B. L. and Perkins, T. A. (2019). Spatial sorting as the spatial analogue of natural selection. *Theoretical Ecology*, 12:155–163.

Phillips, B. L. and Puschendorf, R. (2013). Do pathogens become more virulent as they spread? Evidence from the amphibian declines in Central America. *Proceedings of the Royal Society B-Biological Sciences*, 280(1766):20131290.

Phillips, B. L., Shine, R., and Tingley, R. (2016). The genetic backburn: Using evolution to halt invasions. *Proceedings of the Royal Society B-Biological Sciences*, 283:20153037.

Pianka, E. R. (1970). On r- and K-selection. *The American Naturalist*, 104:592–597.

Pizzato, L. and Shine, R. (2008). The behavioural ecology of cannibalism in cane toads (Bufo marinus). *Behavioural Ecology and Sociobiology*, in press.

Pizzatto, L. and Shine, R. (2011). The effects of experimentally infecting Australian tree frogs with lungworms (Rhabdias pseudosphaerocephala) from invasive cane toads. *International Journal for Parasitology*, 41(9):943–949.

Polechová, J., Barton, N., and Marion, G. (2009). Species' range: Adaptation in space and time. *The American Naturalist*, 174(5):E186–204.

Poloni, S. and Lutscher, F. (2023). Integrodifference models for evolutionary processes in biological invasions. *Journal of Mathematical Biology*, 87(1):10.

Price, G. R. (1970). Selection and covariance. *Nature*, 227(5257):520–521.

Prowse, T. A. A., Cassey, P., Ross, J. V., Pfitzner, Chandran, Wittmann, Talia A., and Thomas, Paul (2017). Dodging silver bullets: Good CRISPR gene-drive design is critical for eradicating exotic vertebrates. *Proceedings of the Royal Society B: Biological Sciences*, 284(1860):20170799.

Pujol, B., Zhou, S.-R., Vilas, J. S., and Pannell, J. R. (2009). Reduced inbreeding depression after species range expansion. *Proceedings of the National Academy of Sciences*, 106(36):15379–15383.

Ramanantoanina, A., Ouhinou, A., and Hui, C. (2014). Spatial assortment of mixed propagules explains the acceleration of range expansion. *PLOS ONE*, 9(8):e103409.

Reznick, D., Bryant, M. J., and Bashey, F. (2002). R- and K-selection revisited: The role of population regulation in life-history evolution. *Ecology*, 83(6):1509–1520.

Reznick, D. and Travis, J. (1996). The empirical study of adaptations in natural populations. In Rose, M. R. and Lauder, G. V., editors, *Adaptation*, pages 243–289. Academic Press, San Diego.

Rode, N. O., Estoup, A., Bourguet, D., Courtier-Orgogozo, V., and Débarre, F. (2019). Population management using gene drive: Molecular design, models of spread dynamics and assessment of ecological risks. *Conservation Genetics*.

Roff, D. A. (1993). *The Evolution of Life Histories: Theory and Analysis*. Springer, New York.

Roques, L., Garnier, J., Hamel, F., and Klein, E. K. (2012). Allee effect promotes diversity in traveling waves of colonization. *Proceedings of the National Academy of Sciences*, 109(23):8828–8833.

Roques, L., Hosono, Y., Bonnefon, O., and Boivin, T. (2015). The effect of competition on the neutral intraspecific diversity of invasive species. *Journal of Mathematical Biology*, 71(2):465–489.

Royle, J. A. and Dorazio, R. M. (2008). *Hierarchical Modeling and Inference in Ecology: The Analysis of Data from Populations, Metapopulations and Communities*. Elsevier.

Sabath, M. D., Boughton, W. C., and Easteal, S. (1981). Expansion of the range of the introduced toad Bufo marinus in Australia 1935-1974. *Copeia*, 1981(3):676–680.

Sax, D. F., Stachowicz, J. J., and Gaines, S. D., editors (2005). *Species Invasions: Insights into Ecology, Evolution, and Biogeography*. Sinauer Associates, Sunderland.

Schliekelman, P., Ellner, S., and Gould, F. (2005). Pest control by genetic manipulation of sex ratio. *Journal of Economic Entomology*, 98(1):18–34.

Schoener, T. W. (1974). Some methods for calculating competition coefficients from resource-utilization spectra. *The American Naturalist*, 108(961):332–340.

Schoener, T. W. (1983). Field experiments on interspecific competition. *The American Naturalist*, 122(2):240–285.

Schreiber, S. J. and Beckman, N. G. (2020). Individual variation in dispersal and fecundity increases rates of spatial spread. *AoB PLANTS*, 12(3):plaa001.

Schreiber, S. J. and Ryan, M. E. (2011). Invasion speeds for structured populations in fluctuating environments. *Theoretical Ecology*, 4(4):423–434.

Scott, J. and Marusyk, A. (2017). Somatic clonal evolution: A selection-centric perspective. *Biochimica et Biophysica Acta (BBA) - Reviews on Cancer*, 1867(2):139–150.

Shaw, A. K., and Kokko, H. (2015). Dispersal evolution in the presence of Allee effects can speed up or slow down invasions. *The American Naturalist*, 185(5):631–639.

Shigesada, N. and Kawasaki, K. (1997). *Biological Invasions: Theory and Practice*. Oxford Series in Ecology and Evolution. Oxford University Press, Oxford.

Shine, R. (2010). The ecological impact of invasive cane toads (Bufo marinus) in Australia. *Quarterly Review of Biology*, 85:253–291.

Shine, R. (2018). *Cane Toad Wars*. University of California Press, Oakland, California.

Shine, R., Brown, G. P., and Phillips, B. L. (2011). An evolutionary process that assembles phenotypes through space rather than through time. *Proceedings of the National Academy of Sciences*, 108(14):5708–5711.

Simmons, A. D. and Thomas, C. D. (2004). Changes in dispersal during species' range expansions. *American Naturalist*, 164(3):378–395.

Skellam, J. G. (1951). Random dispersal in theoretical populations. *Biometrika*, 38(1/2):196–218.

Slatkin, M. (1974). Competition and regional coexistence. *Ecology*, 55(1): 128–134.

Slatkin, M. and Excoffier, L. (2012). Serial founder effects during range expansion: A spatial analog of genetic drift. *Genetics*, 191(1):171–81.

Song, L., Zhang, S., Yu, S., Ma, F., Wang, B., Zhang, C., Sun, J., Mao, X., and Wei, L. (2020). Cellular heterogeneity landscape in laryngeal squamous cell carcinoma. *International Journal of Cancer*, 147(10):2879–2890.

Southwell, D., Tingley, R., Bode, M., Nicholson, E., and Phillips, B. (2017). Cost and feasibility of a barrier to halt the spread of invasive cane toads in arid Australia: Incorporating expert knowledge into model-based decision-making. *Journal of Applied Ecology*, 54(1):216–224.

Stephens, P. A. and Sutherland, W. J. (1999). Consequences of the Allee effect for behaviour, ecology and conservation. *Trends in Ecology & Evolution*, 14(10):401–405.

Strauss, S. Y., Lau, J. A., and Carroll, S. P. (2006). Evolutionary responses of natives to introduced species: What do introductions tell us about natural communities? *Ecology Letters*, 9(3):357–371.

Streicher, J. W., McEntee, J. P., Drzich, L. C., Card, D. C., Schield, D. R., Smart, U., Parkinson, C. L., Jezkova, T., Smith, E. N., and Castoe, T. A. (2016). Genetic surfing, not allopatric divergence, explains spatial sorting of mitochondrial haplotypes in venomous coralsnakes. *Evolution*, 70(7): 1435–1449.

Strogatz, S. H. (2012). *Sync: How Order Emerges from Chaos in the Universe, Nature, and Daily Life*. Hachette UK.

Subramanian, S. (2012). The abundance of deleterious polymorphisms in humans. *Genetics*, 190(4):1579–1583.

Szűcs, M., Vahsen, M. L., Melbourne, B. A., Hoover, C., Weiss-Lehman, C., and Hufbauer, R. A. (2017). Rapid adaptive evolution in novel environments acts as an architect of population range expansion. *Proceedings of the National Academy of Sciences*, 114(51):13501–13506.

Tanaka, H., Stone, H. A., and Nelson, D. R. (2017). Spatial gene drives and pushed genetic waves. *Proceedings of the National Academy of Sciences*, 114(32):8452–8457.

Tang, L. and Chen, S. (2022). Traveling wave solutions for the diffusive Lotka–Volterra equations with boundary problems. *Applied Mathematics and Computation*, 413:126599.

Tingley, R., Phillips, B. L., Letnic, M., Brown, G. P., Shine, R., and Baird, S. J. E. (2013). Identifying optimal barriers to halt the invasion of cane toads Rhinella marina in arid Australia. *Journal of Applied Ecology*, 50:129–137.

Tingley, R., Ward-Fear, G., Schwarzkopf, L., Greenlees, M. J., Phillips, B. L., Brown, G., Clulow, S., Webb, J., Capon, R., Sheppard, A., Strive, T., Tizard,

M., and Shine, R. (2017). New weapons in the toad toolkit: A review of methods to control and mitigate the biodiversity impacts of invasive cane toads (Rhinella marina). *The Quarterly Review of Biology*, 92(2):123–149.

Tobler, W. R. (1970). A computer movie simulating urban growth in the Detroit region. *Economic Geography*, 46(sup1):234–240.

Torchin, M. E. and Mitchell, C. E. (2004). Parasites, pathogens, and invasions by plants and animals. *Frontiers in Ecology and the Environment*, 2(4):183–190.

Travis, J. M. J. and Dytham, C. (2002). Dispersal evolution during invasions. *Evolutionary Ecology Research*, 4:1119–1129.

Travis, J. M. J., Münkemüller, T., Burton, O. J., Best, A., Dytham, C., and Johst, K. (2007). Deleterious mutations can surf to high densities on the wave front of an expanding population. *Molecular Biology and Evolution*, 24(10):2334–2343.

Travis, J. M. J., Mustin, K., Benton, T. G., and Dytham, C. (2009). Accelerating invasion rates result from the evolution of density-dependent dispersal. *Journal of Theoretical Biology*, 259:151–158.

Turchin, P. (1998). *Quantitative Analysis of Movement: Measuring and Modeling Population Redistribution in Animals and Plants*, volume 1. Sinauer Associates Sunderland.

Urban, M. C., Phillips, B. L., Skelly, D. K., and Shine, R. (2008). A toad more travelled: The heterogeneous invasion dynamics of cane toads in Australia. *The American Naturalist*, 171(3):E134–E148.

van Ditmarsch, D., Boyle, K. E., Sakhtah, H., Oyler, J. E., Nadell, C. D., Déziel, É., Dietrich, L. E., and Xavier, J. B. (2013). Convergent evolution of hyperswarming leads to impaired biofilm formation in pathogenic bacteria. *Cell Reports*, 4:697–708.

Van Petegem, K., Moerman, F., Dahirel, M., Fronhofer, E. A., Vandegehuchte, M. L., Leeuwen, T. V., Wybouw, N., Stoks, R., and Bonte, D. (2018). Kin competition accelerates experimental range expansion in an arthropod herbivore. *Ecology Letters*, 21(2):225–234.

van Saarloos, W. (2003). Front propagation into unstable states. *Physics Reports*, 386(2):29–222.

Van Valen, L. (1971). Group selection and the evolution of dispersal. *Evolution*, 25:591–598.

Visher, E., Evensen, C., Guth, S., Lai, E., Norfolk, M., Rozins, C., Sokolov, N. A., Sui, M., and Boots, M. (2021). The three Ts of virulence evolution during zoonotic emergence. *Proceedings of the Royal Society B: Biological Sciences*, 288(1956):20210900.

Volpert, A. I., Volpert, V. A., and Volpert, V. A. (1994). *Traveling Wave Solutions of Parabolic Systems*, volume 140 of *Transl. Math. Monographs*. American Mathematical Soc., Providence, RI.

Volterra, V. (1926). Fluctuations in the Abundance of a Species considered Mathematically1. *Nature*, 118(2972):558–560.

Wang, M.-H., Kot, M., and Neubert, M. G. (2002). Integrodifference equations, Allee effects, and invasions. *Journal of Mathematical Biology*, 44(2):150–168.

Weinberger, H. F. (1982). Long-time behavior of a class of biological models. *SIAM Journal on Mathematical Analysis*, 13(3):353–396.

Weiss-Lehman, C., Hufbauer, R. A., and Melbourne, B. A. (2017). Rapid trait evolution drives increased speed and variance in experimental range expansions. *Nature Communications*, 8:ncomms14303.

Williams, J. L., Hufbauer, R. A., and Miller, T. E. (2019). How evolution modifies the variability of range expansion. *Trends in Ecology & Evolution*, 34(10):903–913.

Williams, J. L., Kendall, B. E., and Levine, J. M. (2016a). Rapid evolution accelerates plant population spread in fragmented experimental landscapes. *Science*, 353(6298):482–485.

Williams, J. L., Snyder, R. E., and Levine, J. M. (2016b). The influence of evolution on population spread through patchy landscapes. *The American Naturalist*, 188(1):15–26.

Windbichler, N., Menichelli, M., Papathanos, P. A., Thyme, S. B., Li, H., Ulge, U. Y., Hovde, B. T., Baker, D., Monnat, R. J., Burt, A., and Crisanti, A. (2011). A synthetic homing endonuclease-based gene drive system in the human malaria mosquito. *Nature*, 473(7346):212–215.

Wright, S. (1931). Evolution in Mendelian populations. *Genetics*, 16(2):97–159.

Yu, Q., Gralka, M., Duvernoy, M.-C., Sousa, M., Harpak, A., and Hallatschek, O. (2021). Mutability of demographic noise in microbial range expansions. *The ISME Journal*, 15(9):2643–2654.

Zahir, N., Sun, R., Gallahan, D., Gatenby, R. A., and Curtis, C. (2020). Characterizing the ecological and evolutionary dynamics of cancer. *Nature Genetics*, 52(8):759–767.

Zhao, B., Sedlak, J. C., Srinivas, R., Creixell, P., Pritchard, J. R., Tidor, B., Lauffenburger, D. A., and Hemann, M. T. (2016). Exploiting temporal collateral sensitivity in tumor clonal evolution. *Cell*, 165(1):234–246.

Zhao, Y., Fu, X., Lopez, J. I., Rowan, A., Au, L., Fendler, A., Hazell, S., Xu, H., Horswell, S., Shepherd, S. T. C., Spain, L., Byrne, F., Stamp, G., O'Brien, T., Nicol, D., Augustine, M., Chandra, A., Rudman, S., Toncheva, A., Pickering, L., Sahai, E., Larkin, J., Bates, P. A., Swanton, C., Turajlic, S., and Litchfield, K. (2021). Selection of metastasis competent subclones in the tumour interior. *Nature Ecology & Evolution*, 5(7):1033–1045.

Name Index

Tables, figures, and boxes are indicated by an italic t, f, and b following the page number.

Subject Index

Tables and figures are indicated by an italic *t*, *f*, and *b* following the page number.